CAMBRIDGE MONOGRAPHS ON PHYSICS

.

GENERAL EDITORS

A. HERZENBERG, PH.D.
Reader in Theoretical Physics in the University of Manchester

J. M. ZIMAN, D.PHIL., F.R.S.
Professor of Theoretical Physics in the University of Bristol

DEFECTS AND RADIATION DAMAGE IN METALS

DEFECTS
AND RADIATION
DAMAGE IN
METALS

M. W. THOMPSON

Professor of Experimental Physics
University of Sussex

CAMBRIDGE
AT THE UNIVERSITY PRESS
1969

Published by the Syndics of the Cambridge University Press
Bentley House, 200 Euston Road, London N.W.1
American Branch: 32 East 57th Street, New York, N.Y.10022

© Cambridge University Press 1969

Library of Congress Catalogue Card Number: 69-10434

Standard Book Number: 521 07068 6

Printed in Great Britain
at the University Printing House, Cambridge
(Brooke Crutchley, University Printer)

CONTENTS

PREFACE

All fields of science grow out of human curiosity, but in some cases this is stimulated by the demands of our technology. Radiation damage in solids is such a field, receiving its major stimuli from the nuclear energy and space exploration programmes.

In writing a book on the subject one has either to restrict the scope to considering the basic atomic mechanisms which give rise to the main effects, as I have done, or else to face the near-impossible task of cataloguing all the radiation-induced effects. A full appreciation of these basic mechanisms leads through a deeper understanding of the solid state to a means of predicting events in unfamiliar situations. Thus, this book should offer something both to the general student of science and to the specialist in materials technology. The further restriction to metals underlines this intention to concentrate on basic mechanisms, for it is in these relatively simple structures that the fundamental processes can be most easily identified. If one were to broaden the scope to include other classes of solid, much of the material presented here would still apply, but extra factors would have to be introduced and for many of these the present state of knowledge might not permit a satisfying treatment.

The book falls naturally into two parts, the first describing the nature of lattice defects in metal crystals and the second going into the mechanisms by which defects are produced under irradiation, emphasizing the special types found in this case. The aim of the first part is to provide a general background of information at a relatively low level of sophistication and much of it would be suitable reading for an undergraduate. Indeed, the book is closely related to lecture courses I have given at the University of Sussex. Of course some excellent books on this subject already exist, but here I have made a special effort to present the material in a way that leads on naturally into radiation damage and recognizes the important contribution that the study of irradiated metals has made to our general knowledge of defects. The level reached in the second part of the book is considerably higher than that of the first, and whilst some of it is suitable for the general student, it should also provide the specialist with a good starting point for his research and the technologist with the background information for solving problems.

Recent years have seen considerable developments in our understanding of collision cascades, of the clustering of point defects and the behaviour of impurities introduced by irradiation. Although other books exist with similar titles, this one includes all these modern developments and probably has a wider coverage.

In the early stages of writing, whilst at Harwell, I received immense help from my colleagues there whose distinguished work covers most of the important topics in radiation damage and forms a significant part of the subject matter. The extent of my debt can be gauged roughly by the number of illustrations they have provided for this book. It is not possible to name all the individuals, but I wish to express my particular thanks to R. S. Barnes, M. J. Makin, D. J. Mazey, R. S. Nelson and J. H. O. Varley. During my past two years at the University of Sussex my students' comments have helped to link the individual topics together and, I hope, make the book more easily comprehensible. I must thank M. Lucas and J. A. Venables for their critical reading of chapters 6 and 7 and Miss N. Kingan for her help in preparing the manuscript, with its many illustrations. Finally, the most important acknowledgement of all is addressed to my wife Sybil for her active encouragement over a much longer period than either of us anticipated when the writing began.

M. W. THOMPSON

Sussex, 1967

CHAPTER I

INTRODUCTION

Perhaps the most striking characteristic of crystalline solids is the extraordinary perfection of the crystal lattice. Thus by knowing the position of a unit cell on one side of a good crystal the position of atoms even 1 cm away can be specified to a precision of better than 10^{-8} cm. By using X-ray diffraction to exploit this perfection the science of crystallography has made remarkable advances during the past half century in unravelling the lattice structure of most common crystals.

Following this knowledge came the application of quantum mechanics to the behaviour of valence electrons in crystals, when it was found that internal diffraction of the electrons placed severe restraints on their momentum and led to the formation of energy bands. This branch of solid state physics has made spectacular advances since the nineteen thirties and led to a fair understanding of electrical, magnetic and cohesive properties of crystals. However, many important characteristics of solids were left unexplained. For example the mechanism of crystal growth was completely unknown, as demonstrated by a ratio between observed and predicted rates of growth of 10^{40}, said to be the largest discrepancy in the history of science!

Similarly one of the most useful properties of metals, their plasticity or malleability, upon which our whole industrial civilization was built, was never satisfactorily explained until the nineteen forties. Another long unexplained effect was the interdiffusion of atoms from solid to solid that takes place far below the melting point.

None of these effects can be described in terms of a perfect crystal lattice, but if one postulates the presence of lattice defects many things become clear. Thus if some lattice sites are vacant, diffusion becomes possible. The presence of certain dislocations in the lattice can introduce spiral terraces on the crystal surfaces allowing it to grow more rapidly and similar defects make it possible for neighbouring atomic planes to slide over one another fairly easily.

In this branch of solid state physics one adopts a special philosophy in which the lattice of atoms and their electrons are partly forgotten, one peers into the crystal looking only for defects in the structure and concentrates on their properties and interactions. In many ways the approach is similar to that of the chemist who neglects the water molecules in his aqueous solution and thinks only of the relatively few radical ions that for him are going to determine its properties.

Knowing that the defect population determines many important properties of solids it is clear that any process which alters their concentration is going to have an effect on physical properties, and irradiation with nuclear particles is such a process. Here is the underlying reason for studying radiation damage, for the technologies of nuclear power and aerospace call for materials that can perform their task under irradiation.

As a typical case take the irradiation of a crystal by neutrons from nuclear fission. These will occasionally collide with the nuclei of atoms and transfer kinetic energy. If this recoil energy is large enough the atom concerned will be knocked out of its lattice site and will travel through the crystal, colliding with its neighbours and maybe displacing these also from their sites. A cascade of atomic collisions is thus generated by the original neutron hit.

The end product of the cascade will be a number of vacant lattice sites and an equal number of displaced atoms wedged into the interstices of the lattice. These defects are usually referred to as vacancies and interstitials. The total number of them, and their distribution relative to one another depends on the way the cascade spreads through the crystal. First one must know how the fast-moving recoil atom loses energy. Some will go in ionization and excitation of electrons, the rest into the kinetic energy of the atoms it collides with.

By restricting ourselves to metals we shall avoid the permanent consequences of ionization, for here the displaced electrons are always able to return to their normal state in a very short time. In insulators effects such as coloration result from these excited electrons finding new sites—where they exhibit different properties.

The lattice pattern can exert some influence on the spreading of the collision cascade. For instance some moving atoms can move in the open channels that exist between the atomic rows of the lattice, and find an easy route through the crystal. Along the rows of atoms themselves one can have energy passed from atom to atom in a sequence of collisions, in the manner of a goods train shunting.

Having produced our interstitials and vacancies, what happens next? If the temperature is sufficiently high, the thermal vibration of the crystal lattice will cause these so-called point defects to migrate, hopping from site to site in the crystal. Occasionally two defects will meet and, if they are unlike, annihilation will result. If they are like there may be a tendency to stick together and clusters of interstitials and vacancies will eventually be formed, like a precipitate. During migration other things may happen, the point defects may attach themselves to defects that were present before irradiation, such as impurity atoms or dislocations, or they may escape to the surface of the crystal.

In the simplest type of irradiation experiment the bombarding particle is chosen so that it only displaces one atom for every primary collision event. It so happens that electrons near 1 MeV are in this category and can be used to show that a minimum energy of about 25 eV must be transferred to an atom, for permanent displacement to occur. Next, one should irradiate at a low enough temperature that the defects cannot migrate. Then by progressively raising the temperature after irradiation, their migration can be controlled and studied carefully, by using a property like the electrical resistance as an index of their presence.

The defect characteristics that one is trying to discover are its spatial configuration of atoms, its formation energy, or the amount by which the energy of the crystal is raised by its presence, and the amount of energy it requires in order to migrate from one site to the next. In metals it appears that interstitials require more formation energy than vacancies, but require less energy for migration. Thus in a progressive warm-up after irradiation, interstitials usually migrate first, sometimes near 10 °K; and vacancies migrate at higher temperatures, often near 300 °K. There appears to be a direct correlation between the melting temperature of the metal and the temperature at which its defects migrate. Presumably the melting temperature is an index of the strength of atomic bonding.

The defect clusters that form in metals take several forms. The simplest is the small group of vacancies probably formed together at the core of a single collision cascade, which form a small void. As this grows larger it apparently becomes unstable and collapses first into a pancake shaped disc then the faces of the disc join up leaving a toroidal ring of lattice defect. On a larger scale such rings are known as dislocation loops and certain types are the defects responsible for plasticity of crystals. It is not surprising therefore that the radiation-produced loops can interact with the pre-existing system of dislocations, drastically affecting the plasticity.

Clusters of interstitials have a very similar configuration, they often form a pancake of extra atoms that lie between the planes of the crystal, forcing them apart. This increases the external dimensions of the crystal, though it is partly offset by the collapsing of vacancy clusters.

Besides radiation damage caused by displacement of atoms there is another form known as *impurity* damage due to the introduction of foreign atoms. This happens either as a result of nuclear transmutation reactions, like $^{10}B(n, \alpha)^{7}Li$ nuclear fission; or by irradiating with ions like alpha particles which are retained in the target as a different chemical species.

A very important class of impurity damage is caused by introduction of the inert gases because they are highly insoluble and precipitate out of solid solution to form small bubbles. In doing this they combine with large numbers of vacancies, and often cause a macroscopic swelling of the crystal.

In the chapters that follow we shall look first at the general nature of lattice defects, then at the collision processes by which they are formed under irradiation, and finally attempt to describe their behaviour and the effects they cause in irradiated metals.

CHAPTER 2

POINT DEFECTS

2.1. Introduction

In this chapter we shall first consider some quite general aspects of defects in the crystalline state. Secondly, those defects which enter the discussion of radiation damage in later chapters will be catalogued, with their basic properties. The treatment of particular defects will mainly be from the theoretical point of view but where a defect is familiar in other fields the relevant experimental information will be surveyed. This procedure may wrongly give the impression that the study of radiation damage has not contributed to our understanding of defects. It is hoped that later chapters will correct any such view.

2.2. The thermodynamics of defective crystals

2.2.1. *Defects in equilibrium.* Even in the absence of radiation a crystal cannot exist at finite temperatures in a state of complete perfection. The vibrations of the lattice constitute one form of imperfection. These may be represented as a statistical distribution of thermal energy amongst the atoms of the crystal and in any such distribution there is always a finite probability of sufficient energy being concentrated, by local fluctuations, on to a group of atoms to form a defect in the crystal lattice.

From thermodynamic considerations one may find the configuration of defects in the lattice that produces the minimum value of the appropriate thermodynamic energy function, and hence determine the best possible degree of perfection. The thermodynamic system is taken as the crystal as a whole, in equilibrium with its surroundings at a constant temperature. For most purposes it is a fair approximation to take the volume of the crystal as constant, when the appropriate energy function is the Helmholtz free energy $F = U - TS$, which applies to systems at constant volume and temperature. The components of the system are the N atoms, and the internal energy U is the sum of their kinetic and potential energies. The entropy S represents the disorder in the system and if w is the number of different configurations of atoms possible when the crystal is in a particular state, then the entropy in that state is $S = k \log w$, with Boltzmann's constant $k = 0{\cdot}86 \times 10^{-4}\,\text{eV degC}^{-1}$.

Suppose one has a crystal with n_r defects of a particular type and N_r sites available for them. The increase in free energy of the crystal, due to their presence, is:

$$F_f^r = n_r\, U_f^r - TS_r \qquad (2.1)$$

U_f^r is the amount by which the internal energy increases when such a defect is introduced and is referred to as the *energy of formation*. S_r is the change in total entropy associated with the introduction of the n_r defects and may be estimated as follows.

For one defect there are N_r sites available and hence N_r possible configurations in the one-defect state of the system. For n_r defects there are N_r for the first, $(N_r - 1)$ for the second, $(N_r - 2)$ for the third, and so on until $(N_r - n_r + 1)$ for the n_rth. This leads to $N(N-1)(N-2)\ldots(N_r - n_r + 1)$ configurations in all. But these cannot all be counted as genuinely different configurations, because the individual defects are indistinguishable from each other and the number above allows for the $n_r!$ ways of distributing n_r defects amongst the n_r chosen sites of a particular configuration. Hence the number of different configurations is:

$$\left.\begin{aligned} w &= \frac{N_r(N_r-1)(N_r-2)\ldots(N_r-n_r+1)}{n_r!} \\[1ex] \text{or}\qquad w &= \frac{N_r!}{n_r!\,(N_r-n_r)!} \end{aligned}\right\} \qquad (2.2)$$

then the entropy is:

$$k[\log N_r! - \log n_r! - \log(N_r - n_r)!]$$

and with Stirling's approximation $\log x! \simeq x \log x$ for large x this becomes

$$k[N_r \log N_r - n_r \log n_r - (N_r - n_r)\log(N_r - n_r)]. \qquad (2.3)$$

In addition to this contribution to S_r one must take account of the alteration of the vibrational disorder, or entropy, brought about by the presence of defects. In the Einstein model of lattice vibration, the atoms are represented as $3N$ independent linear harmonic oscillators and the entropy associated with each atom is:

$$3k \log \left\{ \frac{kT}{h\nu_E} \right\} \qquad (2.4)$$

where ν_E is the natural frequency of the oscillators, given in terms of the Einstein characteristic temperature Θ_E by $h\nu_E = k\Theta_E$. The factor 3 arises from the three degrees of vibrational freedom (see Dekker, 1958).

Suppose each defect affects the vibration of z neighbouring atoms, changing their frequency, on average, to ν_r. Then the entropy of these z atoms will be:

$$3kz \log \left\{ \frac{kT}{h\nu_r} \right\} = 3kz \left[\log \left\{ \frac{kT}{h\nu_E} \right\} + \log \frac{\nu_E}{\nu_r} \right].$$

But the original entropy of these atoms was $3kz \log \{kT/h\nu_E\}$, hence the change in entropy due to the defect is given by the second term. For n_r defects the total change in entropy due to changes in frequency is

$$3n_r kz \log \frac{\nu_E}{\nu_r}. \qquad (2.5)$$

Then taking both components of entropy change, (2.3) and (2.5), and putting these in the free energy expression (2.1) one has

$$F_f^r = n_r U_f^r - kT[N_r \log N_r - n_r \log n_r \\ - (N_r - n_r) \log (N_r - n_r) + n_r \log \left\{ \frac{\nu_E}{\nu_r} \right\}^{3z}]. \qquad (2.6)$$

In equilibrium n_r will satisfy $(dF_f^r/dn_r) = 0$ and applying this to (2.6) leads to

$$\frac{U_f^r}{kT} = \log \left[\frac{N_r - n_r}{n_r} \left\{ \frac{\nu_E}{\nu_r} \right\}^{3z} \right].$$

Assuming $N_r \gg n_r$ and putting the concentration C_r for n_r/N_r one has:

$$C_r = \left\{ \frac{\nu_E}{\nu_r} \right\}^{3z} \exp \left\{ -\frac{U_f^r}{kT} \right\}. \qquad (2.7)$$

Because U_f^r appears in the exponent the concentration is very sensitive to this quantity. For instance if one postulates a defect with $U_f^r = 1$ eV and another with $U_f^r = 10$ eV and calculates their respective concentrations at 1000 °K, assuming $\nu_E = \nu_r$, one finds

10^{-5} for the 1 eV case and 10^{-50} in the other. This shows clearly that the equilibrium concentration of defects having formation energies of more than a few eV can generally be neglected. The effect of temperature is equally severe.

The influence of the factor $(\nu_E/\nu_r)^{3z}$ with a vibrational origin, is relatively small. The percentage change in frequency is unlikely to be more than 10 % and with $z = 10$ one has a factor 17·5 or 0·06 according to whether the frequency decreases or increases. In general one expects an increase, and hence a factor less than unity, if the defect causes neighbouring atoms to occupy smaller spaces. Conversely one expects a decrease and a factor greater than unity if the neighbouring atoms occupy greater spaces.

One can derive (2.7) in a rather different form by introducing the *entropy of formation* ΔS_f^r, which refers only to the entropy change due to effects on the vibrational frequencies. Writing this in (2.6) instead of $3n_r kz \log(\nu_r/\nu_E)$ one obtains

$$C_r = \exp\left\{+\frac{\Delta S_f^r}{k}\right\} \exp\left\{-\frac{U_f^r}{kT}\right\} \qquad (2.8)$$

which is a rather more general form in which no assumptions are made regarding the vibrational model. The factor $\exp\{\Delta S_f^r/k\}$, easily identified with $(\nu_E/\nu_r)^{3z}$, is often referred to as the *entropy factor*. We have seen above how to estimate its magnitude. Experiments which measure the equilibrium concentration of a defect generally yield a quantity known as the *free energy of formation* defined as $(U_f^r - T\Delta S_f^r)$. Note that this is not the difference in free energy between the crystal with and without the defect, since ΔS_f^r is only a part of the entropy change.

Another point worth emphasizing is that one could not expect to produce a given defect by supplying exactly the formation energy from a collision event during irradiation. Under such violent conditions considerably more energy would be required, the excess being dissipated irreversibly to lattice vibrations, etc. Thus, although the formation energy of an interstitial-vacancy pair in Cu is about 5 eV the mean displacement energy is over 20 eV.

2.2.2. *The approach to equilibrium.* Radiation damage consists of defects in a structure introduced at concentrations in excess of the equilibrium values. In order to reduce their concentration the defects must either migrate from site to site in search of a sink at which they are removed or modified, or else they must undergo an internal change of configuration by which they disappear in a single stage. As a simple example of the first process one could take an interstitial atom migrating in jumps from one site to another until it reaches the surface where it is absorbed into a natural step. The second process is typified by a closely spaced interstitial and vacancy. By a single jump of either the vacancy or the interstitial mutual annihilation occurs. The difference between the two cases is not fundamental and lies simply in the number of jumps.

If one were to move a defect very slowly between two neighbouring sites, the free energy of the crystal would rise to a maximum as some intermediate configuration was reached and then return to its original value. The excess energy at the maximum is the free energy of activation for migration: F_m^r. Under normal conditions this can only be supplied by the crystal by random fluctuations in the local density of vibrational energy. Assuming this energy to be distributed according to a Maxwell–Boltzmann law, the probability of an excess free energy F_m^r being available for a defect during a particular period of oscillation is $\exp(-F_m^r/kT)$. If the oscillation frequency of the defect is ν, given in order of magnitude by ν_E, then in an interval dt there are $\nu\,dt$ periods, and the number of jumps dn is $\nu \exp(-F_m/kT)\,dt$. Hence the rate of jumping is

$$\frac{dn}{dt} = \nu \exp(-F_m^r/kT). \qquad (2.9)$$

From the calculations in the previous section it will be clear that significant rates are only possible when F_m is less than a few eV, and that migration rates are strongly dependent on temperature.

The free energy F_m^r may be expressed as:

$$F_m^r = U_m^r - T\Delta S_m^r. \qquad (2.10)$$

U_m^r is easily identified as the difference in potential energy between the crystal with a defect in a stable site and one with the defect in the intermediate configuration between sites. ΔS_m^r is the difference in entropy between these two states and, as in the case of formation entropy, is due to the different effect that the two configurations have on the lattice vibrations.

It is generally more difficult to calculate ΔS_m^r than U_m^r and one often finds equation (2.9) written

$$\left.\begin{array}{c} \dfrac{dn}{dt} = A \exp\left(-U_m^r/kT\right) \\[2mm] A = \nu \exp\left(\Delta S_m^r/k\right). \end{array}\right\} \qquad (2.11)$$

with

The uncertainty in ν and S_m^r is then lumped together, allowing A to be treated as an empirical constant, often referred to as the *frequency factor*, which one expects to be approximately independent of temperature.

The rate at which a damaged crystal approaches equilibrium, or at least an intermediate state with lower free energy, will depend on the rate of jumping and on the number of jumps j required to reach a sink. Suppose one has an excess concentration C of a single type of defect which disappears completely by migrating to sinks, and assume that the number of sinks greatly exceeds the number of defects. In the interval dt, during which C changes by $-dC$, each defect makes $\nu \exp\left(-F_m^r/kt\right) dt$ jumps and its chance of reaching a sink is $\nu \exp\left(-F_m^r/kT\right) dt/j$.

Hence

$$dC = -\frac{C\nu}{j} \exp\left(-F_m^r/kT\right) dt \qquad (2.12)$$

which leads to:

$$\left.\begin{array}{c} C = C_0 \exp\left(-t/\tau\right), \\[2mm] 1/\tau = \nu \exp\left(-F_m^r/kT\right)/j. \end{array}\right\} \qquad (2.13)$$

with

These equations are familiar in the theory of reaction rates and represent a simple *first-order* reaction. They show clearly how increasing temperature strongly enhances the rate of approach to equilibrium.

If in (2.12) the concentration had appeared to the power x, we should refer to an *xth order* reaction. An example of a second-order reaction would be one in which two similar defects coalesced on meeting. Then $j \propto C^{-1}$ making $dC \propto C^2$. In radiation damage the reactions are often too complex for the rate to be proportional to a simple power of the concentration C. A more general form of (2.12) is then useful:

$$\frac{dC}{dt} = F(C) \exp\left(-U_m^r/kT\right) \qquad (2.14)$$

with $F(C)$ some undetermined function of C which also includes the factor $\nu \exp\left(\Delta S_m^r/k\right)$.

2.3. Vacancies

2.3.1. *The bond model.* It is common to classify defects by their geometrical symmetry. In §§ 2.3, 2.4 and 2.5 we consider the *point* defects: *interstitials, vacancies* and *impurity atoms*. The simplest of all defects is the vacancy (vacant lattice site). A rough assessment of its formation energy U_f^v is made by supposing an atom to be removed from the interior of the crystal, where it is surrounded by z neighbours, replacing it on the surface, and calculating the net expenditure of energy. Using the concept of interatomic bonds, which is best justified for covalent crystals, one breaks z bonds in removing the atom and restores an average of $\frac{1}{2}z$ in replacing it. Then if the energy per bond is w, the work done $U_f^v \simeq \frac{1}{2}zw$.

To determine w; the energy required to break all the bonds in a crystal of N atoms is NL_s, where L_s is the latent heat of sublimation per atom. But there are z bonds to each atom, each shared by 2 atoms, hence there are $\frac{1}{2}zN$ bonds in the crystal and the energy to break them all is $\frac{1}{2}zwN$.

Hence
$$L_s = \tfrac{1}{2}zw$$

and
$$U_f^v \simeq L_s. \tag{2.15}$$

L_s in typical solids is a few electron volts but this must give an overestimate of U_f^v because we have neglected the energy that will be restored if the broken bonds around the vacant site regroup amongst themselves. For instance if z is an even number and they linked up in pairs, releasing a bond energy of, say, $\frac{1}{2}w$ per pair, then $\frac{1}{4}zw = \frac{1}{2}L_s$ would be released. In such a case the formation energy would be $\frac{1}{2}L_s$. The bond picture really serves to provide an upper limit to U_f^v and its use is best justified for covalent crystals. See table 2.3.

It is instructive to treat small clusters of vacancies by this method. The simplest is a neighbouring pair of vacant sites; the *divacancy*. Suppose one starts with two isolated vacancies with a total of $2z$ broken bonds. When these occupy adjacent sites only $(2z - 1)$ bonds are removed.

Hence
$$U_f^{2v} = (2z - 1)w - 2 \times \tfrac{1}{2}zw$$

and it follows that
$$U_f^{2v} = 2U_f^v(1 - 1/z). \tag{2.16}$$

A useful concept is that of a binding energy U_b between defects, this being the energy required to separate the cluster into its constituent defects. In the case of the divacancy:

$$U_f^{2v} = 2U_f^v - U_b^{vv}. \tag{2.17}$$

By comparing this with (2.16) one obtains

$$U_b^{vv} = 2U_f^v/z. \tag{2.18}$$

Since z ranges from 4 in the diamond structure to 12 in f.c.c., U_b^{vv} could range from 50 to 10 % of U_f^v. One might therefore expect that $U_b^{vv} < U_f^v$ with the difference being greatest in closely packed structures.

2.3.2. *A cavity in a rigid continuum.* In metals the valence electrons are not localized in bonds and an alternative treatment must be found. A model which regards the crystal as a continuous rigid medium has been found to give surprisingly good results and has the advantage that it is easy to extend to a cluster of vacancies in the form of a void (see §8.2.1). Suppose one creates a small cavity with volume $\Omega = \frac{4}{3}\pi r_a^3$ equal to that occupied by one atom in the crystal, Ω and r_a being the *atomic volume* and *radius*, by definition. The material removed is spread uniformly over the external surface of the crystal. If we take a spherical crystal of original radius R which increases to $R + \Delta R$, then since the medium is rigid volume is conserved and:

$$\left. \begin{array}{l} 4\pi R^2 \Delta R = \frac{4}{3}\pi r_a^3 \\ \Delta R = r_a^3/3R^2. \end{array} \right\} \tag{2.19}$$

and

In this model the formation energy U_f^v is the difference in surface energy of the crystal with and without the cavity, and if γ is the surface energy per unit area it is simple to show that

$$U_f^v \simeq 4\pi r_a^2 \gamma(1 + 2r_a/3R)$$

neglecting terms in $(\Delta R)^2$ and using equation (2.19).

Then since $r_a \ll R$

$$U_f^v = 4\pi r_a^2 \gamma. \tag{2.20}$$

In most metals $\gamma \sim 0.1$ eV.Å^{-2} and $r_a \sim 1.5$ Å, hence one expects $U_f^v \sim 2$ eV. This will be something of an overestimate since by postulating rigidity we have neglected the tendency of the cavity to collapse under surface tension, which will be resisted by elastic strains generated in the surrounding medium. In this relaxed condition the potential energy of the defective crystal is reduced, and from the above energy balance U_f^v will emerge smaller. We go on to consider this in the next section.

Expressing (2.20) in terms of Ω rather than r_a gives:

$$U_f^v = (6\sqrt{\pi\Omega})^{\frac{2}{3}} \gamma.$$

If one takes a spherical void composed of n vacancies the volume Ω in this expression is simply replaced by $n\Omega$ and it follows that

$$U_f^{nv} = n^{\frac{2}{3}} U_f^v \qquad (2.21)$$

and the binding energy,

$$U_b^{nv} = n U_f^v (1 - n^{-\frac{1}{3}}). \qquad (2.22)$$

For the divacancy $n = 2$, $U_f^{2v} = 1\cdot6 U_f^v$, and $U_b^{2v} = 0\cdot4 U_f^v$, which is in general agreement with the results of the bond picture.

2.3.3. *The elastic continuum model.* Suppose we have an isotropic elastic medium, of infinite extent, with a centre of dilatation at the origin. That is to say that at some small radius r_0 a radial displacement δ_0 of the medium has occurred. At a greater radius r this displacement is reduced to $\delta(r)$.

Then our vacancy will be a cavity of radius r_0, the surface tension (if positive) causing an inward (negative) displacement $-\delta_0$. The model is particularly convenient since the interstitial atom or oversize impurity atom can be represented, at least qualitatively, by a positive dilatation.

We first determine the function $\delta(r)$. A spherical surface at r moves to $r + \delta(r)$, hence its circumference changes by $2\pi\delta(r)$. The strain around the circumference, or *tangential strain*, is thus $+2\pi\delta(r)/2\pi r$ or $\delta(r)/r$.

A second surface at $(r+a)$, where $a \ll r$, moves by a distance $\delta(r+a)$ given by

$$\delta(r+a) = \delta(r) + a\frac{d}{dr}.\delta(r).$$

Thus the separation between the two surfaces changes from a to $[a + a(d/dr).\delta(r)]$. Hence there is a *radial strain* of $(d/dr).\delta(r)$.

For a first approximation we assume that any element of the medium suffers only a change of shape without any change of volume. Then an element which, before dilatation is an approximate cube ($a \times a \times a$) contained between our spherical surfaces at r and ($r + a$), changes its shape to a parallelepiped as shown in fig. 1.

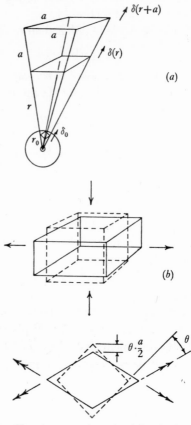

Fig. 1. Showing how dilatation around a point defect leads to a distortion of the volume element (b). In (c) this distortion is seen to be equivalent to a shear strain since no change in volume is involved, only one of shape.

For the volume change to be zero the two tangential strains must exactly cancel the single radial strain, i.e.

$$(\text{radial strain}) = -2 \, (\text{tangential strain})$$

or

$$\frac{\mathrm{d}}{\mathrm{d}r} \cdot \delta(r) = -\frac{2\delta(r)}{r},$$

with

$$\delta(r_0) = \delta_0. \qquad (2.23)$$

Solving this differential equation with its boundary condition gives

$$\delta(r) = (r_0/r)^2 \delta_0. \tag{2.24}$$

The sign of $(\mathrm{d}/\mathrm{d}r).\delta(r)$ then indicates that for positive δ_0 (expansion) the medium is compressed along the radius and hence extended tangentially, as illustrated in fig. 1.

In more sophisticated terms this distortion of the volume element without change of volume can be expressed:

$$\mathbf{div}\,\boldsymbol{\delta}(\mathbf{r}) = 0.$$

Then since the displacement vector $\boldsymbol{\delta}(\mathbf{r})$ is always directed along the position vector $\mathbf{r}$ and angular terms in $\mathbf{div}$ disappear giving

$$\frac{1}{r^2}\frac{\mathrm{d}}{\mathrm{d}r}(r^2\delta(r)) = 0$$

which leads again to the result above.

This distortion with no net volume change is best represented as a pure shear strain θ, as illustrated in fig. 1(c). From this it is clear that θ is equivalent to a linear strain $+\theta$ in one direction and $-\theta$ in the perpendicular one.

We can now think of the distortion of our cube $a \times a \times a$ as two simultaneous shearing strains of magnitude $r_0^2\delta_0/r^3$. Each produces a radial strain of $-r_0^2\delta_0/r^3$ which add up to give $-2r_0^2\delta_0/r^3$, and each produces a separate component of tangential extension $r_0^2\delta_0/r^3$.

Now W the energy stored per unit volume due to shear strain θ is given by

$$W = \tfrac{1}{2}\mu\theta^2, \tag{2.25}$$

where μ is the shear modulus of elasticity. Hence with $\theta = r_0^2\delta_0/r^3$ for each component the stored energy per unit volume at radius r is

$$W(r) = \tfrac{1}{2}\mu(2r_0^2\delta_0/r^3)^2 \tag{2.26}$$

the factor 2 appearing because there are two components of shear strain. The total stored elastic energy is

$$U_{\text{elastic}} = \int_{r_0}^{\infty} 4\pi r^2 W(r)\,\mathrm{d}r$$

and evaluating this with $W(r)$ given by (2.26) gives:

$$U_{\text{elastic}} \simeq \tfrac{8}{3}\pi\mu r_0\,\delta_0^2. \tag{2.27}$$

This is sometimes referred to simply as the *strain energy*.

Now if we represent the crystal with a vacancy in it by an elastic medium with a cavity of radius $r_0 + \delta_0$ the increase of internal energy is:

$$U = 4\pi(r_0 + \delta_0)^2 \gamma + \tfrac{8}{3}\pi\mu r_0 \delta_0^2.$$

At absolute zero, where the term TS in the free energy disappears, we can minimize U with respect to δ_0 in order to find the equilibrium dilatation. Then differentiating with respect to δ_0

$$\frac{\partial U}{\partial \delta_0} = 8\pi r_0^2 \gamma \left[1 + \frac{\delta_0}{r_0} \frac{(1 + 2\mu r_0)}{3\gamma} \right].$$

Equating this to zero should give the equilibrium situation where free energy is a minimum. Hence one finds:

$$\delta_0 = - \frac{r_0}{1 + \dfrac{2\mu r_0}{3\gamma}}. \tag{2.28}$$

As expected, displacement is inwards when γ is positive. Let us now examine the magnitudes, for a typical metal with $\gamma \sim 0 \cdot 1$ eV/Å², $\mu \sim 1$ eV/Å³ and $r_0 \sim 1$ Å

$$\delta_0 = - \frac{1}{1 + 7} \sim -0 \cdot 1 \text{ Å}.$$

It is clear that the first term in the denominator of (2.28) can be omitted without serious error, and the approximation gets better for larger values of r_0, i.e.

$$\delta_0 \simeq - \frac{3\gamma}{2\mu}. \tag{2.29}$$

This is a most interesting result for it is independent of r_0 and shows that the fractional contraction of a cavity under surface tension forces gets less in proportion to $1/r_0$ as r_0 increases. Thus one expects the contraction around a spherical vacancy cluster to be less as the cluster grows. This may be put in a slightly different way, the magnitude of the *strain field* around a cavity decreases with its size.

The formation energy of a vacancy is now obtained by inserting the equilibrium value of δ_0 from (2.29) into the expression for U and neglecting small terms, i.e.

$$U_f^{\text{v}} = 4\pi r_0^2 \gamma - 12\pi r_0 \frac{\gamma^2}{\mu} + 6\pi r_0 \frac{\gamma^2}{\mu}.$$

The first term is the surface energy of the undilated cavity.

The second is the reduction in surface energy due to contraction by the surface tension and the third is the elastic energy stored in the solid. Note that the third term is always positive no matter what sign δ_0 takes, since it depends on δ_0^2. The relative magnitudes of the terms for a vacancy can be seen by taking the values above, whence

$$U_f^v \simeq 1{\cdot}2 - 0{\cdot}4 + 0{\cdot}2 \,\mathrm{eV}.$$

It is now clear why the rigid continuum model succeeds for cavities, the decrease in surface energy due to inward relaxation is partly offset by an increase in elastic energy and

$$U_f^v = 4\pi r_0^2 \gamma - 6\pi r_0 \frac{\gamma^2}{\mu}. \tag{2.30a}$$

It is also possible to estimate the vacancy migration energy using the above model but neglecting the relaxation term. We identify U_m^v with the increase in internal energy when an atom is mid-way between two vacant sites, as illustrated in figure $3(b)$ (see p. 21). In this position we suppose that there are two cavities of radius r_1 whose total volume equals that of the original vacancy, i.e.

$$2r_1^3 = r_0^3.$$

Then
$$\left.\begin{aligned}
U_m^v &= 2 \times 4\pi r_1^2 \gamma - 4\pi r_0^2 \gamma \\
&= 4\pi r_0^2 \gamma (2^{\frac{1}{3}} - 1), \\
U_m^v &= 0{\cdot}3 U_f^v.
\end{aligned}\right\} \tag{2.30}$$

For vacancies the main value of the elastic continuum model is in illustrating the relaxation processes that occur around the defect. In the case of the interstitials however we shall see that the dilatations are large enough to make the elastic energy dominate the formation energy.

2.3.4. *Configuration and energies in an atomic model.* Further improvements in the calculation of U_f^v in metals can only come from an atomic model, which has the added advantage that one also obtains information about the disposition of atoms around the defect. Furthermore, it is possible to calculate U_m^v, the migration energy, by calculating the energy of the crystal when the defect is in its intermediate configuration between two stable sites. One may also adapt the calculations to interstitial atoms. Of course, the technique demands a detailed knowledge of interatomic forces and the electronic behaviour and for most crystals neither are accurately known.

It is well known from the theory of metallic cohesion that in the noble metals Cu, Ag and Au the main forces holding the atoms apart are those of closed shell repulsion, and these contribute a large term to the cohesive energy (see Mott & Jones, 1937). In alkali metals this is not the case and electrostatic terms, from the interactions involving positive ion cores and free electrons, are far more important. Thus in these metals one cannot use the picture of atoms held apart by forces between neighbouring pairs of atoms. This is demonstrated by the b.c.c. structure of alkali metals, which cannot be generated by packing smooth spheres into a box, whereas f.c.c. can. Under these conditions it is not obvious that the vacancy configuration should be a vacant site with neighbouring atoms relaxed slightly inwards, and some collapsed configuration is entirely possible. The b.c.c. and h.c.p. transition metals are not well understood, but as in the noble metals closed-shell repulsion must contribute an important term to the cohesive energy.

It is possible to observe individual atomic sites on the surface of some metals by the *ion emission microscope*. This technique will be described in Chapter 6 and in fig. 106 where vacant sites are seen on the surface of a W crystal after α-particle bombardment. Their appearance suggests that in W the surface vacancy is not a collapsed configuration and relaxations of neighbours are indeed small. One might reasonably infer that if this is the case on the surface, the internal vacancy should also be an open configuration.

It is only in the case of Cu, Ag, Au and Fe that much progress has been made in calculating defect configurations. Cu is the one most extensively studied and for which the assumption of a central repulsive force is most easily supported by the observed elastic constants. The treatment below will illustrate the method developed by Huntington & Seitz (1942 *a*, *b*), Eshelby (1954), Tewordt (1958), Seeger & Mann (1960), Johnson & Brown (1962), Huntington (1953) and others.

The crystal is assumed to be at absolute zero of temperature. Zero-point motion can be neglected, because quantum rules forbid any changes that involve energy being taken from such motion. One supposes an ion core to be removed from the interior and replaced on the surface. The formation energy U_f^v is then the net energy supplied in so doing. Two separate contributions can be identified; that due to the change in total repulsive energy, when atoms surrounding the vacancy relax inwards, and that due to the changed energy of the free electrons.

The first is obtained by finding the new equilibrium position of each atom. Those close to the vacancy are treated individually, assuming a two-body closed-shell repulsive potential of the Born–Mayer form (see Chapter 4).

$$V(r) = A \exp(-r/b).$$

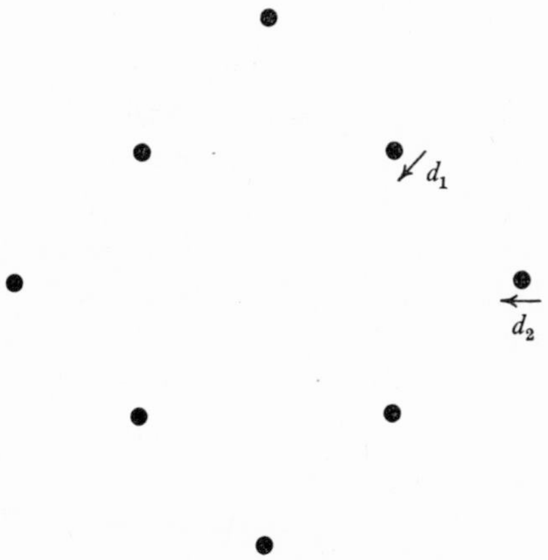

Fig. 2. Atomic positions around a vacancy in Cu, after Tewordt (1958). See table I for values of d_1 and d_2.

More distant atoms are treated by replacing the crystal, in the calculation, with an elastic continuum containing a centre of contraction. The inward displacement in the two regions is made to coincide at some suitable interface where elastic conditions can be shown to prevail (i.e. force $\propto$ displacement). A variational procedure is then carried through to determine the set of inner displacements that results in the minimum energy; this set is taken as the equilibrium configuration.

For Cu a potential with $A = 2\cdot2 \times 10^4\,\mathrm{eV}$ and $b = D/13$ has been very successful in making predictions agree with observations. Tewordt's results are given in fig. 2 and table I.

It will be seen that the relaxations are small and lead to a very

TABLE I

$\dfrac{d_1}{a}$	$\dfrac{d_2}{a}$	$\dfrac{\Delta v}{\Omega}$	Electronic energy (eV)	Repulsive energy (eV)	U_f^v (eV)
0·016	0·001	+0·47	0·9	> −0·1	0·9

small (< 0.1 eV) decrease in the repulsive energy of the crystal. This bears out the prediction of the elastic continuum model that only a small fraction of the formation energy is associated with dilatation, although in the atomic model we find a negative sign. The electronic energy can be identified with surface energy.

This change in energy due to redistribution of the valence electrons has been estimated in the following way. Because the crystal has increased in size the same number of electrons occupy a slightly larger volume and their energy is slightly reduced according to the free electron theory of metals. However, the region around the vacancy has an excess negative charge since an ion core has been removed. Electrons in this locality, being repelled, have a higher total energy and it is this that corresponds to surface energy. Taken together, the two effects give a net increase of 0·9 eV. Since this is at least ten times greater than the change in repulsive energy, it approximates to the formation energy U_f^v.

In order for a vacancy to migrate, one of the neighbouring atoms must jump into it, making the neighbouring site vacant. The process is illustrated in fig. 3. The migration energy U_m^v is easily identified as the energy of the crystal when the atom is at the saddle point between two sites. For Cu many authors have made the calculation, obtaining values between 0·4 and 1·4 eV. The majority of the energy comes from increased closed-shell repulsion as the jumping atom comes into closer proximity to its neighbours. In table II (see p. 37) a value of 1·0 eV is quoted to summarize the situation.

In the case of the divacancy, calculations suggest a lower migration energy than for the single vacancy (Lomer 1959; Bartlett & Dienes, 1953). This is intuitively reasonable since the presence of the second vacancy makes more room for the jumping atom to displace its neighbours. Table II contains a summary of calculations.

Some vacancy calculations for alkali metals have been made by Fumi (1955) who gives rough estimates of U_f^v as Li: 0·55 eV, Na: 0·53 eV, K: 0·36 eV, Rb: 0·31 eV, Cs: 0·26 eV.

A very different method of calculation has been employed by Vineyard and his colleagues at Brookhaven National Laboratory (Gibson, Goland, Milgram & Vineyard, 1960) which is applicable

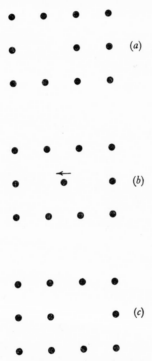

Fig. 3. The migration of a vacancy.

to a wide range of problems in radiation damage. Crystal behaviour is directly simulated by suitably programming a digital computer to solve the simultaneous equations of motion of up to 1000 atoms.

The basic theoretical model takes a small crystallite, built from these atoms, in the form of a rectangular parallelepiped which is supposed to be a part of an infinite crystal of the same structure. The atoms interact with one another according to a potential between pairs of the form $V(r)$, which is a function of separation only. A distributed energy, which depends on the volume of the crystal, is also required to simulate the cohesive effect of electrons in a noble metal according to the Wigner–Seitz model. This is represented by a constant inward force on the outermost atoms of the crystallite, (i.e. a hydrostatic pressure). These boundary atoms are also acted on by Hooke's-law restoring forces and a viscous drag force, proportional to velocity. These are required to simulate the effect of the surrounding crystal in which the crystallite is embedded. Fortunately, it turns out that the results of the calculation are insensitive to the choice of the boundary forces. The only thing that matters is $V(r)$.

In equilibrium the hydrostatic pressure balances the repulsive force between atoms. Each atom is assigned co-ordinates of position and velocity, an equation of motion is set up for each one, and the computer programmed to solve all the equations of motion simultaneously. An iterative method is used which, given a starting configuration at $t = 0$, calculates the subsequent configurations at a series of small time intervals.

Suppose one atom is missing and all the rest are on their lattice sites at $t = 0$. The calculation will follow the relaxation processes that eventually result in the equilibrium vacancy configuration. Alternatively an extra atom can be introduced and the interstitial configuration found. Later on, we shall see how the method has also been applied to follow high-energy collision events with phenomenal success.

The correct choice of $V(r)$ is, of course, vital to the whole technique, and the first requirement is that the chosen lattice shall be stable. Provided $V(r)$ is repulsive (always positive) and extends only to nearest neighbours a closely packed lattice, either f.c.c. or h.c.p. should be stable. The more open b.c.c. lattice is not so easy, and requires that second nearest neighbours shall be affected by $V(r)$, and possibly that $V(r)$ becomes negative (attractive force) near the equilibrium separation. Next, we require that the potential used shall be compatible with the observed elastic moduli, according to the well-known relations. But this is not enough, since a realistic $V(r)$ is characterized by at least two parameters (e.g. A and b in the Born–Mayer form: $A \exp(-r/b)$). Hence one really requires two experimentally known quantities of a different type to fit predictions to. Fortunately the threshold energy for producing radiation damage can be found experimentally and can also be predicted by this model, as we shall see in §5.5.2. On this basis suitable potentials were arrived at for f.c.c. Cu and b.c.c. Fe. In the case of Cu the following Born–Mayer function gave good results:

$$V(r) = 2 \cdot 2 \times 10^4 \exp\left(-13r/D\right) \mathrm{eV} \qquad (2.31)$$

(D is nearest neighbour spacing).

For Fe, the potential cannot be expressed as a simple function and the reader is referred to the original paper for detailed information (Erginsoy, Vineyard & Englert, 1964). For closer separations than $D/2$ it is approximately of the Born–Mayer form, but for both first and second nearest neighbours it is negative by a few tenths of an eV with a minimum value midway between.

One can only be impressed by the elegance of this method but it is important to realize its limitations at the outset, for although it will be quoted extensively throughout this book it may not be based on a good theoretical model in some circumstances. The trouble lies in the behaviour of valence electrons. The inner closed shells probably repel one another with a central force expressible as a positive $V(r)$ and in the perfect crystal the use of such a potential, possibly with a negative tail, and a hydrostatic cohesive pressure probably represents a metal rather well. But when a defect is present the valence electrons must be redistributed locally and changes should be made in $V(r)$ for the atoms in this vicinity, certainly in the tail near the nearest neighbour separation. At the present time *no* theoretical model accounts for this situation adequately and the computer method at least has the advantage of being an accurate solution of the *mathematical* problem. The method is really at its best when dealing with high energy collision events, where the tail of $V(r)$ and the cohesive forces are quite unimportant.

For the simple vacancy in Cu the computer model finds a slightly different configuration to that in fig. 2 and table I. The nearest neighbours relax inwards with $d_1 = 0 \cdot 016a$ as in other calculations, but the second nearest atoms move *outwards* by a very small amount. When the first neighbours move inwards they in fact move *closer* to the site of second neighbours and this leads to the effect. It has an important consequence, for it shows the importance of the anisotropy of elastic effects in a lattice and calls into question an elastic continuum model which assumes isotropy of the medium. We are almost led back to the cavity in a rigid continuum as the best model for a vacancy!

2.3.5. *Changes in volume and lattice parameter.* The formation of a vacancy must cause a change in volume of the crystal. First there is an increase, roughly equal to the volume Ω per lattice site of the crystal, due to the replacement of the removed atom on the surface. Then there may be an overall contraction or expansion if there is a dilatation around the vacancy. We have seen that in an infinite and isotropic elastic medium the displacement at distance r from a centre of dilatation is

$$\delta(r) = (r_0/r)^2 \delta_0.$$

If one takes a spherical reference surface at radius R from the centre, the effect of the dilatation is to move a volume $4\pi R^2 \delta(R)$ across this surface. One might expect that this would be the same as the volume change of a finite spherical region of radius R, leading, with (2.23), to the result: $4\pi r_0^2 \delta(r_0)$. But in a finite region there is an image effect associated with the outer surface, analogous to the image charges that are introduced into the theory of electrostatics, and Eshelby (1954) has shown this to lead to a volume change:

$$\Delta v' = 4\pi r_0^2 \delta_0 \frac{3(1-\sigma)}{(1+\sigma)} \tag{2.32}$$

with σ = Poisson's ratio.

Suppose our crystal behaves like a finite region of elastic continuum. If a vacancy is formed, introducing a centre of dilatation with negative $\delta(r_0)$, there will be an initial negative change in volume $\Delta v'$, before the material from the vacancy is replaced on the surface. The magnitude of the effect may be found by putting $r = a/\sqrt{2}$ and $\delta(r_0) = -0.016a$ from table I. Then since $\Omega = a^3/4$ in a f.c.c. lattice we have $\Delta v' = -0.2\Omega$. This procedure cannot strictly be justified since at r_0 the displacement is too large for the crystal to behave elastically.

In a crystal of N atoms and V vacant sites the vacancy concentration is defined as $C_v = V/(N+V) = V/N$ for $V \ll N$. If one neglects interaction between vacancies, and Eshelby finds this to be reasonable for small concentrations, one can sum the contribution $\Delta v'$ from each vacancy to the initial displacement of outer surface. Then the crystal which would have occupied a volume $(N+V)\Omega$ in the absence of the elastic displacement will occupy a volume

$$(N+V)\Omega + V\Delta v'.$$

Thus the new volume per lattice site is given by

$$\left.\begin{array}{c} \Omega + \Delta\Omega = \Omega + V\Delta v'/(N+V), \\ \Delta\Omega = C_v\Delta v'. \end{array}\right\} \quad (2.32)$$

or

In an X-ray diffraction experiment one measures an average spacing between atomic planes, giving the X-ray lattice constants. These are simply related to $\Omega(\Omega = a^3/4$ in f.c.c.) and a change in Ω will cause a change in X-ray lattice constant given by:

$$\frac{\Delta\Omega}{\Omega} = \frac{3\Delta a}{a}.$$

Thus

$$\frac{\Delta a}{a} = \frac{1}{3}C_v\frac{\Delta v'}{\Omega}. \quad (2.34)$$

With $\Delta v' = -0.2\Omega$ and 1 % of vacant sites $(C_v = 10^{-2})$ one has $\Delta a/a = -7 \times 10^{-4}$ which is rather a small effect since the limit of detection in typical experiments is $|\Delta a/a| \sim 10^{-5}$.

The overall change in volume when a vacancy is formed by removing an atom from an internal site and replacing it on the surface is $(\Delta v' + \Omega)$. A rough value for Cu would therefore be $+0.8\Omega$.

For V vacancies the change is:

$$\Delta v = V(\Delta v' + \Omega).$$

The original volume v is $N\Omega$, hence:

$$\frac{\Delta v}{v} = \left(1 + \frac{\Delta v'}{\Omega}\right)C_v \quad (2.35)$$

This is the volume change that might be observed by measuring the external dimensions of a crystal, or its density. The volume change is related to the length change Δl by

$$3\frac{\Delta l}{l} = \frac{\Delta v}{v},$$

hence

$$\frac{\Delta l}{l} = \frac{1}{3}\left(1 + \frac{\Delta v'}{\Omega}\right)C_v \quad (2.36)$$

2.3.6. *Measurements of the vacancy concentration in equilibrium.*
Subtracting equation (2.34) from (2.36) gives

$$\tfrac{1}{3}C_{\mathrm{v}} = \frac{\Delta l}{l} - \frac{\Delta a}{a}. \tag{2.37}$$

This suggested a method of measuring the equilibrium concentration of vacancies in a crystal by comparing changes in macroscopic length and X-ray lattice parameter. Feder & Nowick (1958) and

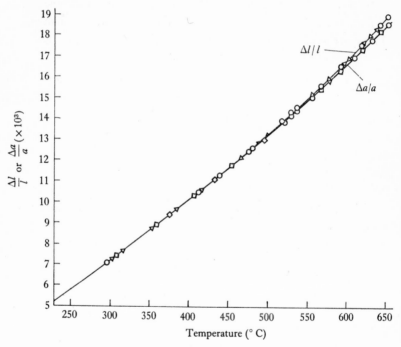

Fig. 4. Changes in length l and lattice parameter a as a function of T for Al. (From Simmons & Balluffi 1960*a*.)

Simmons & Balluffi (1962*a, b*) measured these quantities, to an accuracy of a few parts in 10^5, in single crystal samples of Al, Cu, Ag and Au held at various temperatures up to the melting point. If l and a are the length and lattice parameter at temperature T, and l_0 and a_0 the same quantities at 0 °C then a graph of either $(l-l_0)/l_0$ or $(a-a_0)/a_0$ *versus* temperature should show a steady increase due to thermal expansion. At low temperatures the two graphs behaved identically, but at high temperatures l increased faster than a, showing some form of porosity to be present. In fig. 4 their curves for Al are reproduced. Assuming the porosity to be due to an equilibrium concentration of vacancies, (2.37) leads one directly to C_v as a function of T. Then from (2.8) one deduces U_f^v from the slope of this function and ΔS_f^v from its magnitude.

Although the assumption that vacancies are present in the single state is a fair first approximation, one must allow for the possibility that some are present in small clusters of two, three or more. The estimates of U_f^{2v} in (2.16) and (2.21) suggest that $U_f^{2v} \sim 1\cdot7 U_f^v$, hence that $C_{2v} \sim (C_v)^{1\cdot7}$ and when $C_v \sim 10^{-4} C_{2v} \sim 10^{-7}$. A small correction is necessary to allow for this. Finally Simmons and Balluffi found the values of U_f^v in table II, and that the entropy ΔS_f^v lies between k and $2k$. Thus the entropy factor is between 3 and 8.

It is difficult to interpret these experiments by postulating any other defects than the vacancy and they are generally accepted as evidence that significant concentrations of interstitials are not produced by thermal activation in these f.c.c. metals. In the §2.4 we shall see this is confirmed by calculations of U_f^i for the interstitial which is several times greater than U_f^v.

2.3.7. *Electrical resistivity of point defects.* In metals, an important property of a lattice defect is the contribution it makes to electrical resistivity. Resistance to current is caused by the scattering of electrons from crystal imperfections such as phonons, impurity atoms or point defects. The resistivity ρ depends on the electrons' mean collision free path λ, and is given by

$$\rho = \frac{m_0 v_F}{nq\epsilon\lambda} \qquad (2.38)$$

(see Seitz, 1940). Where v_F is the velocity of a free electron at the top of the Fermi distribution, m_0 and ϵ are the rest mass and charge of the electron, n the density of atoms and q the number of free electrons per atom.

Suppose one has a pure metal containing some vacancies at a low enough temperature for electron-phonon scattering to be disregarded. There are nC_v vacancies per unit volume and if each has a total scattering cross-section σ the mean free path is

$$\left.\begin{array}{l} \lambda = (n\sigma C_{\mathrm{v}})^{-1} \\[2mm] \rho_0 = \dfrac{m_0 v_{\mathrm{F}} \sigma}{q e^2} C_{\mathrm{v}}. \end{array}\right\} \qquad (2.39)$$

and

The electron sees the vacancy as a local anomaly in the normal periodic potential within the crystal with radial dimensions of the order r_{a}. Then

$$\sigma \sim \pi r_{\mathrm{a}}^2.$$

Inserting this in (2.39) with typical numerical values for a noble metal one finds $\rho_0 \sim 3$ micro ohm cm for 1% of vacancies (i.e. $C_{\mathrm{v}} = 10^{-2}$). This is a very crude treatment and is included only to show the physical processes involved. Because the total cross-section has been used without considering the wave nature of the electron it overestimates the effect. More refined calculations are available for Cu which predict values in the range 1 to 2 micro ohm cm per 1% vacancies. The extent to which these agree with experiment will be seen in later sections and table II.

An important feature of the free electron model is seen when one considers the situation when several types of defect are present together. Let the rth type have a cross-section σ_r and a concentration C_r. Then the mean cross-section is $\Sigma_r C_r \sigma_r / \Sigma_r C_r$ and the total defect concentration $\Sigma_r C_r$. Putting these quantities for σ and C_{v} in equation (2.39) gives:

$$\rho_0 = \frac{m_0 v_{\mathrm{F}}}{e^2} \Sigma_r C_r \sigma_r. \qquad (2.40)$$

Thus the contributions of the individual defects are additive. Experiments show this to be approximately true for noble metals and the relationship expressed by (2.40) is generally referred to as Matthiessen's Law.

The characteristic increase of metallic resistivity with temperature is due to electron-phonon scattering, whose $C_r \sigma_r$ increases with temperature. From the simple treatment above one would expect scattering by point defects to be independent of temperature and this behaviour is generally observed, i.e.

$$\rho = \rho_0 + \rho(T).$$

By Matthiessen's Law one would expect $\rho(T)$ to be unaffected by defects, therefore the introduction of extra point defects should simply shift the resistivity *versus* temperature curve upwards by $\Delta\rho_0$ without changing its shape. At very low temperatures the resistivity of most metals approaches the residual value ρ_0 which is decided by the defect concentration. This is illustrated in fig. 5.

Resistivity is therefore a convenient index of defect concentration and has the added advantage of being relatively easy to measure with high precision. Note that this is only true of metals. In semiconductors the number of charge carriers (q in (2.39)) may depend on the concentration of defects. In many such cases the effect of radiation damage on q outweighs the effect on λ.

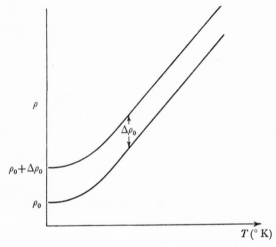

Fig. 5. The resistivity of a metal as a function of temperature.

2.3.8. *Quenching and annealing experiments.* In a *quenching* experiment a sample is heated to a steady temperature T_q, where it contains the equilibrium concentration of vacancies given by (2.8). It is then cooled rapidly at $\sim 10^4$ degC sec^{-1}, to a lower temperature T_0. Such a rate is sufficiently fast to prevent significant movement of the vacancies during the quench and if T_0 is low enough the equilibrium concentration of vacancies will be frozen into the sample.

Quenching experiments vary in detail but are broadly similar to the classical one of Bauerle & Koehler (1957) which will be described here. They heated a thin gold wire to T_q by passing a current through it. Quenching was by rapidly immersing the wire in a bath of water at temperature T_0. The cooling rate could be followed by displaying the potential drop across the sample, with constant current flowing, on an oscilloscope. With materials that are more chemically active than gold it is necessary to carry out the above operations in inert media, and reliable experiments are extremely difficult.

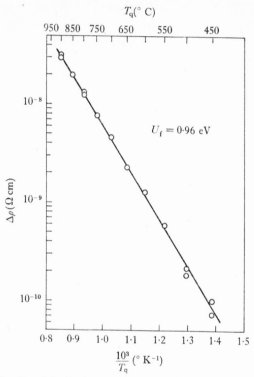

Fig. 6. Semilogarithmic plot of quench-in resistivity versus reciprocal of the absolute quench temperature. (From Bauerle & Koehler, 1957.)

The residual resistance of the wire was measured before and after the quench, immersing it in liquid nitrogen to reduce $\rho(T)$ in comparison with ρ_0 and so improving the accuracy. Many subsequent workers have used liquid He for the measuring bath, thus eliminating $\rho(T)$ completely.

It was found that quenching introduced a resistivity increment $\Delta\rho_0$, and that a graph of $\log(\Delta\rho_0)$ *versus* $1/T_q$ was a straight line as shown in fig. 6. If $\Delta\rho_0$ can be attributed to the equilibrium concentration of vacancies at T_q then $\Delta\rho_0 \propto \exp(-U_f^v/kT_q)$ from (2.8) and

(2.39) and U_f^v can be deduced from the slope of the line. For Au, Bauerle and Koehler deduced a value of 0.98 ± 0.03 eV. Using expression (2.8), giving C_v as a function of T, with the entropy factor (~ 5) found by Simmons and Balluffi, it is possible to use $\Delta\rho_0$ to deduce the resistivity per vacancy as 1.5 ± 0.3 micro ohm cm (see table II).

In the *annealing* experiment the quenched sample is heated rapidly from T_0 to a constant temperature T_1, held there for a measured time, cooled rapidly to T_0, after which the resistivity is

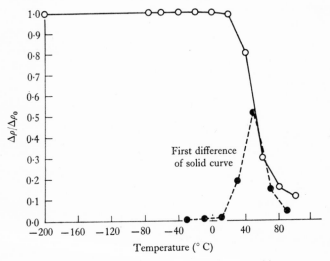

First difference
of solid curve

Fig. 7. Isochronal annealing of Au after quenching.
(From Baurle & Koehler, 1957.)

again measured. If the annealing temperature T_1 is high enough for vacancies to migrate, $\Delta\rho_0$ is reduced. In an *isochronal* annealing study, one carries out a sequence of anneals, maintaining a constant annealing period and using a higher value of T_1 for each. From the resulting graph of $\Delta\rho_0$ versus T_1 it is easy to identify the temperature ranges in which the resistivity recovers towards its original value. Fig. 7 shows the isochronal annealing of Au found by Bauerle and Koehler showing resistivity to recover in a single stage near 40 °C. Such a graph is sometimes known as a *recovery* curve. Its first derivative with respect to T_1 shows recovery stages as peaks and is known as a *recovery spectrum*. Because of the behaviour implied by equation (2.14) the position of a recovery stage on such a graph will depend on the chosen period of annealing, longer periods tending to shift a particular stage to lower temperatures.

In *isothermal annealing* one subjects the sample to many anneals at the same temperature, when a steady decrease in $\Delta\rho_0$ with time is observed. By (2.14) and (2.40) the rate of annealing at T_1 is given by:

$$\frac{d}{dt}(\Delta\rho_0) = G(\Delta\rho_0)\exp(-U_m^r/kT_1), \qquad (2.41)$$

where $G(\Delta\rho_0)$ is an undetermined function, in general. Suppose one anneals isothermally, first at T_1, and then at a slightly higher temperature T_2. At the intersection of the two isothermals where T_1 changes to T_2, one has the same value of $\Delta\rho_0$ and hence the same $G(\Delta\rho_0)$. Then the ratio of the slopes gives:

$$\frac{d}{dt}(\Delta\rho_0)_1 \bigg/ \frac{d}{dt}(\Delta\rho_0)_2 = \exp\frac{U_m^r}{k}\left(\frac{1}{T_2}-\frac{1}{T_1}\right) \qquad (2.42)$$

from which U_m^r may be determined. Bauerle and Koehler obtained the pair of isotherms shown in fig. 8 by annealing a gold sample quenched from 700 °C. They are plotted as $\log(\Delta\rho_0)$ *versus* t and since they appear as straight lines one infers from (2.12) that $x = 1$ and recovery is due to a first order reaction. The ratio of slopes gives an activation energy of 0·8 eV. The results are consistent with the migration of vacancies to sinks whose concentration is effectively constant, and if this is the case then $U_m^v = 0·8$ eV for gold.

Such simple results were not obtained when T_q exceeded 700 °C and isotherms for $T_q = 700$, 800 and 900 °C are compared in fig. 9. Whereas the first is behaving according to first-order reaction kinetics it is clear that the other two are not, and there appears to be an initial delay time before rapid annealing sets in. Koehler, Seitz & Bauerle (1957) suggested that the different behaviour was due to a higher T_q giving large enough concentrations of single vacancies for an appreciable chance of their meeting during the anneal, thus forming divacancies with a lower migration energy. On this hypothesis they deduced a migration energy $U_m^{2v} = 0·65 \pm 0·05$ eV.

In general, where one has several recovery processes with differing values of U_m^r, several stages will occur in the recovery spectrum. This is well illustrated by the complex spectra observed in the recovery

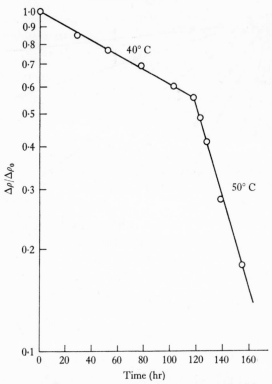

Fig. 8. A pair of isotherms at 40 and 50 °C for Au quenched from 700 °C. (From Bauerle & Koehler, 1957.)

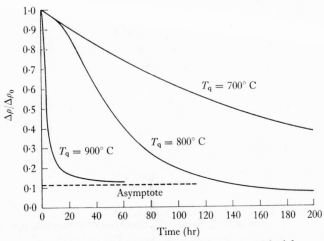

Fig. 9. Comparison of the isotherm at 40 °C for Au quenched from 700, 800 and 900 °C. (From Bauerle & Koehler, 1957.)

TDA

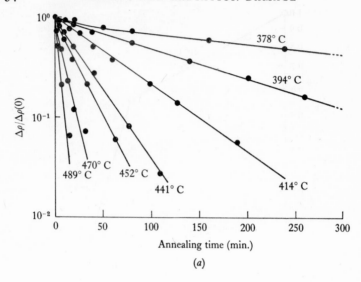

(a)

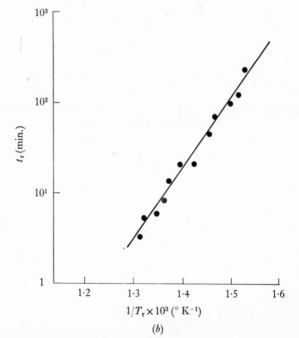

(b)

Fig. 10. (a) Multiple isotherms for quenched Pt.
(b) Plot of log t_r versus $1/T_r$. (Both from Germagnoli, 1962.)

of radiation damage (see fig. 110). In determining U_m^r for a particular stage by isothermal annealing the isotherms must be at temperatures within the stage and as close together as the precision of measurement permits.

An alternative method for determining U_m^r by isothermal annealing is to take a number of samples that have been subjected to the same quench (or irradiation) and to anneal each of them at a different temperature. Fig. 10a shows a set of isotherms obtained in this way for Pt quenched from 1050 °C by Germagnoli (1962). A line drawn horizontally at constant $\Delta\rho_0$ finds each sample in the same state, and hence with the same value of $G(\Delta\rho_0)$ in equation (2.41). Integrating equation (2.41) gives:

$$\int \frac{\mathrm{d}(\Delta\rho_0)}{G(\Delta\rho_0)} = t_r \exp\left(-U_m^r/kT_r\right),$$

where t_r is the time required to reach the chosen value of $\Delta\rho_0$ at temperature T_r. The left-hand side of this equation is the same for all the isotherms at a constant value of $\Delta\rho_0$, hence by plotting $\log t_r$ versus $1/kT_r$ one obtains a straight line with slope U_m as in fig. 10b. The disadvantages of this technique are that it requires a large number of samples in an identical starting condition, and that the range of temperatures covered by the isotherms may be difficult to accommodate in a single recovery stage.

The above annealing experiments have been described in some detail as they illustrate the different techniques that can be applied to determining migration energies.

2.3.9. *Self-diffusion experiments.* The long range movement of the constituent atoms of a crystal, or *self-diffusion*, requires that atoms can change sites. In f.c.c. and transition metals at least, this can occur with the least amount of activation energy when vacancies are present since U_f^v is much less than U_f^i. Referring to fig. 11, when a vacancy moves from V_1 to V_2 the atom originally at V_2 moves to V_1.

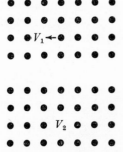

Fig. 11. Self-diffusion by the vacancy mechanism.

In a self-diffusion experiment one measures the rate at which atoms change their sites, often by following the spread of a radio-active isotope of the element concerned. The temperature dependence of this rate depends on two factors. First, the probability of an atom having an adjacent site vacant, which is proportional to the vacancy concentration $\exp(-U_f^v/kT)$. Second, the probability per unit time of an atom jumping into an adjacent vacant site, which is proportional to $\exp(-U_m^v/kT)$. Then the rate of jumping, and hence of self-diffusion, depends on the product of these two probabilities: $\exp-(U_f^v+U_m^v)/kT$. The observed self-diffusion is a thermally activated rate process of the type discussed in §2.2.2. The activation energy U_{sd} observed is clearly given by

$$U_{sd} = U_f^v + U_m^v \qquad (2.43)$$

and measurements of this quantity provide further information about the vacancy.

An interesting correlation exists between the absolute melting points of metals, T_m, and the temperature at which self-diffusion first appears. It is even more striking to compare the ration U_{sd}/T_m in different metals, which in all cases falls close to 0·15, with the f.c.c. family slightly above the b.c.c. If a vacancy mechanism holds in each case this appears reasonable, since U_f^v and U_m^v should both depend on the strength of atomic bonding, which is also reflected in T_m. (See Shewman, 1962.)

2.3.10.   *Summary of vacancy properties.*

(See tables II and III on pages 37–38.)

The prefix C denotes a calculation, E equilibrium experiments, Q quenching experiments, E+Q comparison of concentrations in equilibrium with property changes after quenching. Figures in brackets refer to the original papers listed below.

	Al	Cu	Ag	Au
U_f^v (eV)	C Q 0·77±0·03 (20, 21, 22, 23, 40) E 0·75±0·07 (9, 44)	C 1·2 (1, 2, 3, 4, 5, 35, 36) Q 1·1±0·1 (29, 37, 38) E 1·17±0·11 (12)	C 1·1 (35, 36) Q 1·08±0·04 (18, 19, 39) E 1·09±0·01 (9)	C 1·0 (35, 36) Q 0·98±0·03 (17, 34) E 0·94±0·09 (15)
U_f^{2v} (eV)	C 1·3±0·1 Q $2U_f^v$−0·17 (40)	C $2U_f^v$−0·15 (34) Q $2U_f^v$−(0·12±0·05) (34)	C 2·1 (34) Q 1·8±0·05 (19, 24)	C 1·85 (35, 36) Q 1·86±0·04 (17)
U_m^v (eV)	C Q 0·63±0·05 (20, 21, 40)	C 1·0 (2, 3, 7, 35, 36) Q 1·08±0·03 (30, 34)	C 0·86 (35) Q 0·85±0·05 (19, 35, 36, 42)	C Q 0·83±0·05 (17, 34)
	$U_{sd}-U_f^v = 0·72$	$U_{sd}-U_f^v = 0·94$	$U_{sd}-U_f^v$ 0·82	$U_{sd}-U_f^v = 0·85$
U_m^{2v} (eV)	C Q 0·46 (40)	C 0·6±0·4 (8, 3) Q 0·67±0·05 (33, 34)	C 0·52 (34) Q 0·57±0·03 (19, 24, 34, 42, 43)	C Q 0·67±0·04 (25, 17, 34)
U_{sd} (eV)	1·48±0·1 (10, 11, 41)	2·11±0·02 (13)	1·91±0·01 (14)	1·81±0·02 (16)
$\Delta\rho_0$ for 1 % (μΩ cm)	E+Q 2·2±0·7 (26, 20, 21, 27)	C 1·6 (6, 34)	C 1·7 (34) E+Q 1·3±0·7 (9, 18)	C 1·7 (34) E+Q 1·5±0·3 (15, 17)
$\frac{\Delta a}{a}$ for 1 %		C −(0·11±0·05) × 10⁻² (4, 5)		E+Q −(0·18±0·05) × 10⁻²
$\frac{\Delta l}{l}$ for 1 %		C +(0·22±0·05) × 10⁻² (4, 5)		E+Q +(0·15±0·05 × 10⁻² (17, 28, 15)

(1) Huntington & Seitz (1942a)
(2) Huntington & Seitz (1942b)
(3) Lomer (1959)
(4) Teworde (1958)
(5) Seegar & Mann (1960)
(6) Blatt (1957)
(7) Johnson & Brown (1962)
(8) Bartlett & Dienes (1953)
(9) Simmons & Balluffi (1960c)
(10) Spokas & Slichter (1959)
(11) Lundy & Murdock (1961)
(12) Simmons & Balluffi (1962b)
(13) Kuper et al. (1955)
(14) Tomizuka & Sonder (1956)
(15) Simmons & Balluffi (1962a)
(16) Makin, Rowe & Le Claire (1957)
(17) Bauerle & Koehler (1957)
(18) Doyama & Koehler (1960)
(19) Doyama & Koehler (1962)
(20) de Sorbo & Turnbull (1959a)
(21) de Sorbo & Turnbull (1959b)
(22) Bradshaw & Pearson (1957)
(23) Panseri & Frederighi (1958)
(24) Palmer & Koehler (1958)
(25) de Jong & Koehler (1963)
(26) Simmons & Balluffi (1960a)
(27) Simmons & Balluffi (1960b)
(28) Takamura (1961)
(29) Airoldi, Bachella & Germagnoli (1959)
(30) Schür et al. (1961)
(31) Gibson et al. (1960)
(32) Benneman (1961a, b)
(33) A. Seeger et al. (1963)
(34) Seegar & Schumaker (1965)
(35) Schottky, Seegar & Schmid (1964a)
(36) Schottky, Seegar & Schmid (1964b)
(37) Budin et al. (1963)
(38) Lucasson et al. (1963)
(39) Quéré (1960)
(40) Frederighi (1965)
(41) Lundy & Murdock (1962)
(42) Quéré (1961)
(43) Cuddy & Machlin (1962)
(44) Feder & Nowick (1958)

TABLE III. *Vacancy properties in some other solids*

The prefix C denotes a calculation, E equilibrium experiments, Q quenching experiments, SD self-diffusion.

	U_f^v (eV)	U_m^v (eV)	U_{sd} (eV)	L_s (eV/atom)
Graphite	Q 3·5 (1)	Q 3·2 (1)	SD 7·1 (2)	7·4 (4)
Diamond	C 4.2 (7)	C 2·0 (7)		
Silicon	C 2·1 (7)	C 1·1 (7)		
	Q 2·0 (8, 9)			
Germanium	C 2·2 (7, 11)	C 1·0 (7)		
	Q 2·0 (8, 9)	Q 1·3 (10)	SD 3·0 (3)	3·9 (5)
Molybdenum	C 5·4 (6)		SD 4·2 (13)	
Tungsten	Q 3·3 (16)	Q 1·9 (16)	SD 5·2 (17)	
Sodium	C 0·53 (12)		SD 0·44 (14, 18)	
	E 0·42 (15)	0·02 (18)		

(1) Baker & Kelly (1962)
(2) Kanter (1957)
(3) Letaw, Portnye & Slifkin (1956)
(4) Glockler (1954)
(5) Kubaschewski & Evans (1958)
(6) Girifalco & Streetman (1958)
(7) Swalin (1961)
(8) Mayburg (1954)
(9) Logan (1956)
(10) Hiraka & Suita (1962)
(11) Scholz (1963)
(12) Fumi (1955)
(13) Askill & Tomlin (1963)
(14) Nachtrieb (1952)
(15) Feder & Charbnau (1966)
(16) Schultz (1965)
(17) Danneburg (1961)
(18) Gugan (1966)

2.4. Interstitial atoms

It appeared from the discussion of the vacancy in § 2.3.3 that the energy stored in the elastic strain field around a point defect is proportional to the square of the local expansion δ_0. Now in the case of the vacancy, δ_0 was due only to the relatively weak surface tension forces and was less than $\frac{1}{10}$th the atomic radius. In that case we saw that the strain energy contributed only about 0·1 eV to the formation energy of a vacancy U_f^v. But now consider the case of an interstitial atom, which we suppose to be formed by taking an atom from the surface and inserting it into the solid, making space for it by a large outward movement of the atoms surrounding its site. Unfortunately one cannot strictly apply the elastic continuum model to this situation because δ_0 must approach the atomic radius r_a and the approximation is only valid for $\delta_0 \ll r_a$. However it is clear that since the expansion is an order of magnitude greater than in the case of a vacancy, so the strain energy must be *two* orders of magnitude greater, i.e. ~ 10 eV, according to the indications of equation (2.27). This is as far as one can go with the elastic continuum model but the general prediction is quite clear: that the formation energy of an interstitial U_f^i is dominated by the strain energy associated with a large outward dilatation.

Using various atomic models many attempts have been made to calculate interstitial properties, particularly in Cu and Fe (see the references of table IV). Experimental information about the interstitial comes almost exclusively from radiation damage experiments but discussion of these must be left until Chapters 6 and 7. Here, we shall only consider the theoretical predictions.

The formation energy U_f^i is found by supposing an atom to be removed from the surface and forced into one of the natural interstices of the lattice. The resulting reduction in local interatomic separation causes a large increase in repulsive energy, probably an order of magnitude greater than the change in electronic energy if we make the reasonable assumption that this is roughly the same (~ 1 eV) as for the vacancy. Unlike the vacancy, it appears that there are several configurations with almost the same formation energy and it is really impossible to decide, on theoretical grounds, which is the stable one. The trouble is, that although the repulsive energy can be found with fair precision the electron energy cannot, and the decision rests with small differences between large energies. The three most probable types are shown in figs. 12, 13 and 14. In the first of these, often referred to as the *dumb-bell*, the extra atom is accommodated by sharing a site with another atom,

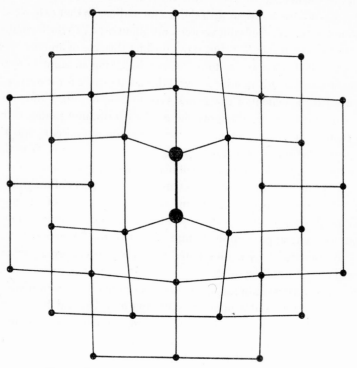

Fig. 12. The ⟨100⟩ dumb-bell interstitial configuration.

the axis of the pair lying along a ⟨100⟩ direction. The second is the *body-centred* interstitial, where the extra atom occupies the largest open space in the f.c.c. unit cell. In the third configuration, shown in figure 2.14, and known as the *crowdion*, a long-range relaxation occurs along a ⟨110⟩ close packed row. Neighbouring rows are little affected.

The calculations are generally agreed that the first two have the lowest formation energy with the crowdion requiring considerably more. Johnson & Brown (1962), for instance, found the dumb-bell to be the stable configuration, the body-centred to require an extra 0·084 eV, and the crowdion an extra 0·45 eV. It should be pointed out that, because possible differences in electronic energy are not taken into account here, the choice should remain open, certainly between the dumb-bell and the body-centred interstitial.

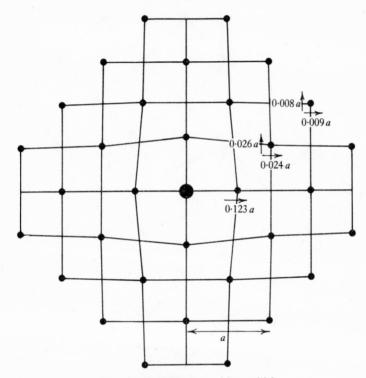

Fig. 13. The body-centred interstitial.

It is important to consider whether or not there are metastable configurations which, though requiring a higher U_f^i, must be supplied with activation energy before they transform into the stable state. Some interpretations of radiation damage experiments require such a property, and both the $\langle 111 \rangle$ split interstitial and the crowdion have been suggested as such metastable defects. However, although the majority of calculations do not support these hypotheses a final decision can only come from experiment.

The migration energy of interstitials has been generally found by calculating the repulsive energy at the saddle point configuration and neglecting electronic changes, thereby introducing the uncertainties encountered above. For the dumb-bell this is shown in fig. 15b and Johnson & Brown (1962) find U_m^i in this case to be 0·05 eV. The crowdion and body-centred interstitial require

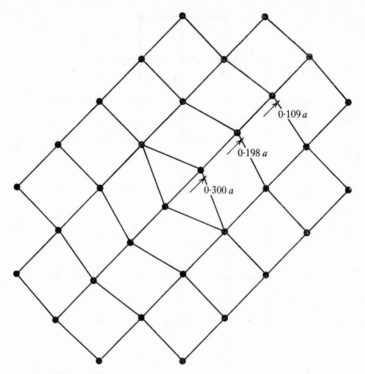

0·109 a

0·198 a

0·300 a

Fig. 14. The ⟨110⟩ crowdion interstitial configuration.

similarly low activation energies. Because it is spread out along a close-packed row, extending through several lattice cells, the crowdion has the interesting property of being constrained to move in a straight line. This has a great influence on the number of jumps required in annealing (see Chapter 6).

Di-interstitials have been examined and are found by most authors to be stable (i.e. with positive binding energy). The stable configuration is somewhat in doubt, but it appears from Johnson and Brown's calculations that a dumb-bell pair with ⟨100⟩ axes parallel is stable with a binding energy of 0·6 eV; the migration energy is not greater than 0·26 eV.

Table IV contains a summary of calculated interstitial properties in Cu. Included in this are estimates of $\Delta a/a$ and $\Delta \rho_0$ due to inter-

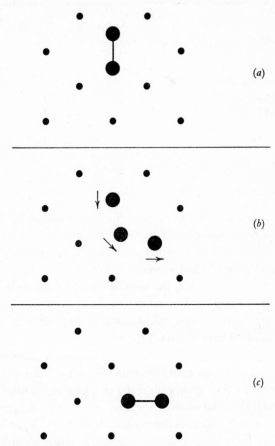

Fig. 15. The migration route for a dumb-bell interstitial.

stitials. Because the relaxation of neighbouring atoms is outwards, δ_0 in equation (2.23) is positive, leading to an increase in lattice parameter just as the negative δ_0 of the vacancy lead to a decrease in a in § 2.3.5. It will be seen that the larger dilatation leads to a larger $\Delta a/a$ per interstitial than in the case of vacancies, as one should expect from the elastic continuum model.

TABLE IV. *Calculated properties of the interstitial in* Cu

(References are to the list in table II, see p. 37.)

	Body-centred	Dumb-bell	Crowdion
U_f^i (eV)	4·0±0·5 (1, 2, 4, 5, 7, 31)	3·9±0·5 (1, 2, 5, 7, 31, 32)	4·7±0·1 (4, 5, 7)
U_m^i (eV)	0·05±0·05 (1, 2, 7)	0·05±0·05 (1, 2, 7)	0·25 (7)
Difference in U_f^i from dumb-bell (7)	+0·084 (7)	0	+0·449 (7)
U_b^{2i} (eV)		0·6 (7)	
U_m^{2i} (eV)		<0·26 (7)	
$\dfrac{\Delta a}{a}$ for 1 %		0·5±0·2 (4, 7)	0·5±0·3 (4, 7)
$\Delta\rho_0$ for 1 % micro ohm. cm.		2·9±2·1 (6)	

It is clear from table IV that the magnitude of U_f^i relative to U_f^v rules out any question of significant concentrations of interstitials being produced by thermal activation. The ratio C_v/C_i is of the order 10^{13} for Cu and a similar result could be expected for other f.c.c. metals. This was the original justification for assuming that self-diffusion in f.c.c. metals proceeds by a vacancy mechanism.

Apart from Cu few calculations of interstitial properties have been made. Iwata, Fukita & Suzuki (1961) have set up a theoretical model of graphite in which the basal planes are supposed to behave like elastic membranes between which the interstitial is forced. They find a formation energy U_f^i of 2·5 eV and a migration energy of 0·016 eV. The sum of these is very small compared with the observed self-diffusion energy and there is clearly an inconsistency here as their results would suggest an interstitial self-diffusion process. On the other hand the relaxation they calculate is consistent with the observed $\Delta a/a$.

2.5. Interstitial-vacancy pairs and their stability

In radiation damage, interstitials and vacancies are produced in equal numbers and in many cases the interstitial is associated with a nearby vacancy, the two having been formed in the same collision event. It is often useful to think of this interstitial-vacancy pair, or *Frenkel* pair, as a single defect. From the continuum model one would expect that because their dilatations are of opposite sign an attractive force should exist between the members of the pair. Under thermal activation there should then be a high probability of any migration leading to annihilation, rather than the more mobile partner escaping. Taking the spacing even closer might result in a spontane-

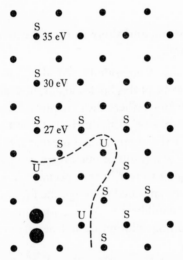

Fig. 16. Stability of Frenkel pairs in {100} plane of Cu. Split interstitial is at lower left. Dotted line separates stable from unstable sites for a vacancy. Approximate threshold energies for dynamic production of three particular pairs are indicated. S, stable; U, unstable. (From Gibson *et al.* 1960.)

ously unstable defect, annihilation occurring even in the absence of thermal activation.

Such behaviour can be simulated by the Brookhaven computer method. Fig. 16 shows the stable and unstable sites for a vacancy relative to a $\langle 100 \rangle$ dumb-bell interstitial in the {100} plane of Cu. It will be seen that the unstable zone extends furthest along the most closely-packed row of atoms. This is intuitively reasonable since in an atomic model the strain field cannot be isotropic, as in the continuum, and should spread out furthest along the directions in which atoms interact most strongly with their neighbours. The size of the unstable zone is quite large, affecting about 50 sites and it is clear therefore that the attractive influence of the interstitial on its vacancy partner must affect a considerably larger number of sites making direct annihilation the most likely result of thermally activated migration, even if the migrating defect is outside the unstable zone.

Similar results were obtained in b.c.c. Fe, the unstable zone extending along the closely-packed $\langle 111 \rangle$ directions.

2.6.   Foreign atoms and their interaction with interstitials and vacancies

2.6.1.   *Introduction.* Compounds and alloys are specifically excluded from the scope of this book and our interest in foreign atoms is confined to situations where they are present in small concentrations. Two cases may be distinguished; those that are present as natural impurity in the sample, and those that are introduced during irradiation, either as a result of nuclear transmutation, or by bombarding ions coming to rest in the specimen. Since one generally starts out with well-annealed material, the first case will concern a solid solution in equilibrium. Where foreign atoms are forcibly injected, the resulting solution cannot be assumed to be in equilibrium.

The equilibrium concentration depends on the *energy of solution* U_s, defined as the increase in internal energy of the crystal for one foreign atom. If U_s is negative the amount in solution is limited only by the breakdown of the small-concentration approximations in § 2.2, and for our purposes any concentration can exist in equilibrium. The energy of solution is not the only factor in determining the state of solid solution when $U_s < 0$, for it is possible that solute atoms will group together, releasing a binding energy and reducing the crystal's energy still further.

Where U_s is positive the atom is insoluble and equilibrium concentrations will be of the order of $\exp(-U_s/kT)$, by equation (2.8). If concentrations higher than equilibrium are produced, by quenching or by injection as ions, the approach to equilibrium will depend on some thermally activated migration process, which could involve other defects.

2.6.2. *Dilute alloys.* It is difficult to calculate even the sign of U_s from a detailed atomic model, although some progress has been made in this direction. For one specific type of solid solution, alloys, an empirical approach has been developed by W. Hume-Rothery. It takes the form of three conditions that must be met for high solubility to occur with atoms of a metallic solvent B replaced on their lattice sites by atoms of the solute element A. They therefore tell us when to expect both U_s and U_b^{2s} to be negative (i.e. solution favourable; binding unfavourable), and are as follows:

(1) *A and B must come from columns in the periodic table that are not far separated.* If this is not fulfilled there will be a tendency for ionic bonding to occur, forming molecules AB and a precipitated phase. For instance, if one adds an electronegative element like oxygen to a metal, an oxide is formed which generally exists as a second phase rather than a solution of oxygen in metal.

(2) *The valency of solute A should be greater than that of B.* Using the bond picture this is reasonable, for if A is successfully to occupy the site of a B atom, it must have at least as many bonds available. Thus, although small amounts of trivalent Al dissolve in the monovalent Cu, when Cu is added to pure Al a precipitate forms.

(3) *The atomic radii of A and B, defined from the atomic volume in the pure elements, must not differ by more than* 15 %. In the preceding sections we have seen how the strain in the lattice depends on dilatation around a point defect. It is obviously necessary for the dilatation to be small to avoid a large positive contribution to U_s. For example, divalent Be dissolves in monovalent Cu up to 17 at. %, but in Ag up to only 3·5 %. Conditions (1) and (2) are satisfied, but whereas the radii differ by 12·9 % in Cu, they differ by 22·9 % in Ag. If the radii differ by more than 50 % and conditions (1) and (2) are met, one frequently finds a structure with A occupying *interstitial* positions in B's lattice. Hydrogen and carbon occupy such positions in Fe.

These rules give little direct information about the atomic forces involved, nor do they differentiate between the effects of U_s and U_b^{2s}.

In the context of this book an important feature of solid solutions is the interaction between solute atoms and point defects. One can separate the interaction energy into two terms, that due to changes in elastic strain energy as the solute atom and defect come together, and the energy change due to electron redistribution.

The first term can be discussed qualitatively in terms of the elastic continuum model and it can be argued that there will be an elastic binding between vacancy and solute atom because the void space enables the dilatation around the solute atom to be accommodated locally. For interstitials one might expect elastic binding to undersized solute atoms with negative dilatation rather than to large ones. There is no satisfactory means at present for estimating even the sign of the electronic term owing to inadequate understanding of the mechanisms involved.

It proves possible to estimate binding energies U_b^{vs} between vacancy and solute atom by careful quenching experiments with dilute alloys (see review by Takamura, 1965 and Hasiguti, 1965). Table V lists some values for various solute atoms. It is interesting that they are all greater than the divacancy binding energy U_b^{2v}, which is 0·17 eV in Al, and that there is no systematic variation with difference in atomic radii. At the present time there is no adequate theory and we shall not try to explain the data. Interstitial-solute atom binding will be considered later in the context of irradiated metals.

TABLE V. *Vacancy–Solute binding energies in* Al.
The atomic radius of Al *is* 1·43 Å
(From Takamura 1965 and Hasiguti, 1965.)

Solute	Si	Ge	Cu	Zn	Sn	Ag	In	Cd	Mg
Atomic radius (Å)	1·18	1·22	1·27	1·31	1·40	1·44	1·45	1·49	1·62
U_b^{vs} (eV)	0·26	0·33	0·20	0·19	0·43	0·25	0·39	0·32	0·18

2.6.3. *Inert gas atoms in metals.* The inert gases present a class of foreign atom of especial importance in radiation damage, for irradiation with nucleons often results in a nuclear reaction having an inert gas as a product. A list of some of the more favourable reactions is given in table VI. In addition to such transmutation processes, the bombarding particle is often an inert gas ion, such as an α-particle, which, if it comes to rest in the sample, constitutes a foreign atom.

TABLE VI. *Some nuclear reactions with thermal neutrons that result in formation of inert gases*

Reaction	Cross-section (barns)	Inert gas
^6Li (n, α) ^3H	950	He
^{10}B (n, α) ^7Li	3990	He
^{25}Mg (n, α) ^{22}Ne	0·27	He+Ne
^{235}U $(n,$ fission$)$	$\begin{cases} 14 \\ 46 \end{cases}$	$\begin{cases} Kr \\ Xe \end{cases}$

The fact that the first ionization potentials of inert gases lie in the range 10–25 eV shows that the outer electrons are much more tightly bound than in other atoms, and results in their chemical inertness. One might therefore expect that in a metal such an atom would retain all its electrons and have zero valency.

In the periodic table they occupy the last column, whereas the metals are in the early columns. Thus the first Hume-Rothery condition is not met. With the possible exception of He and Ne their atomic radii are much greater than metals. This is shown in table VII. Hence, one or two of the Hume-Rothery conditions are not fulfilled and one should expect very low solubility.

TABLE VII. *Atomic radii of inert gases and metals*

Inert gases ...	He	Ne	A	Kr	Xe
r_a (Å)	—	1·60	1·92	1·96	2·19
Metals ...	Al	Cu	Ag	Au	U
r_a (Å)	1·43	1·28	1·45	1·44	1·49

For Cu it is possible to improve on this empirical approach and Rimmer & Cottrell (1957) have estimated U_s from an atomic model. In calculating the increased repulsive energy, two situations are considered; first with the gas atom in a body-centred interstitial position, secondly, substitutional in a Cu site. Large dilatations occur in the first case and lead to a correspondingly high value for the increase in repulsive energy: 21 eV in the case of Xe.

The electronic term is calculated by assuming the atom to occupy a vacancy, whose formation we have already seen to require an increase of 0·9 eV in electronic energy. But, because the dilatation is so much greater than around the normal vacancy, some further energy is required, making a total electronic term of about 6 eV for interstitial Xe. The details of the calculation are summarized in table VIII.

TABLE VIII. *Energies of solution of inert gases in* Cu

U_{s1} refers to interstitial and U_{s2} to substitutional solution, respectively. Data is taken from Rimmer & Cottrell (1957), table II and table IV.

		He	Ne	A	Kr	Xe
Interstitial gas atom	repulsive energy	1·7	3·4	9·2	12·9	2·1
	electronic energy	0·8	1·2	2·7	3·8	5·7
	U_{s1}	2·5	4·6	11·9	16·7	26·7
Substitutional gas atom	repulsive energy	0	0·3	2·0	3·3	6·8
	electronic energy	0·9	1·0	1·5	2·0	2·9
	U_{s2}	0·9	1·3	3·5	5·3	9·7
$U_{s2} + U_f^i$		4·8	5·2	7·4	9·2	13·6
$(U_{s2} + U_f^i) - U_{s1}$		+2·3	+0·6	−4·5	−7·5	−13·1

Consider now whether it is energetically favourable for an interstitial gas atom to remain in this state or to replace a Cu atom on the lattice, creating a Cu interstitial in the process. If $(U_{s2} + U_f^i) - U_{s1}$ is *positive* the hypothetical process could not occur and the *interstitial* solution is the equilibrium state; if *negative, substitutional* solution follows. The values in table VIII show that whereas A, Kr and Xe should exist in substitional solution, He and Ne are expected to occupy interstitial sites.

The calculations show clearly that in all cases U_s is positive and therefore the equilibrium concentrations at temperature T will be approximately $\exp(-U_s/kT)$. Thus, even at its melting point the equilibrium concentration of heavy inert gases in Cu is negligibly small ($\sim 10^{-40}$), although for He one might dissolve appreciable concentrations ($\sim 10^{-4}$).

Because the gas atom causes a large positive dilatation and the vacancy is a void able to reduce this, one expects a large positive contribution from atom-atom repulsive energy to the binding energy U_b^{vg} of gas atom and vacancy.

From the figures in table VIII it seems likely that the decrease in repulsive energy will far outweigh any electronic term and lead to a positive binding energy. It is intuitively reasonable that a large atom, causing dilatation of the lattice, should be able to relax some strain by absorbing a small void. For the heavier gases several vacancies might become associated with one gas atom. The migration of such gas atoms should most easily occur by movement of the vacancies surrounding them with activation energy not far removed from U_m^v. However, when non-equilibrium concentrations of gas atoms are produced by irradiation, even this form of migration will not permit the approach to equilibrium unless sufficient vacancies are present. Such vacancies could stem either from the collision cascades, or from thermal activation with U_f^v. We shall see in Chapter 8 that these predictions are born out by experiment.

CHAPTER 3

EXTENDED DEFECTS

3.1. Dislocations

3.1.1. *Geometrical properties.* It is not possible here to present more than a brief account of dislocations in crystals and the reader is referred to one of the comprehensive works on the subject (e.g. Cottrell, 1953, *Dislocations and Plastic Flow in Crystals*, O.U.P. or

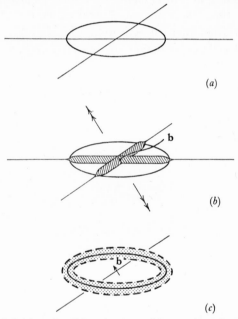

Fig. 17. The definition of a dislocation loop: (*a*) a cut is made within a closed loop; (*b*) stress is applied to separate the faces of the cut by **b**; (*c*) extra material is added to fill the gap and the external stress is removed.

Read, 1953, *Dislocations in Crystals*, McGraw-Hill). For the purpose of definition one supposes the formation of a dislocation to proceed as follows. On a chosen plane within the crystal a line is drawn in the form of a closed loop and a cut made through the enclosed area (fig. 17*a*), stresses are applied to separate the plane faces of the cut and extra material is added to fill the lenticular cavity so created

[52]

(fig. 17*b*). The applied stress is removed and relaxation allowed to occur.

One expects the resulting strain to be maximum in a toroidal region around the original line. For a large enough loop the defect extends along a line and we call this a *dislocation line*. The vector joining points on the faces of the cut that were initially in contact is known as the *Burgers Vector* **b** of the dislocation loop. Its direction is fixed relative to the plane of the loop, but with respect to the line

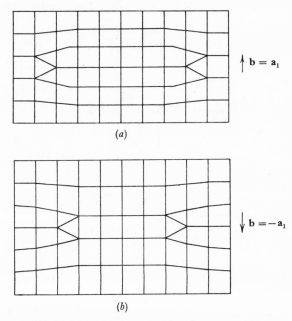

(*a*)

(*b*)

Fig. 18. (*a*) A loop with **b** normal to its plane and equal to a lattice vector $\mathbf{a}_1$. (*b*) A loop with $\mathbf{b} = -\mathbf{a}_1$, here material has been removed in order for the cut surfaces to be translated through one another.

its orientation changes as one passes around the loop. If **b** corresponds to one of the primitive translation vectors of the lattice, the crystal at the centre of a large loop is not greatly distorted, and loops with such a vector are likely to have the lowest formation energy.

Some simple examples of dislocation loops are shown in figs. 18, 19, 20. Although the simple cubic lattice is drawn, the same schematic representation would apply where each basic cube represents the primitive (Bravais) lattice cell in a more complex structure.

In fig. 18*a*, **b** is perpendicular to the loop plane and equal in magnitude to one lattice vector a_1. A part of a new plane of atoms has therefore been introduced. In fig. 18*b*, **b** = −**a**, and the faces of the cut have been translated *through* one another by one lattice constant, and part of an atomic plane *removed*. Since **b** is normal to the loop plane it is everywhere normal to the dislocation line in either of these cases.

Another type of dislocation loop is shown in fig. 19, where **b** lies in the loop plane and **b** = a_2. Within the loop the crystal is simply

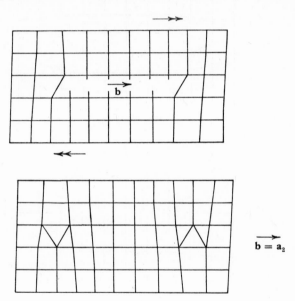

Fig. 19. A loop formed by shearing the faces of the cut by a_2, no extra material is involved.

sheared across its plane and no extra material is required. In the section shown in fig. 19, **b** is perpendicular to the line and parallel to the loop plane. The atomic configuration approximates to an extra half plane of atoms on one side of the loop plane and this segment of the loop is known as an *edge dislocation*.

In an alternative section across the loop, at right angles to that in fig. 19, the configuration is quite different, since **b** is parallel to both dislocation line and loop plane. This segment of line is illustrated in fig. 21. It will be seen that parallel planes of the crystal are turned into a helical surface when pierced by the dislocation line. For this reason such a dislocation is known as a *screw dislocation*. The remainder of the loop in fig. 19 is made up of a hybrid line which has

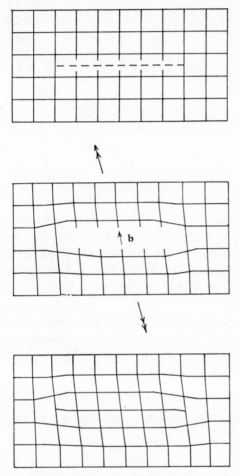

Fig. 20. A loop for which **b** is not a primitive lattice vector resulting in a region of faulted lattice within the loop.

some properties in common with both screw and edge. A pure screw dislocation has **b** parallel to the line, in a pure edge **b** is perpendicular to the line. The hybrid can be described in terms of its edge and screw components; i.e. by resolving **b** into components perpendicular and parallel to the line.

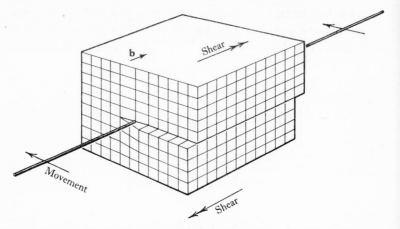

Fig. 21. A segment of screw dislocation line. Movement of the line extends the region of sheared crystal.

3.1.2. *Formation energy.* It is easy to make a rough calculation of the formation energy U_f^d for a segment of screw dislocation line. The crystal is taken in the form of a cylinder of radius R and length L, with the dislocation line along its axis. One assumes the material to behave as an elastic continuum and considers the strain in a cylindrical shell between r and $r + dr$ from the axis. As illustrated in fig. 22 this shell is sheared by **b**. By unwrapping the shell the shear strain is seen to be $\theta = b/2\pi r$. If the shear modulus of elasticity is μ, the shear stress is $\mu\theta$, and the elastic energy stored per unit volume is $\frac{1}{2}\mu\theta^2$. Therefore the energy stored between r and $r + dr$ is

$$(\mu b^2 l \, dr/4\pi r).$$

Then for r between limits r_0 and R one has an integrated energy:

$$\frac{\mu b^2 l}{4\pi} \log\left(\frac{R}{r_0}\right). \tag{3.1}$$

The inner limit r_0 is the radius at which the elastic approximation breaks down. From the results of previous sections we may expect $r_0 \sim 10$ Å. Since r_0 appears only in the logarithm its exact value is not too important for an order of magnitude calculation. The energy stored inside r_0 is generally neglected as a first approximation (see Cottrell, 1953 or Read, 1953). $R/r_0 \sim 10^6$ for a typical single crystal sample and hence

$$U_f^d \sim \mu b^2 l. \tag{3.2}$$

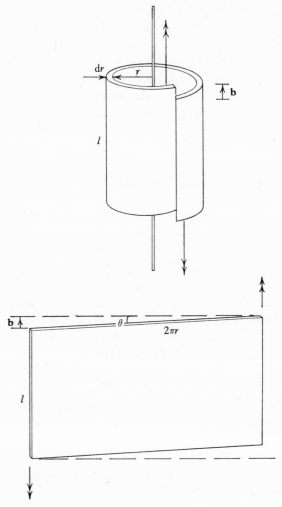

Fig. 22. A cylindrical shell of material at radius r from a screw dislocation contains a shear strain θ.

μb^2 is then the energy per unit length of line, and because l will tend to a minimum value it can be regarded as a *line tension* force. For most metals $\mu b^2 \sim 3 \, \text{eV} \, \text{Å}^{-1}$ and even in the most minute crystal U_f^d is too great for dislocations to be created by direct thermal activation.

The logarithmic dependence of expression (3.1) on R shows that the strain around a dislocation extends over a great distance. It is this fact that leads to the long range of the interaction force between dislocations.

The calculation of strain energy for an edge dislocation is complicated by its non-cylindrical symmetry. The formation energy is still of the order $\mu b^2 l$ but the interaction force between edge dislocations is strongly dependent on the relative orientation of the dislocations involved.

The entropy associated with a dislocation is very small because of its linear properties. Suppose we have a crystal with a single dislocation line running through it. Each atomic plane contains something analogous to a point defect where the line penetrates it. If there were no correlation between the position of these defects in adjacent planes the entropy increase would be roughly the same as if the crystal contained the one point defect per plane. But because the choice of sites is limited by the restriction that they must follow a line, the number of arrangements of the system and hence its increase in entropy is much smaller than in the point defect case. This effectively means that we can always neglect the contribution of entropy to the free energy of a crystal containing a dislocation. Thus in (2.6) the second group of terms can be omitted and it becomes:

$$F_f^d \simeq n_d\, U_f^d.$$

Thus the minimum free energy occurs when there are no dislocations present. In spite of this, crystals nearly always contain some dislocations and the reasons for this will be considered in § 3.1.5.

3.1.3. *Mechanical properties of dislocations.* Although they are not of direct concern here, the mechanical properties are important because, historically, they are the reason for the discovery of dislocations and much of the nomenclature is built around them. Also, a number of experiments in radiation damage involve changes in the mechanical properties of crystals.

The stress-strain curve of most crystals initially shows a steeply rising linear elastic portion where Hooke's law is obeyed and over which the strain is completely recovered when the stress is removed (i.e. deformation is reversible). At a critical point, known as the elastic limit, this gives way to plastic deformation, which is irreversible. Finally, fracture occurs. The stress at which the elastic limit is reached in a well-annealed metal crystal may be as low as 10^8 dyne . cm^{-2}.

One might expect it to be easier for plastic deformation to occur by atomic planes sliding over one another than by pulling apart in tension. Experimentally it is observed that single crystal samples

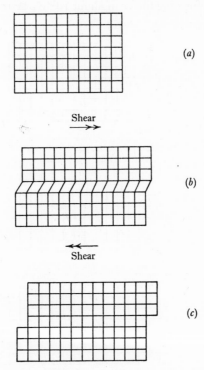

(a)

Shear
$\longrightarrow\!\!\!\rightarrow$

(b)

$\twoheadleftarrow\!\!\!-$
Shear

(c)

Fig. 23. Shearing of a crystal by simultaneous slip between a pair of lattice planes.

deform by shearing across planes of low index. In most cases these are planes in which the atoms are the most closely packed and which therefore have the largest spaces between. For instance f.c.c. crystals generally shear on {111} planes and b.c.c. on {110}. Let us consider some possible mechanisms by which sliding between neighbouring planes, or *slip*, could occur. First suppose the movement to occur simultaneously over the whole slip plane as illustrated in fig. 23 and calculate the stress at the elastic limit. In the intermediate state (fig. 23 *b*) one has a severe disruption of the crystal structure over a region extending for one atomic spacing either side of the slip plane. The internal energy of this region will be raised by an amount similar to that found in a phase transformation (i.e. ~ 1 eV per atom). Then for the whole slip plane in a 1 cm cube of crystal, where there are $\sim 10^{15}$ atoms, there has to be an energy input from the deforming forces of about 10^{15} eV $\triangleq 10^{4}$ erg. The distance

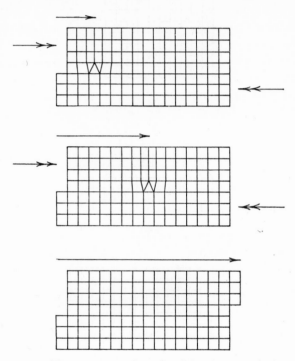

Fig. 24. The movement of an edge dislocation extends the
region of slip through a crystal.

moved by the force in one atomic unit of slip is about 10^{-8} cm, hence the force is $\sim 10^{11}$ dyne. Since this acts over 1 cm² of slip plane the critical shear stress is 10^{11} dyne cm⁻². The fact that the observed stress at the elastic limit can be three orders of magnitude smaller than this rules out simultaneous slip as a possible mechanism.

Consider now the situation illustrated in fig. 24, where a crystal under a shear stress is penetrated by a segment of edge dislocation with **b** in the slip plane and in the same direction as the shear. If the dislocation moves from left to right the region of slipped crystal increases until the entire top half has moved by an amount **b**.

In moving from one site to the next the dislocation requires very little energy, because for each bond on one side of the dislocation that shortens there is a corresponding one on the opposite side that lengthens, and the changes in energy tend to cancel out. Also, there are far fewer atoms raised to a higher energy configuration than in the case of simultaneous slip.

A screw dislocation is also able to generate slip by moving across the slip plane, as shown in fig. 21. Because **b** is directed along the line, the line moves transverse to the direction of shear and deformation is by a tearing action.

If one has a complete loop, of the type illustrated in fig. 19, lying in the slip plane, rather than a segment stretched across the crystal, a moment's reflection, with fig. 21, 24 and 19, will show that an applied shear stress will cause the loop to either expand or contract according to whether **b** is directed with the shear or against it. If a loop changes its area by a on a slip plane of unit area the amount of slip is ba, hence the shear strain in a unit volume is given by:

$$\epsilon = ba. \tag{3.3}$$

In making calculations of the dislocation movements induced by stress it is advantageous to think of a force per unit length (dF/ds) acting on the line and perpendicular to it. Suppose one has a loop of radius R on a slip plane of area A. In expanding from R to $R+b$ the increase in area of the slipped region is $2\pi Rb$ and the amount dy by which the slip increases, averaged over the crystal as a whole, is $(2\pi Rb/A)\,b$. The applied stress σ comes from a force σA and the work done by this is $\sigma A\,dy = 2\pi\sigma b^2 R$. The hypothetical force (dF/ds) does work $2\pi Rb(dF/ds)$ during the expansion and if one neglects any energy dissipation, which is reasonable for metals, one has by equating these:

$$dF/ds = \sigma b. \tag{3.4}$$

The increase in line length increases its formation energy U_f^d, since this is proportional to l, and if we take the energy per unit length as μb^2 from (3.2) the energy increase in going from R to $R+b$ is $2\pi\mu b^3$. Under stress the loop will only expand if the work done by the applied stress, in a small expansion, can exceed this increase in line energy (again neglecting any other energy loss). For example,

$$2\pi\sigma b^2 R \geqslant 2\pi\mu b^3$$

or

$$\sigma \geqslant \mu b/R. \qquad (3.5)$$

With $R = 10^{-3}$ cm, $\sigma \geqslant 10^8$ dyne/cm^2 from this expression.

Thus for a given radius of loop there is a minimum stress $\mu b/R$ that must be applied before expansion can occur. Alternatively, let us consider the case of a line segment, anchored at each end and initially straight with $R = \infty$. When stress σ is applied it should bow out into a shape that approximates an arc of circle, reducing R to the value $\mu b/\sigma$ given by the equality in expression (3.5). This will be considered further in later sections.

Note that movement under stress can only occur if **b** lies in the slip plane, and when this condition is satisfied the loop or line segment concerned is said to be *glissile* (i.e. able to slip or glide). Otherwise, the term *sessile* is applied. In some situations although the loop as a whole is sessile its Burgers vector may be directed parallel to another slip plane of the crystal. Then those segments of the dislocation line that lie in this plane will be able to move in it, bowing out to a radius R given by (3.5) and from its originally planar geometry.

An impressive bulk of experiments have confirmed that plastic deformation of crystals occurs by dislocation movement. In plate I*a* the points at which the ends of a half-loop emerged from a LiF crystal during deformation are marked by etch pits. The stress was increased in steps and after each new stage the crystal was etched further so that the expansion of the loop may be seen as a sequence of etch pits (Johnston & Gilman, 1959).

The observed radius of curvature R and stress σ are consistent in magnitude with expression 3.5.

Plate I*b* shows dislocations on the slip planes of a thin Au crystal, seen in the electron microscope. Stressing the crystal would cause these to move providing further clear evidence of the rôle of dislocations in the slip process.

3.1.4. *Dissociation, partial dislocations and stacking faults.* A dislocation with a Burgers vector that is the vector sum of one or more whole lattice translation vectors is known as *full* or *perfect*. Where

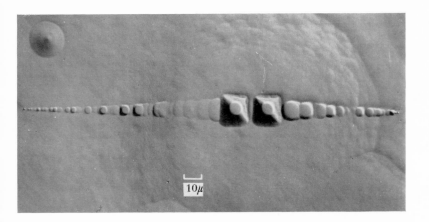

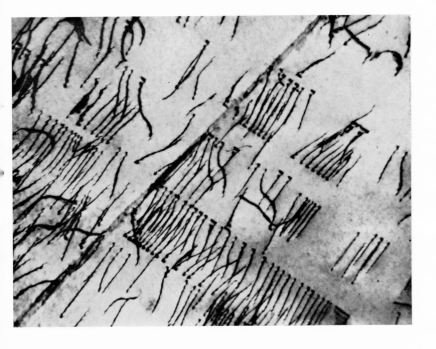

Plate I (a) Etch pits at the ends of a half-loop of dislocation emerging from the surface of a LiF crystal, formed between successive stages of deformation. The largest pits were the earliest and the sequence shows how the half-loop expanded under stress (from Johnston & Gilman, 1959). (b) Electron micrograph of dislocations on the slip planes of a thin Au crystal (by courtesy of J. A. Venables).

(Facing p. 62)

more than one lattice vectors are involved it is possible for this to dissociate into a number of other full dislocations, subject to the condition that

$$\mathbf{b} = \mathbf{b}_1 + \mathbf{b}_2 + \dots \qquad (3.6)$$

To a first approximation we may take $U_f^d = \mu b^2 l$ and find that such dissociation is energetically favourable if

$$b^2 > b_1^2 + b_2^2 + \dots$$

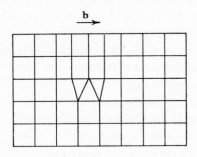

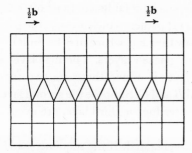

Fig. 25. Dissociation of an edge dislocation into two partials and a region of stacking fault.

This will be the case if $\mathbf{b}$, $\mathbf{b}_1$, $\mathbf{b}_2$, ..., etc. have the same direction. If they form a polygon it is not necessarily so, but it may be possible to find some combinations that favour dissociation. Thus one might expect that dislocations will generally have Burgers vectors that are primitive lattice translations or very simple combinations of these. This prediction is borne out by experiment.

Dissociation may be carried a stage further. Consider the situation illustrated in fig. 25, in which an edge dislocation with $\mathbf{b} = \mathbf{a}$ dissociates into two with $\mathbf{b} = \frac{1}{2}\mathbf{a}$ known as *partial* dislocations as their Burgers vector is some fraction of the translation vector **a**. Between the two lies a planar defect in the lattice, known as a *stacking fault* where slip has proceeded to some metastable position, half way to that produced by a full dislocation. In the structure illustrated we show only one metastable position. Here, slip would proceed in *two* substages and there would be *two* partials. With more complex structures there are often several metastable positions and a full dislocation dissociates into as many partials as there are substages in one unit of slip. These are separated by stacking faults that differ in character from one another.

Returning to fig. 25, the separation of the partials is determined by a balance between the force of repulsion between the partials, which fall off as the separation increases, and the increase in surface energy of the stacking fault, which acts as a surface tension force, pulling the partials together. In metals the separation can vary from a few Å, in open structures like Al where atomic displacements near the dislocation core can be easily accommodated, to ~ 10 Å in full metals like Au where they cannot.

In f.c.c. metals stacking faults are most often found between the closely packed {111} planes, which are also the principal slip planes. Fig. 26 shows how the f.c.c. lattice may be built up by stacking hard spheres in closely packed layers. The first layer is labelled A and when the next layer is added there are two types of site for the spheres, labelled B and C. Suppose the second layer is put in B sites. The alternative sites for the third layer are over A or C on the first layer. If it occupies C sites and one repeats the sequence ABC, the f.c.c. structure is generated. If the sequence had been $ABAB...$the h.c.p. structure would result.

A stacking fault arises whenever a layer is placed to interrupt the sequences ABC, and we shall now show that this is the same type as found between partials. Suppose that an A layer is removed, making $ABCABC.BCABC$. The stacking fault is at the point marked with a dot where a B layer comes directly after a C. If a plane is added, one has $ABC.B.ABC$ and two stacking faults result.

We shall now introduce a vector notation that is useful for describing dislocations and stacking faults in cubic crystals. Three edges of the cubic lattice cell are taken as Cartesian co-ordinate axes. The length of the cube edge is a. The vector joining the origin (o, o, o) to a point (ha, ka, la) is written $a(h, k, l)$ or simply $a(hkl)$. Then the letters in the bracket give the Miller indices of the vector's direction and its magnitude is $a\sqrt{(h^2 + k^2 + l^2)}$. In f.c.c. crystals the line joining nearest neighbours along a (100) direction is a(100), and along a (110) direction: $\frac{1}{2}a$(110). Thus the translation vectors of the primitive f.c.c. lattice are of the type $\frac{1}{2}a\langle 110\rangle$.

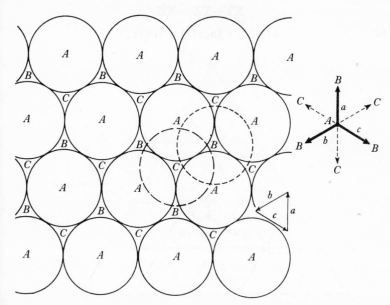

Fig. 26. Generation of a f.c.c. lattice by stacking hard spheres in closely packed {111} layers. The first layer is marked A, the second occupies the position of type B, the third occupies position C. The sequence $ABCABC$ then repeats. (From Read, 1953.)

To illustrate the use of this notation we take the example illustrated in fig. 27 where a dislocation line with Burgers vector $\frac{1}{2}a\langle 110\rangle$ passes over a {111} slip plane of a f.c.c. crystal. Its passage causes slip by an amount $\frac{1}{2}a$(110) and since this is a primitive translation vector no stacking fault is created and the atom originally at X moves to Z, which is an exactly equivalent position. Notice that if it moves first to Y a stacking fault exists between the planes, and the nature of this is defined by giving the *fault vector* $\mathbf{f} = \mathbf{XY} = \frac{1}{6}a$(121) which is the amount by which adjacent planes are sheared at the fault.

5

If the dislocation is dissociated into a pair of partials with stacking fault between, the passage of the first partial moves the atom from X to Y, the second completes the slip by moving from Y to Z. Thus the Burgers vector of the first partial is $\mathbf{XY} = \frac{1}{6}a(121)$ and that of the second is $\mathbf{YZ} = \frac{1}{6}a(21\bar{1})$. The vector triangle $\mathbf{XYZ}$ gives the vector relation

$$\mathbf{XY} + \mathbf{YZ} = \mathbf{XZ}$$

or

$$\tfrac{1}{6}a(121) + \tfrac{1}{6}a(21\bar{1}) = \tfrac{1}{2}a(110).$$

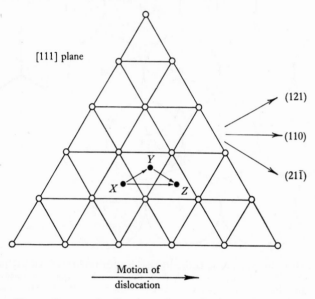

Fig. 27. Vector diagram showing how slip on the {111} plane occurs in two stages $\mathbf{XY}$, $\mathbf{YZ}$ as the two partials of a dislocation with $\mathbf{b} = \mathbf{XZ}$ sweeps from left to right.

The convenience of the notation is obvious for if we take components:

$$x \text{ components:} \quad \tfrac{1}{6}a + 2\tfrac{1}{6}a = \tfrac{1}{2}a,$$
$$y \text{ components:} \quad 2\tfrac{1}{6}a + \tfrac{1}{6}a = \tfrac{1}{2}a,$$
$$z \text{ components:} \quad \tfrac{1}{6}a - \tfrac{1}{6}a = 0.$$

Thus we have a means of adding up the effect of successive dislocations and hence of expressing dislocation reactions. Further examples of this procedure will be found in Chapter 7 later in the book.

3.1.5. *Dislocation networks.* Up to this point, dislocations have been idealized to some extent, but we now turn attention to the situation in real crystals. To express the number of dislocations in a crystal quantitatively one uses the *dislocation density* ρ_d, such that there are ρ_d dislocations per unit area crossing any surface drawn inside the crystal. This is numerically the same as the total length of line per unit volume ($cm/cm^3 = cm^{-2}$).

In spite of the large magnitude of U_f^d, and their very small entropy which should lead to an infinitesimal equilibrium concentration, crystals always contain dislocations. Some of these originate from the condensation of point defects and these will be considered in the next section. Amongst the rest there are those that are 'grown in' and those produced subsequently by plastic deformation.

Take the example of crystals grown by solidification of the liquid phase. For any finite cooling rate it will be impossible for all atoms to find exactly the right position on the liquid/solid interface. It appears from observations that the crystal reduces its free energy by concentrating its defective regions into segments of dislocation line that link up to form a three-dimensional network. A typical example is shown in plate II.

By annealing crystals at high temperature for a long period it is possible to reduce the dislocation density as low as 100 dis. cm^{-2}, but there is a residual network that is practically impossible to remove. Such reductions are brought about by line tension acting to shorten the length of line, or by the dislocations moving out of the crystal as a result of point defects condensing on them. For instance, if a row of vacancies condenses on the edge dislocation in fig. 24, the line moves on to the next plane. Such a process is known as *climb*, and it will receive more detailed attention in later sections.

As a result of the long range interaction forces between dislocations, energy is often reduced by the network forming special arrays. One example is shown in plate III where a set of like edge dislocations form a wall, lying parallel to one another in different slip planes. The result is a *low-angle tilt boundary* in the crystal with a slight rotation of the lattice about an axis in the boundary plane. A boundary formed by a two-dimensional network of screw dislocations is a *low-angle twist* boundary with the rotation axis normal to its plane.

When crystals are grown from dilute liquid solution, or from the vapour, it is sometimes easier to control the rate of arrival of new atoms, and a higher degree of perfection can be attained. Those dislocations that are present often contribute to the growth process by providing special sites on the surface. For example, where a screw dislocation emerges from the surface there is a step, such as is shown in fig. 21. Because the dislocation has turned planes of the crystal into a single helicoid surface, when atoms condense on the step the rotation it executes can continue indefinitely without it being eliminated. This process is responsible for the *spiral growth steps* that are frequently observed on crystal surfaces. A class of crystals which have a high degree of perfection and which grow by this process are the 'whiskers'. Here one often finds a single screw dislocation along the axis which appears to be associated with the growth. An example is shown in plate IV. Such crystals have a tensile strength that approached the theoretical limit set by simultaneous slip, since their growth dislocation is unable to glide in any useful slip plane. This suggests that stress alone is not able to generate dislocations in such a perfect crystal.

When a crystal that already contains a dislocation network is plastically deformed, the dislocation density increases rapidly and may reach a value as high as 10^{12} dis. cm^{-2}. One mechanism by which glissile dislocations can multiply under stress is shown in fig. 28 where a glissile line segment is anchored at its ends. The nature of the anchoring points is unimportant, but they could be nodes in the network where sessile dislocations meet the glissile one. As stress is increased from zero the line bows out to a radius of curvature R given by (3.5). When $R = L/2$, $\sigma \sim 2\mu b/L$, and the bulge will expand without further increase in stress. Eventually the situation in fig. 28c is reached from which a complete glissile loop emerges, leaving an anchored line segment that can repeat the sequence indefinitely. This system is known as the Frank–Read source (Frank & Read, 1950) and has been observed experimentally, as is shown in plate V. Several other mechanisms, such as multiple cross slip (Gilman, 1961) have been suggested by which dislocations

Plate II Dislocation network in Al (courtesy of A. R. Lang).

(*Facing p.* 68)

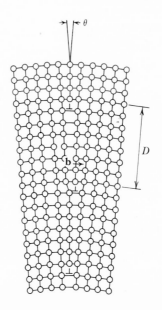

(a)

(b)

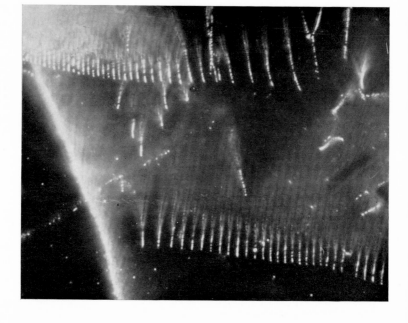

Plate III (a) Showing how a low-angle tilt boundary is composed of a parallel array of edge dislocations. (b) Optical micrograph of edge dislocations forming a low angle tilt boundary in CsBr. The dislocations have been rendered visible by decorating them with particles of Au, diffused in at high temperature. (From Amelinckx, 1962.)

Plate IV Electron micrograph of a 'whisker' crystal of AlN, about 1μ wide, showing the screw dislocation necessary for growth (by courtesy of C. Drum).

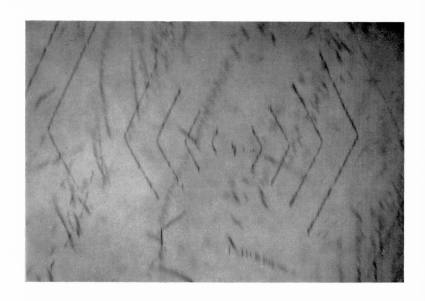

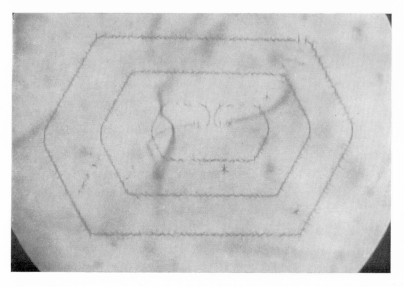

Plate V Optical micrograph of a Frank–Read source in Si decorated with Cu particles (after C. W. Dash). (From Amelinckx, 1962.)

can multiply, and the Frank–Read source is only quoted as an example.

In addition to forming new dislocations, plastic deformation can modify existing dislocations and sometimes produce point defects. As an example, suppose a screw dislocation pierces the slip plane of a second screw dislocation which then passes through it. Then the

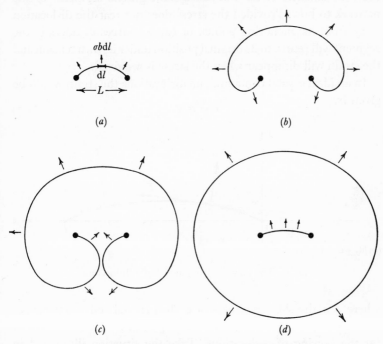

Fig. 28. Stages in the generation of glissile loops by a Frank–Read source.

first line is no longer straight but contains a *jog* corresponding to the Burgers vector of the second. This will impede the motion of the dislocation, since a line of point defects must be formed in the wake of the jog. Whether they are interstitials or vacancies depends on the relative orientation of the two Burgers vectors. Other dislocation processes can also produce lines of point defects. When deformation is at a sufficiently low temperature for some point defects to be retained it is referred to as *cold-working* of the sample. Many attempts have been made to determine point defects properties from annealing experiments with cold-worked samples but results are rather ambiguous since so many different forms of damage are introduced simultaneously.

3.1.6. *Dislocations and elastic properties.* The presence of dislocations can also affect the elastic region of the stress-strain curve, for even the smallest stress will cause some glissile segments of the network to bow. Provided the stress does not tear the dislocation away from its anchoring points, or cause sources to operate, the segment will return to its original position under its own tension and the strain will disappear when the stress is removed.

In an ideal crystal containing no dislocations the strain would be given by

$$\epsilon = \sigma/\mu, \tag{3.7}$$

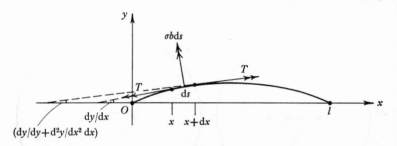

Fig. 29. The bowing string model of a dislocation.

where μ is the shear modulus of perfect crystal and σ is the shear stress. In a real crystal a contribution ϵ_d must be added to account for the bowing of dislocations. Take the situation illustrated in fig. 29 where the bowing line segment is originally of length l. The force on the element ds at (x, y) is $\sigma b\, ds$, and for small bowing we shall assume this to act perpendicular to the axis Ox. The element is also acted on by line tension forces T from either side which make angles $\dfrac{dy}{dx}$ and $\left(\dfrac{dy}{dx} + \dfrac{d^2y}{dx^2}\, dx\right)$ with Ox. For small displacements, $ds \simeq dx$, and there is a resultant restoring force, roughly parallel to Oy, of magnitude $T\dfrac{d^2y}{dx^2}\, dx$, and in equilibrium

$$T\frac{d^2y}{dx^2} = \sigma b,$$

but since $T \sim \mu b^2$, from the text following equation (3.2), we have,

$$\frac{d^2y}{dx^2} = \frac{k\sigma}{\mu b} \qquad (3.8)$$

with k a numerical constant of order unity. Solving (3.8) with the boundary conditions $y = 0$ for $x = 0$ and 1, gives the equation of the bowed line as the parabola:

$$y = \frac{k\sigma}{2\mu b} x(1-x). \qquad (3.9)$$

The area swept out is easily found by integration to be $k\sigma l^3/12\mu b$, and hence from equation (3.3) the strain due to this segment alone is:

$$d\epsilon_d = k\sigma l^3/12\mu. \qquad (3.10)$$

Suppose there are $N(l)\,dl$ segments per unit volume in the range dl at l, all operating in the same manner. Then the total strain from dislocations is:

$$\left.\begin{aligned} \epsilon_d &= \frac{\sigma}{\mu} \int_0^\infty \frac{k l^3}{12} N(l)\,dl \\[2mm] \epsilon_d &= D\sigma/\mu, \end{aligned}\right\} \qquad (3.11)$$

or

with D representing the integral. Adding this to the ideal strain from equation (3.7) gives the total strain:

$$\frac{\sigma}{\mu} + D\frac{\sigma}{\mu}.$$

Defining μ_{eff}, as the effective modulus, this strain can be equated to σ/μ_{eff}, giving:

$$\mu_{\text{eff}} = \frac{\mu}{1+D}. \qquad (3.12)$$

Thus the effective modulus is reduced, as we expected.

The magnitude of the effect depends on $N(l)$ and one can only make reasonable assumptions about this. In the case where one wishes to represent a dislocation network in a well-annealed crystal, where all segments are about the same length and are anchored only at the nodes of the network, one can assume a delta function: for example,

$$N(l) = \frac{\rho_d}{l_0} \delta(l - l_0), \qquad (3.13)$$

where ρ_d is the total length of dislocation per unit volume under the given stress conditions, and l_0 the segment length.

An alternative, of greater relevance in radiation damage, is where pinning is by defects randomly spaced along the dislocation segments in large enough numbers to greatly outnumber the nodes. Then one assumes an exponential function.

$$N(l) = \frac{\rho_d}{l_0^2} e^{-l/l_0}. \tag{3.14}$$

Here, it is easily verified that l_0 is the average segment length. Using the two forms to calculate the integral D, one finds the two corresponding expressions for μ_{eff}:

Delta: $\mu_{\text{eff}} = \mu/(1 + k\rho_d l_0^2/12). \tag{3.15}$

Exponential: $\mu_{\text{eff}} = \mu/(1 + k\rho_d l_0^2/2). \tag{3.16}$

Thus, both assumptions lead to the same general function. We shall use these results when considering irradiation effects in Chapters 6 and 7.

3.1.7. *Internal friction due to dislocations.* It is well known that elastic vibrations in a solid are damped, even in the absence of external forces, and the origin of this damping is referred to in a general way as *internal friction*. Dislocation movement contributes to internal friction by three main processes: *dynamic loss, breakaway* and *relaxation*. Recent reviews which relate this subject to radiation damage have been written by D. K. Holmes (1962), and D. O. Thompson & Pare (1964) and the reader is referred to these for bibliography.

Whenever a dislocation moves through a crystal at finite speed, energy is radiated as phonons and this is the basis of dynamic loss. The assumption is generally made that the resulting force of friction is proportional to velocity. Take the dislocation segment in fig. 29 and suppose the stress to be oscillatory: $\sigma = \sigma_0 \cos \omega t$. The element of segment then has an equation of motion:

$$M\ddot{y} - B\dot{y} - T\frac{\mathrm{d}^2 y}{\mathrm{d}x^2} = \sigma_0 b \cos \omega t,$$

where M is an effective mass per unit length and B is the friction force per unit length for unit velocity. First consider natural vibrations when $B = 0$ and $\sigma_0 = 0$, the solution must be of the standing wave form:

$$y = a \cos \omega_0 t \sin \frac{\pi x}{l}$$

with

$$\omega_0 = \frac{\pi}{l} \sqrt{\frac{T}{M}}$$

the natural frequency, whence

$$\frac{d^2 y}{dx^2} = \frac{-\pi^2}{l^2} y.$$

Provided damping is small the complete solution will also be of this general form and the equation of motion becomes:

$$M\ddot{y} - B\dot{y} + T\frac{\pi}{l^2}y = \sigma_0 b \cos \omega t \qquad (3.17)$$

which will be recognized as the usual equation for forced and damped harmonic motion.

The meaning of M, the mass per unit length, may be seen by noting that ω_0 must be similar to the frequency of transverse acoustic waves with wavelength $2l$, hence

$$\omega_0 = \frac{\pi c}{l},$$

where c is the velocity of sound ($\sim 10^5$ cm/sec). Comparing this with the earlier expression gives

$$Mc^2 = T$$

by the definition of T, or $\quad Mc^2 = U_{\mathrm{f}}^{\mathrm{d}}/l$.

Thus the mass per unit length is directly related to the intrinsic energy of the dislocation, and one notes an interesting analogy with particles in Relativity Physics. The dislocation mass is contributed to by all the atoms that are moved when the dislocation moves, the importance of their contribution depending on their displacement and hence their effect on $U_{\mathrm{f}}^{\mathrm{d}}$.

Fortunately one seldom has to solve (3.17) exactly because for typical segments with $l \sim 10^{-3}$ cm and $c \sim 10^5$ cm/sec the natural frequency of $\sim 10^8$ c/sec is far above the frequency at which most internal friction experiments are conducted. Under these conditions the inertial term, $M\ddot{y}$, can be omitted from the equation and the motion is in phase with the driving force. Neglecting the damping term for the present, the solution is:

$$y(x, t) = \frac{\sigma_0}{2\mu b} x(l-x) \cos \omega t. \qquad (3.18)$$

Inclusion of the damping term makes a slight change in the amplitude that can be neglected in all practical cases. Taking the typical case of Al, $B \sim 10^{-4}$ dyne.cm^{-2}.sec.

Internal friction is generally expressed as a *decrement* Δ, defined as the energy loss per cycle ΔE, divided by twice the maximum elastic strain energy W.

Since
$$\left. \begin{aligned} W &= \tfrac{1}{2}\epsilon\sigma_0 = \frac{\sigma_0^2}{2\mu}, \\ \Delta &= \frac{2\mu\,\Delta E}{\sigma_0^2}. \end{aligned} \right\} \qquad (3.19)$$

The contribution to ΔE by a single segment of length l is calculated by integrating the work done on element ds according to:

$$\mathrm{d}(\Delta E) = \int_0^l \int_0^{2\pi/\omega} B\dot{y}^2 \,\mathrm{d}t\,\mathrm{d}x.$$

Then, calculating $\dot{y}^2$ from equation (3.18) one obtains for the contribution to Δ from a single segment:

$$\mathrm{d}\Delta = \frac{\pi B\omega l^5}{60\mu b^2}. \qquad (3.20)$$

With the delta function form of the segment length distribution from equation (3.13) one obtains the total decrement as:

$$\Delta = \frac{\pi B\rho_d\,\omega l_0^4}{60\mu b^2}. \qquad (3.21)$$

Using the alternative exponential distribution (3.14) gives the same function but without the factor 60 in the denominator.

The dependence on l_0^4 makes Δ very sensitive to small changes in the pinning points, as we shall see in Chapters 6 and 7. The dependence on temperature is mainly through B, which increases linearly with T. Δ is proportional to ω and it is this that generally makes the dynamic loss dominate the others at high frequency ($>$ 10^6 c.sec^{-1}).

Note, however, that dynamic loss is independent of stress, and hence strain amplitude ϵ_0.

This is not the case with *breakaway* loss, which is due to dislocations being pulled away from their pinning points on the forward quarter cycle, catching on the original pins on stress reversal, and pulling away again in the third quarter cycle. The breakaway energy must be supplied twice in a cycle and it is this that causes the decrement. Naturally this loss increases for larger stresses, or strain amplitudes, as the line can be pulled away from an increasing proportion of pins, and its presence is usually detected by an increasing graph of Δ *versus* ϵ_0.

This mechanism is very dependent on temperature since breakaway is assisted by thermal activation. For a given temperature the probability of breakaway per cycle is roughly proportional to the jump frequency from (2.9) multiplied by the period of vibration, i.e. the number of jumps per cycle.

Hence

$$\Delta \propto \frac{\nu}{\omega} e^{-U/kT} \tag{3.22}$$

with U the activation energy for breakaway. This is a very crude treatment and this point must be emphasized, for no satisfactory theory has yet been developed. However, it serves to show how decrement due to breakaway will be negligibly small until a critical temperature range where it will rapidly take over from all other mechanisms. The behaviour with respect to ϵ_0 is very similar. The inverse proportionality to frequency shows that this mechanism is most likely to dominate at low frequencies ($\sim$ kc.sec^{-1}) where dynamic loss is small. At megacycle frequencies the experimental strain amplitude is rarely large enough for breakaway to occur.

Finally we come to *relaxation* loss. Suppose a crystal contains defects with several alternative geometrical configurations, each having the same formation energy and therefore being equally populated states, but being separated by a maximum in free energy. If the applied stress makes one configuration energetically favourable, thermal activation will tend to shift the population in this direction. On reversing, the stress activation will again make the defect change into the newly favoured configuration. The jump frequency j for thermal activation is $\nu \exp(-U/kT)$ from equation (2.9), with U the activation energy to change configurations. If

$\omega \ll j$ the configuration changes follow the stress without phase lag. If $\omega \gg j$ there is no effect at all. But when $\omega = j$ a relaxation phenomenon occurs, and the energy loss rises to a maximum. Experimentally one observes a peak in the graph of Δ *versus* T or ω. The characteristic temperature T_r is related to the relaxation frequency ω_r by

$$\omega_r = \nu \exp\left(-U/kT_r\right). \tag{3.23}$$

Away from the critical frequency this loss is usually swamped by one of the others.

With dislocations, relaxation can occur between two stable positions of pinning. Hutchinson, Rogers & Turkington (1964) for example, suggest that a vacancy can pin by either of the two partials of a dislocation, and that activation energy is required to shift the dislocation from one position to the other.

Certain types of point defect can also cause relaxation loss. One requires that they should possess an axis of symmetry whose rotation is favoured by applied stress. For instance, Seeger (1962) suggests that a relaxation peak observed in Ni at room temperature with 1 c.sec^{-1}, is due to the rotation of $\langle 100 \rangle$ split interstitials.

3.2. Clusters of point defects

3.2.1. *Vacancy clusters.* In §2.3.2 and 2.3.3 it was shown to be energetically favourable for n vacancies to collect together into a spherical void of roughly volume $n\Omega$, for which the formation energy was:

$$U_f^{\text{sphere}} = (6\sqrt{\pi}\, n\Omega)^{\frac{2}{3}} \gamma. \tag{3.24}$$

Although this shows that clustering is a favourable process the spherical void is not necessarily the cluster form with the lowest energy. In this section we shall examine this and other configurations, and in 3.2.2 extend the examination to clusters of interstitials.

Suppose n vacancies condense to form a flat circular void of radius r and equal in thickness to one vacancy ($\sim \Omega^{\frac{1}{3}}$). Using the rigid continuum model its formation energy is:

$$U_f^{\text{disc}} = 2\pi r^2 \gamma \tag{3.25}$$

and since
$$
\left.\begin{aligned}
n\Omega &\simeq \pi r^2 \Omega^{\frac{1}{3}}, \\
U_f^{\text{disc}} &\simeq 2n\Omega^{\frac{2}{3}} \gamma.
\end{aligned}\right\} \tag{3.26}
$$

It is obvious that for a given large n the sphere has the smallest surface area and hence the lowest formation energy. The above expressions give:

$$U_f^{\text{disc}}/U_f^{\text{sphere}} \to 0\cdot 44 n^{\frac{1}{6}}$$

for large n, which clearly shows the tendency.

But if a disc shaped void collapses in its central region to reunite the exposed surfaces, it releases a large amount of surface energy and we are left with a dislocation loop like that in figure 3.2.6. In this process the surface energy has been converted to strain energy. Taking μb^2 as the formation energy per unit length of dislocation line, we have for the loop:

$$U_f^{\text{loop}} \simeq 2\pi r \mu b^2 \tag{3.27}$$

then using (3.25) for r

$$U_f^{\text{loop}} \simeq 2\sqrt{(n\pi)}\,\Omega^{\frac{1}{3}}\mu b^2. \tag{3.28}$$

Comparing (3.27) with (3.25) shows that when r exceeds a critical value r_0, collapse becomes energetically favourable, and

$$r_0 \sim \mu b^2/\gamma.$$

Taking typical values: $\mu b^2 \sim 3\,\text{eV}/\text{Å}$, $\gamma \sim 0\cdot 1\,\text{eV}/\text{Å}^2$ and $r_0 \sim 30\,\text{Å}$. Comparing the loop with the spherical void gives:

$$\frac{U_f^{\text{loop}}}{U_f^{\text{sphere}}} \to \frac{\mu b^2}{\gamma}\,\frac{2\pi^{\frac{1}{6}}}{6^{\frac{2}{3}}n^{\frac{1}{6}}\Omega^{\frac{1}{3}}}$$

showing that for large enough n the loop has the lowest formation energy.

The collapse of the disc need not necessarily join the lattice planes in the correct manner, and the loop will then enclose an area of stacking fault. For instance, in an f.c.c. crystal the disc void may be formed by removing part of a $\{111\}$ A plane. Collapse perpendicular to the plane, by a vector $\frac{1}{3}a(111)$, will lead to a B and C plane in contact. The resulting stacking fault $ABC.BCABC$ in principle could be removed by a shearing in the $\{111\}$ plane by a vector $\frac{1}{6}a(11\bar{2})$. One could regard this operation as the dissociation of a partial from the loop, with Burgers vector $\frac{1}{6}a(11\bar{2})$, which sweeps across the loop to produce the desired shear. The reaction is

$$\tfrac{1}{3}a(111) + \tfrac{1}{6}a(11\bar{2}) \to \tfrac{1}{2}a(110)$$

and if it is energetically favourable will result in a loop of full dislocation ($\mathbf{b} = \frac{1}{2}a(110)$) with no stacking fault enclosed. The process is illustrated in fig. 30.

In the electron microscope* it is possible to observe vacancy clusters in quenched metals, the minimum size resolvable under

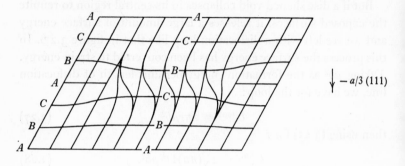

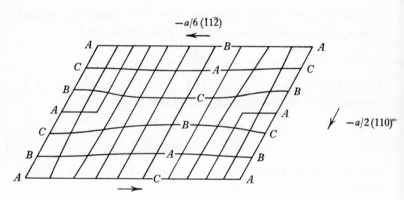

Fig. 30. Showing how a collapsed vacancy loop with $\mathbf{b} = -\frac{1}{3}a(111)$ undergoes shear by $-\frac{1}{6}a(11\bar{2})$ to remove the stacking fault and become $\mathbf{b}-\frac{1}{2}a(110)$.

ideal conditions being of the order of 10 Å. Clusters have been seen in many forms: voids, loops with and without stacking fault, and more complex defects. Plate VI shows an example of voids in quenched Al, first seen by Kiritani & Yoshida (1963, 1964), Kiritani, Shimomura & Yoshida (1965), and Yoshida *et al.*

---

* The detailed techniques of electron microscopy and their theoretical basis are outside the scope of this book and the reader is referred to articles by Amelinckx (1964) and Hirsch (1962), or to the book *Transmission Electron Microscopy of Metals* by G. Thomas (Wiley, 1963). A brief introduction will be found in § 3.2.3 below.

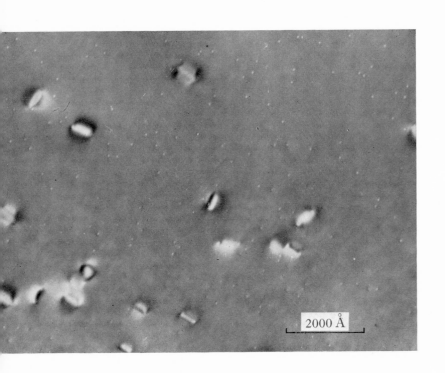

Plate VI Octahedral voids in quenched Al. (From Kiritani, 1964.)

(*Facing p.* 78)

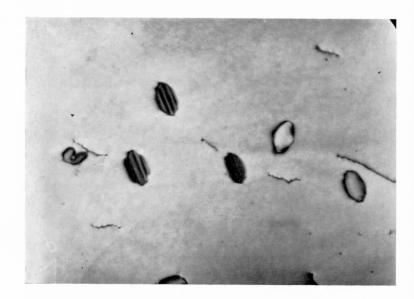

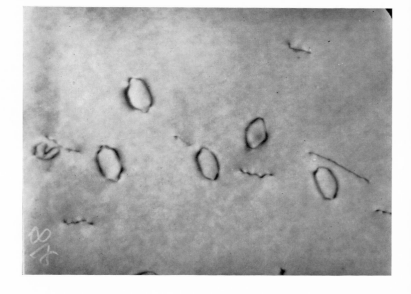

Plate VII Electron micrographs of hexagonal loops about 1000 Å diameter, on {111} planes in quenched Al. In (a) some loops show contrast that indicates the presence of enclosed stacking fault. After slight heating in (b) this disappears by the process of fig. 30 (by courtesy of D. J. Mazey).

(1965 a, b). Loops were also seen and it appears that void formation is favoured by high purity and slow quenching rates ($\sim 10^4$ deg. sec^{-1}). Annealing the Al specimens at about 50 °C produced the large voids seen in plate VI. These have a crystallographic shape and are octohedra formed of {111} planes. Since the {111} planes are the most closely packed they have the lowest value of γ and hence this shape is not surprising.

Historically, the first clusters were loops in Al reported by Hirsch, Smallman *et al.* (1958), and plate VII shows a micrograph of quenched Al obtained by D. J. Mazey in which hexagonal loops are clearly visible. At first they have a dark striped centre, which indicates the presence of an enclosed stacking fault, but this is seen to disappear suddenly on heating or stressing the foil. It therefore seems that some activation energy is needed to form the necessary partial. The orientation of loops can be determined by comparing the micrographs with electron diffraction patterns obtained in the microscope from the same area of foil (see footnote on p. 78).

Then it is found that the hexagon lies in a {111} plane with edges parallel to $\langle 110 \rangle$ directions. There are three other {111} planes in the crystal and each pair of parallel sides of the hexagon lies on one of these. It is observed that in the unfaulted loop, under stress, two pairs of sides are able to bow out on to the other {111} planes and are therefore glissile. This shows that the Burgers vector of the loop is along a $\langle 110 \rangle$ axis, out of the loop plane, consistent with the scheme presented above.

Image contrast of loops in the electron microscope is due to local changes in diffraction conditions near to the dislocation line. In favourable circumstances it is possible to determine the Burgers vector by following the change in contrast as the specimen is tilted. Such techniques are discussed in § 3.2.3. In the f.c.c. metals that have been examined one finds that the loops formed from quenched-in vacancies have $\mathbf{b} = \frac{1}{3}a \langle 110 \rangle$ as expected.

The number of vacancies involved in a regular hexagonal loop with sides of length L is found from:

$$n\Omega = 3L^2(a/\sqrt{3})$$

a is the f.c.c. cell constant hence $a/\sqrt{3}$ is the effective thickness of each $\{111\}$ plane and $\Omega = a^3/4$, then

$$n = 4\sqrt{3}\,L^2/a^2. \tag{3.29}$$

By counting the number of loops in a known volume of foil and determining their size distribution, it is possible to estimate the concentration of vacancies retained in the loops. Order of magnitude agreement is obtained with expression (2.8), showing that a significant fraction of quenched-in vacancies condense to form loops.

Makin & Hudson (1963) find that in some f.c.c. metals the loops often lie on $\{120\}$ planes rather than $\{111\}$. A calculation of the elastic energy associated with a loop by Bullough & Foreman (1964) shows that this is a manifestation of a general result, that the minimum energy occurs when the loop plane is close to $\{120\}$. It seems likely that although loops are initially formed on $\{111\}$ planes they are sometimes able to rotate into the equilibrium position. An ideal situation is illustrated in plate VIII, where a rhombus-shaped loop with $\mathbf{b} = \tfrac{1}{3}a(110)$ has its line segments on two pairs of $\{111\}$ planes forming a prism with axis (110). Because they lie on slip planes containing the Burgers vector all segments are glissile and the loop is free to rotate by slip. Bullough showed that its equilibrium position is close to the $\{120\}$ plane.

Of course, a loop having non-glissile segments will not be able to rotate by slip. For instance, in a hexagonal loop only four out of six segments are glissile. In such cases rotation can only occur if atoms can be redistributed around the loop perimeter. This is thought to occur by a process of *self-climb by pipe diffusion*, in which vacancies formed on one side of the loop find an easy migration path along the dislocation line. Such diffusion is generally easier than normal self-diffusion and occurs at lower temperatures.

The question of interaction forces between loops has also received some theoretical attention. Foreman & Eshelby (1962) have calculated the interaction energy for a pair of like parallel loops with their centres joined by the vector (r, θ), in polar coordinates. This energy has its origin in the overlap of two elastic strain fields, and since these are each very anisotropic the force may be either attractive or repulsive according to relative position.

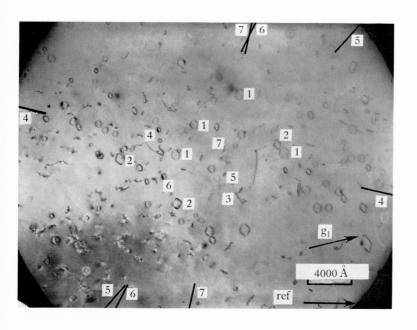

Plate VIII Rhombus-shaped loops in quenched Al–1% Mg.
(From Makin & Hudson, 1963.)

(*Facing p.* 80)

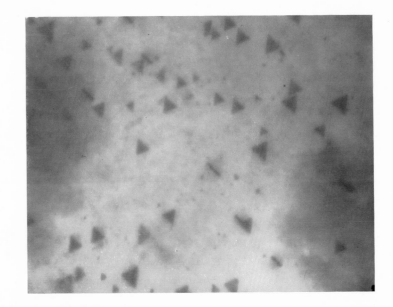

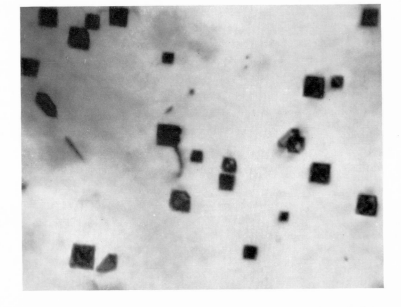

Plate IX Electron micrographs of stacking fault tetrahedra in quenched Au. (a) Viewed along ⟨111⟩ when their projection is triangular and (b) viewed along ⟨100⟩ when it is square. The characteristic striped contrast of stacking fault is visible (by courtesy of P. Hirsch).

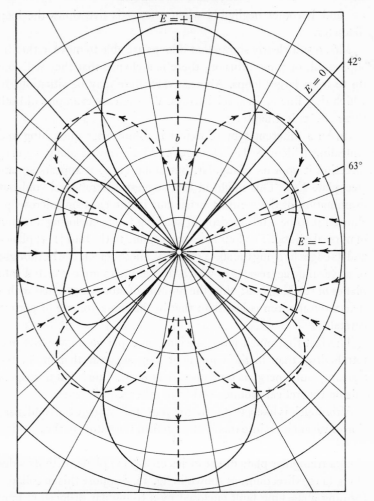

Fig. 31. Showing how two small dislocation loops interact as a function of their relative positions. One is held stationary at the origin whilst the other is moved about with its Burgers vector parallel to the first. Solid lines connect points of equal interaction energy E, whilst lines of force are shown broken. (From Foreman & Eshelby, 1962.)

Fig. 31 shows the contours of constant interaction energy for the case where **b** is normal to the loop plane for each loop. Near $\theta = 0$, when the loops are above one another, energy is positive and the loops repel. Near $\theta = \frac{1}{2}\pi$ the loops approach the same plane and the force is attractive. The interaction energy in each case falls off as

6

r^{-3} and becomes negligible when r exceeds five times the loop diameter.

This result clearly shows that if loops are able to move, either by slip, climb or pipe diffusion, there is a chance that they will join up to form larger loops. Alternatively there may be situations in which they line up in special arrays. We shall see examples of both these effects in Chapter 7.

In Au a different type of defect is found after quenching and annealing (Silcox & Hirsch, 1959). When viewed along a $\langle 111 \rangle$ direction it appears triangular, but along $\langle 110 \rangle$ it appears square (see plate IX). This might at first suggest a tetrahedral void but examination of the contrast behaviour in the electron microscope showed it to be a tetrahedron of stacking faults enclosing a region of normal crystal. The four faces correspond to the four $\{111\}$ planes and its edges lie along the six $\langle 110 \rangle$ axes. These defects can be thought of as vacancy aggregates in which the vacancies are converted into planes of stacking fault, thus the number of atoms associated with a defect tetrahedron is less than that in the same volume of perfect crystal.

The fact that much more stacking fault per vacancy exists in a tetrahedron than in the equivalent loop suggests that in Au the stacking fault energy is much less than in Al. One is led to suspect this on theoretical grounds and from other experiments.

Although it is likely that from the very earliest stage the condensation of vacancies is into this form of defect, the nature of the stacking faults can best be found by considering the formation to proceed from a triangular plate of vacancies on a $\{111\}$ plane, with its sides along $\langle 110 \rangle$ directions, as shown in fig. 32. Suppose this to collapse forming a stacking fault enclosed by a triangular dislocation loop with $\mathbf{b} = \frac{1}{3}a(111)$. A regular tetrahedron can be drawn through this triangle as its base, whose faces form $\{111\}$ planes and whose sides are the six $\langle 110 \rangle$ directions. The triangular dislocation throws off a partial of the type $\frac{1}{6}a\langle 121 \rangle$ onto each of the upright faces of the tetrahedron, leaving a dislocation of the type $\frac{1}{6}a\langle 110 \rangle$ on the base according to:

$$\frac{1}{3}a(111) \to \frac{1}{6}a(101) + \frac{1}{6}a(121).$$

The partials $\frac{1}{6}a\langle 121 \rangle$ bow out on the upright faces and meet one another along the upright edges of the tetrahedron forming new dislocations of the same type, $\frac{1}{6}a\langle110\rangle$, that surround the base:

$$\tfrac{1}{6}a(121)+\tfrac{1}{6}a(\overline{1}\overline{1}2)\rightarrow\tfrac{1}{6}a(01\overline{1}).$$

Then the whole tetrahedron is constructed of $\frac{1}{6}a\langle110\rangle$ type disloca-tion and each face covered with the same type of stacking fault.

To calculate the number n of vacancies involved in a tetrahedron of side L we simply consider the triangular plate that could be formed on its base by reversing the above procedure.

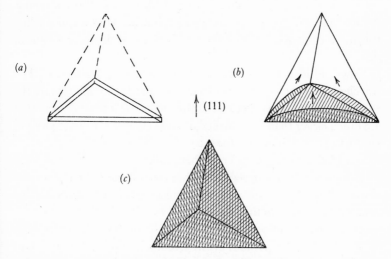

(a)

(b)

$\uparrow$ (111)

(c)

Fig. 32. Illustrating the formation of a stacking fault tetrahedron on $\{111\}$ planes from a triangular plate of vacancies.

Then

$$n\Omega = \tfrac{1}{2}L^2(a/\sqrt{3})$$

and

$$n = 2L^2/\sqrt{3}\,a^2. \tag{3.30}$$

From counts of the number of tetrahedra it appears again that the concentration of condensed vacancies is of the same order of magnitude as that predicted by expression (2.8). The fact that in Au one forms tetrahedra of stacking fault rather than the loops found in Al and Cu suggests that the stacking fault energy must be lower in Au than in the others. Estimates of stacking fault energy from other experiments are in accord with this idea.

It is interesting, at this stage, to reconsider the fate of vacancies in a quenching and annealing experiment, such as described in § 2.3.8.

As the annealing temperature is raised, and vacancies become mobile, some will condense to form loops or tetrahedra. Those near to the edges of a crystal may escape rather than condense, leading to the denuded zones free from clusters that are seen close to grain boundaries or free surfaces. Others may condense on the existing dislocation network of the crystal causing the dislocation lines to move by the process known as *climb*. In the case of a pure edge dislocation the extra half plane of atoms is eaten away by the vacancies and the line climbs perpendicular to its Burgers vector. A pure screw dislocation cannot climb but becomes a spiral when vacancies condense on it. This may be seen in the micrograph of plate X.

Now consider the resistivity decrease during annealing. The wavelength of the conduction electrons is of the same order as the vacancy's dimensions. Because scattering is weak from defects that are much larger than the wavelength, one expects the resistivity per vacancy to fall as these become incorporated into loops or tetrahedra, in accordance with the observed behaviour.

As the annealing proceeds to higher temperatures the dislocation loops will try to boil off vacancies and shrink in the process. The smallest loops will do this most readily since the line tension acts to assist the release of vacancies. Indeed, small loops are often seen to shrink whilst larger ones grow. Eventually the temperature will be sufficiently high for all loops to disappear and this is observed at temperatures approaching those where self-diffusion first becomes appreciable (roughly 40 % of the absolute melting temperature). Tetrahedra in Au are evidently more stable than loops; persisting up to 600 °C, well above the self-diffusion temperature.

3.2.2. *Interstitial clusters.* These have most often been observed in irradiated solids and the experimental evidence must be deferred until Chapter 7. Here, we shall simply make some theoretical remarks about their possible form.

Spherical clusters of interstitials are out of the question since they must obviously require a very large strain energy in the surrounding matrix. Only plates or needle-shaped clusters appear feasible. The plate can be thought of as part of a new plane of atoms surrounded by a dislocation loop. The most likely plane in f.c.c.

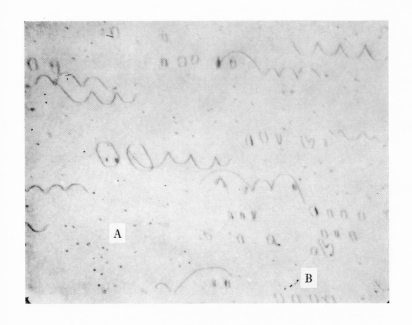

Plate X Electron micrograph of spiral dislocations in quenched Al–1·2% Si.
This behaviour is expected for screw dislocations with vacancies condensed on
them. (From Westmacott *et al.* 1961.)

(a)	(b)	(c)	(d)	(e)	(f)	
0°		20°		26°		

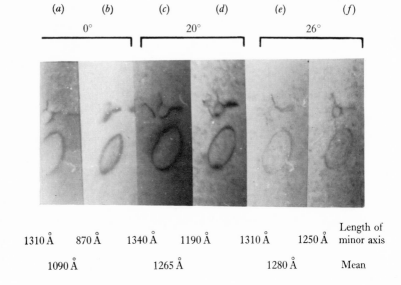

1310 Å	870 Å	1340 Å	1190 Å	1310 Å	1250 Å	Length of minor axis
1090 Å		1265 Å		1280 Å		Mean

Plate XI The change in appearance of loops produced by high temperature deformation of MgO as the specimen is continuously tilted. The pairs of photographs (a) and (b), (c) and (d), (e) and (f) were taken on opposite sides of extinction contours. (From Edmonson & Williamson, 1964.)

metals is the $\{111\}$ plane and we have seen in § 3.1.4 that there is a stacking fault on either side of the extra plate of atoms. In this state one has a loop with $\frac{1}{3}a(111)$ but if this is swept by two partials of the type $\frac{1}{6}a\langle 121\rangle$, one above and one below the plane, the stacking faults can be removed to give a loop with $\frac{1}{2}a(110)$.

$$\frac{1}{3}a(111) + \frac{1}{6}a(2\overline{1}\overline{1}) + \frac{1}{6}a(\overline{1}2\overline{1}) \to \frac{1}{2}a(110).$$

3.2.3. *Observation in the electron microscope.* In quenching experiments one has some justification for assuming that loops seen in the electron microscope are formed by vacancies clustering, although there are cases where mechanical strains during the quench can produce loops. In irradiation experiments no such assumption is justified since both types of point defect are produced. It is clearly of great importance to find an experimental technique for distinguishing between the two types of loop.

In principle it is possible to determine the magnitude and direction of the Burgers vector from a set of measurements in the electron microscope, involving a complete determination of the electron diffraction conditions in relation to the observed contrast behaviour, and the inclination of the loop to the foil (Hirsch, Howie & Whelan, 1960; Howie & Whelan, 1962). Knowing **b** it is a simple matter to decide whether planes of atoms were added or removed in forming the cluster and hence whether interstitials or vacancies were involved. One also knows how many planes were added or removed and what type of stacking fault, if any, is enclosed by the loop. These measurements are often difficult to make and to interpret. The detailed electron diffraction theory is outside the scope of this book and the reader is referred elsewhere for a complete description of them (see footnote in § 3.2.1).

In many cases, however, it is enough to distinguish between interstitial and vacancy clusters and a simple technique suggested by Edmondson & Williamson (1964) provides a good example of the principles involved in such measurements. This need only involve us in a simple introduction to diffraction contrast theory. In the electron microscope, a beam of mono-energetic electrons from the electron gun passes through a thin foil sample which is in the object plane of a system of electron-optical lenses which eventually form a greatly magnified image on a fluorescent screen. The specimen's crystal lattice produces a set of diffracted beams each corresponding to Bragg reflection from a particular lattice plane.

Each diffracted beam is brought to a separate focus in the image plane of the objective lens. At this point a diffraction pattern is formed, and if desired an image of this can be produced on the viewing screen. Normally, however, the lenses are set to produce an image of the specimen on the screen and a small moveable aperture is used in the image plane of the objective to select one or more beams. If one excludes all the diffracted beams and selects only the directly transmitted beam, a *bright field* image of the specimen is formed. In this condition any rotation of the crystal that enhances the diffracted beams does so at the expense of the direct beam, producing a darker field on the screen. Conversely, a rotation that reduces diffraction brightens the image. A defect in the crystal that causes a local rotation will either brighten or darken the image near the defect, according to the sense of rotation, and it is this effect that renders defects, such as dislocations, visible.

In *dark field* conditions one selects one of the diffracted beams with the aperture. Unless the specimen is exactly oriented for the Bragg condition of this beam the field of view appears dark, and then any defect that rotates the lattice locally to produce diffraction will appear bright.

In practice the crystals are often slightly curved, since they are extremely thin ($\sim 10^2$ to 10^3 Å), and it is unusual to find a large area over which the same Bragg diffraction conditions operate. The bright field image is then crossed by dark bands, known as *bend extinction contours*, which mark the regions where the crystal is oriented to make one or more diffracted beams very strong. These contours therefore show the progressive rotation of the lattice across the specimen due to its curvature. If the specimen is rotated as a whole, on a tilting stage, these contours move across the field. In addition there may be a second type of dark band known as *thickness extinction contours* due to variations in thickness of the specimen. These do not sweep across the field when the specimen is tilted.

Fig. 33 shows a schematic section through a dislocation loop formed by interstitial clustering. In regions A and B there are local

rotations of the lattice in opposite senses, indicated by the circu-
lating arrows. If the foil is oriented so that the loop appears just to
one side of a bend extinction contour one set of reflecting planes are
just approaching the Bragg condition. But in one or other of the
regions A and B the local rotation will be in the right direction to

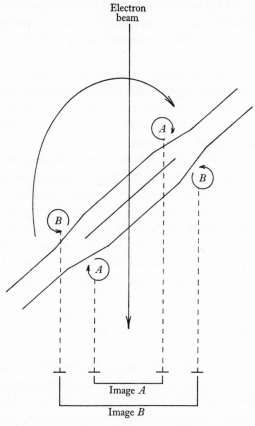

Fig. 33. Observation of the lattice rotation due to different parts of a dislocation
loop and the determination of the sign of the Burger's vector (see text).

enhance the diffraction condition. Suppose this is region A, then
A will appear dark whilst B, where the rotation diminishes the
diffraction, appears brighter. The image will then appear slightly
larger than the loop defined by the dislocation core. If the whole
crystal is now rotated by a small amount in the same sense as A,
just sufficient to move the bend extinction contour through the

loop, the situation will be reversed. In A the local rotation reduces the diffraction and A appears bright, whereas B now appears dark. The loop image will apparently contract, becoming more elliptical, in the situation illustrated. One has next to decide in which sense the loop is inclined to the horizontal, and this is done by observing whether there is a general tendency to become more or less elliptical as a large external rotation is applied in a known direction.

Plate XI shows how these properties can be applied to distinguish between interstitial and vacancy loops. The external rotation is clock-wise so that in a large rotation each image becomes less elliptical as a general trend. Take the situation at (b) and (f). Just before bend extinction contour crosses the loop, regions A will be dark, since they contain clockwise rotations and anticipate the external rotation. After the contour passes, regions B will be dark since their rotation tries to restore the dark extinction condition. From the diagram it is clear that during the passage of the contour the interstitial loop has apparently become *less* elliptical whereas the vacancy loop has become *more* elliptical. The general tendency of the external rotation was to make the loops appear *less* elliptical. One may generalize to say that when the apparent change as the bend extinction contour passes *enhances* the general trend of ellipticity due to external rotation, one has an *interstitial* loop. When the apparent change *reverses* the general trend one has a *vacancy* loop.

An example of this technique is shown in plate XI where the micrographs show the behaviour of a loop in quenched MgO. As external rotation proceeds the general trend for a *less* elliptical loop is always *reversed* when a bend extinction contour passes through. This is therefore a vacancy loop. It must be emphasized that this simple version of the technique is chosen mainly as an illustration of the physical principle involved. If the Burgers vector were inclined at a small angle to the loop plane the loop behaviour might be difficult to interpret since the regions A and B would be shifted from the positions shown in fig. 33. Further, we have neglected the alteration of Bragg conditions by the local change in lattice spacing and have concentrated only on rotations. Hirsch and his colleagues suggest that the second effect is more important in the vast majority of cases, but some care is necessary in applying the above procedure.

CHAPTER 4

THE PRIMARY EVENT

4.1. Introduction

Having discussed the nature of defects, we now consider how they are produced in a crystal by irradiation. The primary event is the interaction between radiation and solid that leads to an atom recoiling from its lattice site. If the recoil is sufficiently energetic, at least one interstitial-vacancy pair is formed, and in many cases the primary recoil energy will be distributed in a cascade of atomic collisions ejecting secondary atoms from their sites. Such cascades are the subject of Chapter 5.

The simplest primary event is the collision between a charged particle and the atomic nucleus. This can be treated as a two-body collision provided that the mean free path between collisions is much greater than the interatomic spacing. The chance of correlation effects due to neighbouring atoms recoiling almost simultaneously is then very small. The systems treated in this chapter will satisfy this condition, but multiple collisions must be considered in Chapter 5.

The momentum of the recoiling atom is the parameter which determines the damage to the solid structure and it will be our aim to calculate it. The interaction of radiation with atoms is generally considered from the viewpoint of the scattered particle, here however we shall be most concerned with the energy of the recoil atom and the angle between its path and that of the incident particle. First let us derive some general relations governing two-body collisions considering only the asymptotic values of momentum at great distances from the collision. There is then no violation of quantum laws in assigning a sharp momentum to a particle since we are not localizing its position along the path. The Principles of Conservation of Momentum and Energy are all that are required to obtain recoil energy as a function of recoil angle. Initially we shall assume that collisions are elastic and, further, that velocities are small enough for non-relativistic mechanics to apply.

Let the mass, velocity and energy of the incident particle be M_1, u_1 and E_1 before collision; the angle through which it is scattered θ_1 and its final velocity v_1. Let M_2 be the mass of the struck particle, v_2 its final velocity and θ_2 the angle of recoil (fig. 34).

Some simplification is brought about by treating the motion relative to the centre of mass G in what will be referred to as G co-ordinates. The co-ordinate system at rest in the laboratory will be called the L system. The velocity $\mathbf{u}_g$ of G in L co-ordinates is found

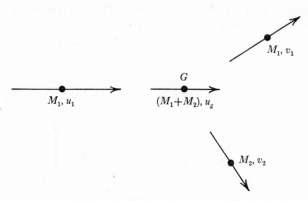

Fig. 34. A two-body collision in laboratory (L) coordinates.

by assuming a particle of mass $(M_1 + M_2)$ to be travelling with $\mathbf{u}_g$ and equating its momentum to that of the incident particle: $M_1 \mathbf{u}_1$.

Hence

$$\mathbf{u}_g = M_1 \mathbf{u}_1/(M_1 + M_2). \tag{4.1}$$

By subtracting this vector from all velocities in the problem one transforms from L to G co-ordinates, as shown in fig. 35 where the meaning of $\mathbf{V}_1$, $\mathbf{V}_2 \mathbf{U}_1$ and ϕ are defined. Because G remains at rest, $\mathbf{V}_1$ and $\mathbf{V}_2$ are in opposite directions. A further simplification arises from conservation of energy and momentum which require:

$$M_1 U_1^2 + M_2 u_g^2 = M_1 V_1^2 + M_2 V_2^2, \tag{4.2}$$

$$M_1 U_1 + M_2 u_g = M_1 V_1 + M_2 V_2. \tag{4.3}$$

In L co-ordinates the equations would include trigonometric functions and would not possess the simple form of this pair. Because (4.2) and (4.3) are similar to the extent that where (velocity)^2 appears in (4.2) the corresponding (velocity) appears in (4.3) they can only be simultaneously satisfied by each particle leaving G with the same speed with which it approached, i.e.

$$U_1 = V_1 \quad \text{and} \quad u_g = V_2. \tag{4.4}$$

Then, calculating V_1 from figure 35 and (4.1):

$$V_1 = M_2 u_1 / (M_1 + M_2) \tag{4.5}$$

the simplified collision is represented in fig. 36.

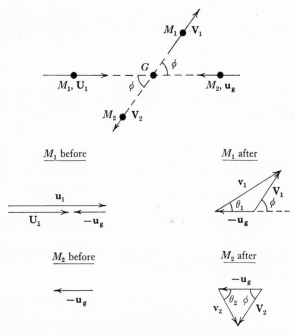

Fig. 35. The transformation to centre-of-gravity (G) coordinates.

An important quantity is the sum kinetic energy, either before or after collision, in the G system. This is easily shown to be:

$$M_2 E_1 / (M_1 + M_2). \tag{4.6}$$

It is easily verified that this, when added to the kinetic energy of the centre of mass $\frac{1}{2}(M_1 + M_2) u_g^2$ is just E_1 confirming that the total energy of the two bodies is unchanged by the new description in G co-ordinates.

We require the recoil energy E_2 in L co-ordinates and must therefore calculate v_2 from the appropriate vector triangle in fig. 35 using the cosine law to give:

$$v_2^2 = 2(1 - \cos \phi) M_1^2 u_1^2 / (M_1 + M_2)^2 \tag{4.7}$$

and hence
$$E_2 = \Lambda E_1 \sin^2 (\phi/2) \tag{4.8}$$
with
$$\Lambda = 4 M_1 M_2 / (M_1 + M_2)^2.$$

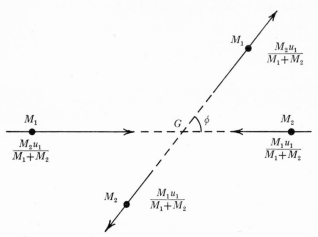

Fig. 36. The two-body collision in G co-ordinates.

Λ has special significance for when $\phi = \pi$ (head-on collision in classical terms) and particles approach and recede along one axis, one has the *maximum possible energy transfer* with:

$$\hat{E}_2 = \Lambda E_1. \tag{4.9}$$

When $\phi = 0$, no scattering occurs and $E_2 = 0$ as one would expect.

In any radiation damage problem there will be a minimum recoil energy, $\breve{E}_2$, which is just capable of producing damage. There is also a minimum bombarding energy $\breve{E}_1$ which has $\breve{E}_2$ as its maximum possible recoil energy, i.e.

$$\breve{E}_2 = \Lambda \breve{E}_1. \tag{4.10}$$

Clearly, we are only concerned with particles for which $E_1 > \breve{E}_1$. Some values of $\breve{E}_1$ for various combinations of bombarding particle and target are given in table IX, assuming the $\breve{E}_2 = 10\,\mathrm{eV}$.

TABLE IX. *Values of E_1^*, the energy to be greatly exceeded by the product $\phi^2 E_1$ in order for classical collision mechanics to apply. Also shown are values of $\check{E}_1$, the threshold energy for various particles to produce damage*

Incident particle	Target atom	$\dfrac{a_0}{a}$	E_1^* (eV)	$\check{E}_1$ (eV)
electron	Li	1·4	—	10^4
electron	U	1·5	—	10^6
proton	Li	1·8	10^{-2}	10^1
proton	U	4·6	10^{-3}	10^3
heavy ion ($M_1 = 100$)	Li	3·8	10^{-4}	10^1
heavy ion ($M_1 = 100$)	U	5·7	10^{-5}	10^3

The relation between θ_2 and ϕ is obtained by considering the components of M_2's velocity parallel and perpendicular to the axis of incidence. The values in L co-ordinates must equal those in G co-ordinates when appropriate components of $\mathbf{u_g}$ are added, i.e.

$$V_2 \cos \phi + v_2 \cos \theta_2 = u_g, \quad V_2 \sin \phi = v_2 \sin \theta_2.$$

Hence using expression (4.4) above for V_2 we find

$$\tan \theta_2 = \sin \phi / (1 - \cos \phi). \tag{4.11}$$

Before being able to calculate the number of recoils in the energy interval dE_2 at E_2 one must know the angular distribution function giving the probability of recoil into $d\phi$ at ϕ. This can only be obtained with a detailed knowledge of the interaction forces between particles M_1 and M_2. Various types of interaction will be dealt with later, but some general methods of calculation can be established. The primary decision is whether a particular problem requires quantum mechanical solution or if a classical approximation would be valid.

Before applying classical mechanics two criteria must be satisfied:

(1) The particle trajectories must be well defined in relation to some linear dimension and that characterizes the range of the forces between particles. Since the particle should be represented by a wave packet with mean wavelength λ, this criterion is expressed as:

$$\lambda \ll a.$$

(2) The deflexion of the incident trajectory must be well defined. The target particle can be thought of as an obstacle of radius $\sim a$ by which incident particle waves are diffracted. It is well known that strong diffraction exists for small angles of the order λ/a and the second criterion is therefore

$$\phi \gg \frac{\lambda}{a}.$$

For a particle of energy E_1 the wavelength is:

$$\lambda = \hbar \sqrt{\left(\frac{2}{M_1 E_1}\right)}.$$

This can be put in a more convenient form if we introduce the Bohr radius a_0 (0·53 Å) as our scale of distance and the Rydberg energy E_R (13.6 eV) as our scale of energy.

For
$$a_0 = \frac{\hbar^2}{m_0 \epsilon^2},$$

where m_0 and ϵ are the mass and charge of the electron and

$$E_R = \frac{\epsilon^2}{2a_0},$$

hence
$$\hbar = a_0 \sqrt{(2m_0 E_R)}$$

and
$$\lambda = a_0 \sqrt{\left(\frac{4E_R m_0}{E_1 M_1}\right)}.$$

Then the criteria become:

(1)
$$E_1 \gg E_1^* \tag{4.12}$$

with
$$E_1^* = 4E_R \frac{m_0}{M_1} \left(\frac{a_0}{a}\right)^2.$$

(2)
$$\phi^2 E_1 \gg E_1^*. \tag{4.13}$$

Since ϕ is always either less than unity, or at the worst, of the same order as unity, the second criterion is sufficient for both. It is clear from (4.13) that in glancing collisions, involving small values of ϕ, the requirement for large E_1 becomes more stringent. Further consideration is given to this point in §4.3.

For charged particles incident on atoms it will be shown in the §4.2.1 that $(a_0/a)^2 \simeq (Z_1 Z_2)^{\frac{1}{3}}$. In nuclear collisions $a \sim 10^{-13}$ cm.

Table 9 shows values of E_1^* for various cases with Li and U atoms chosen as extreme cases of light and heavy metals.

It is plain to see that for charged particles with mass at least as great as the proton's, classical collision approximations are valid in the energy range of interest in radiation damage. Although the electron is charged and has an a value comparable to that of the proton, its mass is very small, and collisions must fall into the quantum category. Further, because its energy must be of the order of $m_0 c^2$ ($= 0.51$ MeV) to be of interest in radiation damage, collisions must be treated by *relativistic* quantum mechanics. Equation (4.8) is not valid in this case and $\breve{E}_1$ and E_q have to be calculated using (4.56) to give $\hat{E}_2$. The neutron, because of the short range of nuclear forces, always requires the use of quantum mechanics to describe its collisions.

4.2. Interatomic Forces

4.2.1. *Theoretical.* This section deals with the most fundamental question in the book, for until one knows the forces between pairs of atoms one cannot give detailed consideration to the collision problems in the primary event, nor to the calculation of defect properties. We are interested both in the forces between unlike atoms, or ions and atoms; and in the forces between like atoms. Suppose we have two atoms with mass ratios M_1 and M_2, nuclear charges $Z_1 \epsilon$ and $Z_2 \epsilon$ respectively, their nuclei separated by a distance r. The force is best described by a potential energy $V(r)$ which arises from many-body interactions involving the electrons and the nuclei. Even in the simplest cases $V(r)$ has never been determined exactly but some simple considerations show that it must be dominated by two distinct contributions in the range of separations that we are interested in.

There are two useful reference points in the scale of separation: the Bohr radius of the hydrogen atom, $a_0 = 0.53$ Å, which gives a rough idea of the position of the atomic electron shells; and D, the spacing between neighbouring atoms in the crystal (typically 2.5 Å). When $r \gg D$ the electrons populate the energy levels of the individual atoms and it follows from the Pauli Exclusion Principle that there is a maximum number that can occupy any set of levels. The lowest levels corresponding to the inner closed shells will all be occupied and there will only be empty levels in the outer valence shells. As our pair of atoms is brought together and these valence shells begin to overlap there may be attractive interactions of the type that form chemical bonds and the relatively weaker van de Waals forces. Neither of these involve energies of more than a few eV and need not concern us in solving the collision problem, although they are very important in the context of defects.

When $a_0 < r \lesssim D$ the closed inner shells begin to overlap and some electrons will find themselves in the same region of space occupying similar energy levels. The Exclusion Principle demands that they change their levels and since all lower levels are occupied they can only move up. The extra energy is supplied by the work done in forcing the atoms together and hence constitutes a positive potential energy of interaction. This effect, known as *closed shell repulsion*, contributes an important term to the total potential function in the region $a_0 < r < D$. Many attempts have been made to calculate this repulsive term for the simplest case: pairs of inert gas atoms, whose electronic shells are all closed (Slater, 1928; Bleick & Mayer, 1934; Abrahamson, 1963, 1964). It is found that the potential is roughly of the form

$$V(r) = A e^{-r/b}. \qquad (4.14)$$

Since this function was first used by Born & Mayer (1932) to represent ion core repulsion in their theory of ionic crystals it is often referred to as the *Born–Mayer* potential. In such crystals, of course, the ions have the same configuration as the next inert gas and one can easily justify this assumption. It is not obvious that the same assumption can still be made for the atom–atom interaction where one is not dealing with ion pairs of the alkali–halide type.

When $r \ll a_0$ and the nuclei become the closest pair of charged particles in the system, their Coulomb potential dominates all other terms in $V(r)$. Then

$$V(r) = \frac{Z_1 Z_2 \epsilon^2}{r}. \qquad (4.15)$$

At larger distances when there is a possibility of electrons entering the internuclear space, there is a reduction of the Coulomb potential because of the electrostatic screening of the nuclear charges by the space charge of the innermost electron shells. One then refers to a *Screened Coulomb* potential. Using a Thomas–Fermi method (see

Schiff, 1955) it is possible to deduce the effective charge density around an atom and to estimate this Screened Coulomb potential between the two nuclei of a pair of atoms (Bohr, 1948; Firsov, 1957a, b, 1958; Brinkman, 1962b; Abrahamson, 1963). There is no simple function to describe it, but a fair approximation for $r < a_0$ is

$$V(r) = \frac{Z_1 Z_2 e^2}{r} e^{-r/a} \qquad (4.16)$$

with
$$a = \frac{Ca_0}{(Z_1 Z_2)^{\frac{1}{6}}}, \qquad (4.17)$$

C is a numerical constant of order unity. Clearly, (4.16) reduces to the Simple Coulomb form (4.15) when $r \ll a$. An alternative to equation (4.7), suggested by Bohr (1948) substitutes $\sqrt{(Z_1^{\frac{2}{3}} + Z_2^{\frac{2}{3}})}$ for the denominator, but the two versions give similar results.

We have now identified two important contributions to the total repulsive potential. At small enough separations the behaviour of $(1/r)$ will make a Screened Coulomb term dominate all others, but for $r > a_0$ the screening effect, roughly represented by the exponential function $e^{-r/a}$, will make this inter-nuclear term effectively vanish. In the region $a_0 < r < D$ one is left with electronic interactions as the main contribution to $V(r)$.

Perhaps the most reliable calculations of $V(r)$ have been made by Abrahamson (1963, 1964) for pairs of inert gas atoms.

The Thomas–Fermi method for obtaining charge densities and hence the screening effect does not take account of inter-electronic interactions. Abrahamson therefore used Fermi–Dirac statistics to treat the electron clouds and his potential contained terms corresponding to the Screened Coulomb and the electronic interactions, having roughly the same form as indicated by equations (4.14) and (4.16). It is often referred to as the T–F–D (Thomas–Fermi–Dirac) potential. His potential for the A—A interaction is shown in fig. 37(a) and it will be seen to vary like Screened Coulomb for $r < a_0/2$ and like Born–Mayer for $r > a_0$.

Abrahamson has computed potentials for all combinations of inert gas atoms, like and unlike, and finds two important features; first that if a Born–Mayer expression like (4.14) is fitted to the region $r > a_0$ the value of b is roughly $a_0/2$ (i.e. 0·27 Å) and that the pre-

TABLE X. *The constants in the Born–Mayer potential,* (4.14)

System	Z_1	Z_2	A (eV)	b (Å)	Method and reference
Ne–Ne	10	10	3×10^3	0·26	Theory of Abrahamson (1963)
Na$^+$–F$^-$	11	9	$2·4 \times 10^2$	0·34	Thermodynamic data, Fumi & Tosi (1962)
A–A	18	18	$7·2 \times 10^3$	0·27	Theory of Abrahamson (1963). Scattering experiments, Berry (1955). Amdur & Bertrand (1962). Amdur & Mason (1954)
K$^+$–Cl$^-$	19	17	$1·9 \times 10^3$	0·34	Thermodynamic data, Fumi & Tosi (1962)
Cu–Cu	29	29	$2·8 \times 10^3$	0·25	Empirical equations (4.18), Brinkman (1962)
			$2·2 \times 10^4$	0·20	Compressibility and displacement energy, Gibson et al. (1960).
Kr–Kr	36	36	$1·5 \times 10^4$	0·29	Theory of Abrahamson (1963)
				0·24	Empirical equations (4.18), Brinkman (1962)
Rb$^+$–Br$^-$	37	35	$4·1 \times 10^3$	0·34	Thermodynamic data, Fumi & Tosi (1962)
Xe–Xe	54	54	$2·7 \times 10^4$	0·23	Theory of Abrahamson (1963)
			$8·3 \times 10^4$	0·21	Empirical equations (4.18), Brinkman (1962)
Cs$^+$–I$^-$	55	53	$1·1 \times 10^4$	0·34	Thermodynamic data, Fumi & Tosi (1962)
Au–Au	79	79	$6·7 \times 10^5$	0·18	Empirical equation (4.18), Brinkman (1962)
			2×10^5	0·20	Compressibility and focusing energy, Thompson (1968)

exponential constant A varies roughly in proportion to $Z^{\frac{3}{2}}$. Values of A and b are shown in table X.

One might reasonably apply these results to the repulsion between ion cores in alkali metals or in alkali halide crystals where the electronic configurations of the cores are similar to inert gases. Noble metals might reasonably be treated in the same way. However, some allowance should be made for the fact that the inert gas shells surround a different nuclear charge than in the case of an inert gas atom. Thus in a metal ion core, where there is a larger nuclear charge than in the corresponding inert gas the shells will be drawn inwards slightly and one should expect the screening length a and the constant b to be rather smaller. In alkali halides the decrease in size of the metal core will be compensated approximately by the growth of the halide ion, thus the inert gas pair should be roughly the same as the alkali-halide pair.

In transition metals, because the inner shells are incomplete it is not easy to justify the use of potentials deduced for inert gas configurations. The error in the Screened Coulomb term for $r < a_0$ should not be serious, because this is mainly determined by the innermost, complete, electron shells. But the electronic term could be appreciably different from that of the Born–Mayer form, and even if this form were valid any estimate of A and b based on an interpolation between inert gas systems could be seriously in error. One should rather attempt to determine A and b from experimental data for the metal in question.

4.2.2. *Experimental verification.* Experimental information about the potential for $r < a_0$ comes from observations of ion–atom scattering and from measurements of the range of ions in solids. The detailed interpretation of such experiments can be deferred until later (see §4.3 and 5.4), but their dependence on $V(r)$ must be obvious. At this stage we simply quote the conclusions. Rutherford's classic interpretation of alpha-particle scattering in 1911 showed that for $r \ll a_0$ the simple Colomb potential is valid. For slightly larger separations a study of collisions between 100 keV inert gas ions and their atoms by Everhart & Lane (1960) is consistent with a screened Coulomb potential of the form in (4.16) and (4.17) with $C = 0.7$. Interpretation of the data on the penetration of heavy ions, with $\sim 10^5$ eV depends on the potential near $r = a_0$. If one assumes the screened Coulomb form, deduced values of C range from 1 to 2. These two pieces of information suggest that the Screened Coulomb potential alone is too small near $r = a_0$ and justify the inclusion of another term such as (4.14) to account for electronic interactions. This conclusion is strongly reinforced by inert gas scattering experiments at relatively low energy (< 1 keV) which give information about $V(r)$ for $D > r > a_0$ (Berry, 1955; Amdur & Mason, 1954;

Amdur & Bertrand, 1962). Fig. 37(a) compares the experimentally deduced potentials for A—A with the theoretical potential functions and it can be seen that the Born–Mayer form of (4.14) is a good approximation to the potential. The same is true for other inert gas pairs that have been studied in this range for separation. It was pointed out earlier that the repulsion between the anions and cations in alkali halide crystals must be very similar to the interaction between pairs of inert gas atoms. For instance in KCl the K^+ and Cl^- ions have the A electron shell configuration and one may expect that the same Born–Mayer potential could describe both the K^+—Cl^- and the A—A repulsion. By analysing the compressibility and thermochemical data for such crystals it is possible to deduce the constants A and b in (4.14) assuming the Born–Mayer potential to be a fair approximation (Fumi & Tosi, 1962). Potentials deduced in this way are only valid near $r = D$ since they are deduced from experiments in which the atoms are very close to the equilibrium separation. Table 10 quotes a number of values and compares them, where possible, with theoretical predictions. The value of b is essentially constant at $0\cdot34$ Å throughout the periodic table, in fair agreement with Abrahamson's result. The Z-dependence of A, however, is much stronger than Abrahamson's $Z^{\frac{3}{2}}$ prediction, though the potential for K^+—Cl^- agrees well with Abrahamson's A—A potential, as shown in fig. 37(a).

Unfortunately, there is much less experimental information about the forces between metal atoms and these are the main concern of this book. There are at present no atom–atom scattering experiments and very few ion penetration experiments. The only significant experimental data, apart from radiation damage, is from the analysis of compressibility and elastic moduli in metals and here it is only the noble metals Cu, Ag and Au that have proved amenable to such treatment. Brinkman (1962b) has suggested empirical formulae for the Born–Mayer constants that are consistent with observed compressibilities of these three metals; these are

$$\left.\begin{aligned} A &= 2\cdot58 \times 10^{-5}(Z_1 Z_2)^{11/4}\,\text{eV}, \\ b &= 1\cdot5 a_0/(Z_1 Z_2)^{\frac{1}{6}} \end{aligned}\right\} \qquad (4.18)$$

(i.e. $b = 1\cdot5a$ when $C = 1$). These are evaluated in table X for the inert gas cases and it will be seen that they consistently give smaller values of b. This is to be expected in the light of the remarks in the previous section where it was suggested that the metal ion core would be smaller than the corresponding inert gas because of the higher nuclear charge, and Brinkman's Born–Mayer constants are deduced for metallic ion cores.

Radiation damage experiments have themselves shed considerable light on $V(r)$. We have already seen in §2.3.4 that Vineyard's computer calculations give a reasonable value for the displacement energy E_d in Cu, when $A = 2 \times 10^4$ eV and $b = 0.196$ Å (Gibson *et al.* 1960). These values are also consistent with compressibility. In Fe, a potential has been deduced in the same way by Erginsoy (1964). In the next chapter another radiation damage parameter will be discussed: the focusing energy. This depends sensitively on $V(r)$ and can be determined by sputtering experiments. A potential for the Au—Au interaction has been deduced in this way, valid for separations in the range $D > r > \frac{1}{2}D$ (Thompson, 1968). All of this data is shown in table X, where it can be seen that although Brinkman's formulae (4.18) are far from accurate, they might serve as a useful guide to the Born–Mayer potential in cases where no other information is available.

4.2.3. *Approximations.* In the case of the inert gas potential there is ample justification for using the sum of a Screened Coulomb term like (4.16) and a Born–Mayer term, like (4.14), to represent the total potential. In the present state of knowledge one can reasonably hope that the same will be true for other types of atom. Brinkman's estimates of A and b in (4.18), because they are compatible with compressibility in noble metals, should give a rough guide to the Born–Mayer term for $Z > 25$. Equation (4.17) with $C = 1$ appears to give a reasonable value of the screening radius a. Thus our first approximate potential may be written

$$V(r) = 2Z_1 Z_2 E_R \frac{a_0}{r} e^{-r/a} + A e^{-r/b}, \qquad (4.19)$$

where the Rydberg energy E_R (13.6 eV) has been written in place of $e^2/2a_0$. Fig. 37 shows how the first term dominates for small separations and the second for large separations.

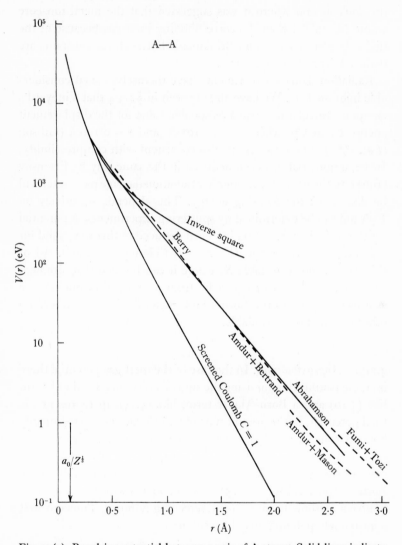

Fig. 37(a). Repulsive potential between a pair of A atoms. Solid lines indicate theoretical curves for: the Screened Coulomb potential of (4.17), the Inverse Square potential of (4.21), and the calculations of Abrahamson (1963). Dotted lines show potentials deduced from: ion–atom scattering experiments by Berry (1955), Amdur & Bertrand (1962), and Amdur & Mason (1954), or KCl thermo-chemical data by Fumi & Tozi (1962).

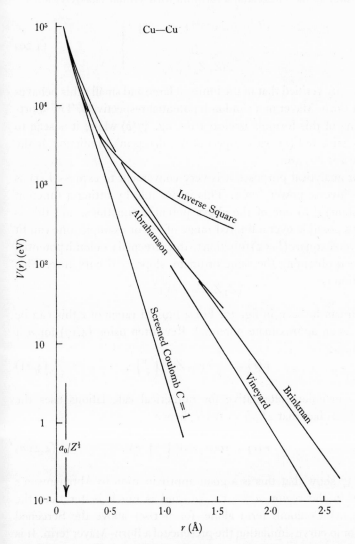

Fig. 37 (b). The Cu–Cu potentials given by the theory of Abrahamson (1963), Screened Coulomb (4.17), Inverse Square (4.21), Brinkman (1962) (4.20) and Vineyard (see Gibson *et al.* 1960).

Brinkman has suggested a formula with similar characteristics:

$$V(r) = A\,e^{-r/b}[1 - e^{-\alpha r/b}]^{-1}$$

with
$$\alpha = \frac{Ab}{2Z_1 Z_2 E_R a_0}.$$

(4.20)

It is easily verified that in the limits of large and small r this behaves like a Born–Mayer or a Coulomb potential respectively. The short-coming of this formula is clear from fig. 37(b) where it is seen to over-estimate $V(r)$, with respect to Abrahamson's prediction, in the range $a < r < 5a$.

For analytical purposes it is very convenient to express $V(r)$ as some inverse power l of r. This is possible by fitting a function (constant)/r^l to one of the above potential functions, and this is always possible over a limited range of r. For example, one can fit an inverse square ($l = 2$) function to the Screened Coulomb potential at $r = a$ obtaining the same ordinate, slope, and curvature. This function is:

$$(Z_1 Z_2 e^2 a\,e^{-1})/r^2$$

which can be seen in fig. 37. For a limited range of r this can be used as an approximate potential. Rewritten using (4.17) for a it becomes:

$$V(r) = \frac{2E_R}{e}(Z_1 Z_2)^{\frac{5}{6}}\left(\frac{a_0}{r}\right)^2.$$

(4.21)

A convenient alternative for numerical calculations uses the fortuitous fact that $2E_R/e \simeq 10\,\text{eV}$, hence

$$V(r) = 10(Z_1 Z_2)^{\frac{5}{6}}\left(\frac{a_0}{r}\right)^2\,\text{eV}.$$

(4.21a)

Fig. 37 show that this is a good approximation to Abrahamson's curve in the region $a/2 < r < 5a$, and is much better than the Screened Coulomb term alone, for it rises above the Screened Coulomb curve simulating the presence of a Born–Mayer term. It is fortunate that this is just the region where the Brinkman formula (4.20) is unreliable.

In choosing an approximate potential for a specific collision problem one considers the range of separations that will be involved by simply equating the available kinetic energy to the potential, and hence obtaining the smallest possible separation. It is then possible to decide the terms that are important and hence which approximation to use.

For example, where one is dealing with interactions between metal atoms where rather low kinetic energies, of say 0·1 to 10^3 eV, are involved the Born–Mayer term can be used alone, with the constants given by (4.18) if no better information is available. In cases of atom–atom collision in the collision cascade where energies from 10^3 to 10^5 eV are involved, the inverse square potential of (4.21) is extremely convenient and a rather good approximation to either Abrahamson's total potential or to (4.19). This potential is also applicable to cases of bombardment by heavy ions with energies from 10^3 to 10^5 eV. In the case of light ions at high energy, such as 5 MeV protons, the Simple Coulomb potential is generally adequate. Further consideration will be given to this matter in §4.3.1 that follows.

4.3. Ions and atoms in collision

4.3.1. *General theory of collision orbits.* For the energies of interest, table IX shows that classical mechanics are appropriate and we may therefore discuss orbits. In fig. 38 collision orbits are shown in G co-ordinates with impact parameter p and particle co-ordinates (r_1, ψ) and (r_2, ψ) for M_1 and M_2 respectively. We require to know the probability of M_2 recoiling in the L system with energy in the interval dE_2 at $E_2 : P(E_2) dE_2$. The first step is to determine the orbits in detail and hence to express ϕ as a function of p.

The radial and transverse velocities of M_1 are $\dot{r}_1$ and $r_1 \dot{\psi}$ respectively, and the resultant velocity is $(\dot{r}_1^2 + r_1^2 \dot{\psi}^2)^{\frac{1}{2}}$. A similar expression holds for M_2 with suffix 1 replaced by 2. In an elastic collision the sum of potential and kinetic energy at any point in the orbit must equal the asymptotic sum of kinetic energies, given by (4.6), i.e.

$$\frac{M_2 E_1}{M_1 + M_2} = V(r_1 + r_2) + \tfrac{1}{2} M_1 (\dot{r}_1^2 + r_1^2 \dot{\psi}^2) + \tfrac{1}{2} M_2 (\dot{r}_2^2 + r_2^2 \dot{\psi}^2).$$

This may be simplified by putting

$$r = r_1 + r_2$$

and

$$r_1 = \frac{M_2}{M_1 + M_2} r, \quad r_2 = \frac{M_1}{M_1 + M_2} r$$

and using the time derivatives of these equations. Then

$$\frac{M_2 E_1}{M_1 + M_2} = V(r) + \frac{1}{2} \frac{M_1 M_2}{M_1 + M_2} (\dot{r}^2 + r^2 \dot{\psi}^2)$$

now

$$\dot{r}^2 + r^2 \dot{\psi}^2 = \left(\frac{dr}{d\psi} \frac{d\psi}{dt}\right)^2 + r^2 \dot{\psi}^2$$

$$= \dot{\psi}^2 \left[\left(\frac{dr}{d\psi}\right)^2 + r^2\right].$$

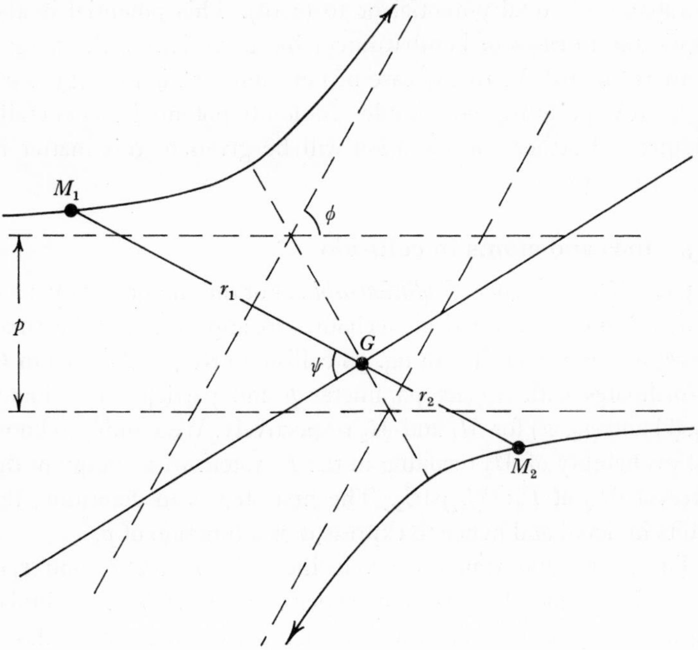

Fig. 38. Collision orbits in G coordinates.

In addition to the conservation of energy there must be conservation of angular momentum, hence the value at any point in the orbit must equal the asymptotic value:

$$\frac{M_1 M_2 u_1 p}{M_1 + M_2} = M_1 r_1^2 \dot{\psi} + M_2 r_2^2 \dot{\psi}$$

from which

$$\dot{\psi} = \frac{p u_1}{r^2} \qquad (4.22)$$

then if in addition one substitutes $u = 1/r$

$$\dot{r}^2 + r^2\dot{\psi}^2 = u_1^2 p^2 \left[\left(\frac{du}{d\psi}\right)^2 + u^2\right].$$

Then the energy equation becomes

$$\frac{du}{d\psi} = \left\{\frac{1}{p^2}\left[1 - \frac{V(u)}{E_1}\frac{M_1 + M_2}{M_2}\right] - u^2\right\}^{\frac{1}{2}}. \tag{4.23}$$

This is no longer time dependent and hence is the equation of the orbit. The scattering angle ϕ is found from it by expressing $d\psi$ as a function of u and du and integrating over the first half of the orbit; ψ from $\phi/2$ to $\pi/2$. Then since

$$\int_{\phi/2}^{\pi/2} d\psi = \tfrac{1}{2}(\pi - \phi),$$

$$\phi = \pi - 2\int_0^{1/\rho}\left\{\frac{1}{p^2}\left[1 - \frac{V(u)}{E_1}\frac{M_1 + M_2}{M_2}\right] - u^2\right\}^{-\frac{1}{2}} du. \tag{4.24}$$

The quantity ρ in the upper limit of u is the value of r when $\psi = \pi/2$ and is hence the distance of closest approach. Since $du/d\psi = 0$ when $\psi = \pi/2$, ρ is given, from (4.23), by

$$\frac{V(\rho)}{1 - p^2/\rho^2} = \frac{M_2 E_1}{M_1 + M_2}. \tag{4.25}$$

(4.24) with (4.25) is the required relation between ϕ and p.

Consider now an atom M_2 being bombarded by ions M_1. Those ions which cross an area $2\pi p\, dp$ enclosed between circles of radii p and $p + dp$ will be scattered into $d\phi$ at ϕ, the relation between dp and $d\phi$ coming from (4.24) by differentiation, thus the *differential cross-section* $d\sigma$ is given by

$$d\sigma = 2\pi p\, dp. \tag{4.26}$$

Knowing $V(r)$ (4.24) enables p^2 to be expressed in terms of ϕ, then using (4.8) this may be put in terms of E_2. Differentiating gives $2p\, dp$ as a function of E_2 and dE_2. Then from (4.26) the differential cross-section for collisions having recoils in dE_2 at E_2 follows immediately.

The *total cross-section* for collisions with E_2 anywhere in the interesting range $\check{E}_2$ to ΛE_1 is

$$\sigma = \int_{\check{E}_2}^{\Lambda E_1} \frac{d\sigma}{dE_2}\, dE_2. \tag{4.27}$$

The probability of recoil in dE_2 at E_2 is clearly the ratio of differential and total cross-sections, i.e.

$$P(E_2)\,dE_2 = \frac{1}{\sigma}\frac{d\sigma}{dE_2}\,dE_2. \qquad (4.28)$$

Explicit evaluation of the integral in (4.24) is only possible for simple potentials such as the Coulomb and Inverse Square. In all other cases numerical methods must be employed. It is therefore profitable to introduce two approximate methods of calculation.

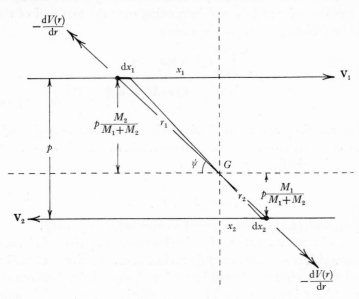

Fig. 39. The Impulse Approximation in G coordinates, valid for $\phi \ll 1$.

When ϕ is small enough, the velocity of either particle does not alter substantially throughout the collision. The orbits are then approximately straight lines running parallel at a separation p as shown in fig. 39. The force between the particles is $-dV(r)/dr$ and in the intervals dx_1 or dx_2 which occupy a time $dx_1/V_1 = dx_2/V_2$ the momentum impulse is

$$-\frac{dV(r)}{dr}\frac{dx_1}{V_1}.$$

Taking components of the impulse and integrating over x_1 from $-\infty$ to $+\infty$ it is clear that, because of symmetry about G, components parallel to the paths will cancel out. On the other hand, the total impulse delivered perpendicular to the paths is:

$$\Delta P = \frac{2}{u_1} \int_p^\infty \left(-\frac{dV(r)}{dr} \right) \frac{p\,dr}{\sqrt{(r^2 - p^2)}}. \tag{4.29}$$

From here the angle ϕ is found

$$\phi = \frac{\Delta P}{M_1 V_1} = \frac{\Delta P}{M_2 V_2} \tag{4.30}$$

and

$$E_2 = \frac{(\Delta P)^2}{2M_2}. \tag{4.31}$$

Note that E_2 is always proportional to E_1^{-1}, irrespective of the potential. This is a characteristic of the approximation. This treatment, known as either the 'Impulse' or 'Momentum' approximation is applicable only to glancing collisions.

It is of importance to establish the validity of this classical treatment since ϕ is necessarily small. If we write $I(p)$ for the integral in (4.29) we have

$$\phi = I(p)/E_1$$

and the criterion (4.13) for classical treatment

$$E_1 \phi^2 \gg E_1^*$$

becomes

$$E_1 \ll \frac{1}{E_1^*} [I(p)^2].$$

Now $I(p)$ is a function only of p and the potential thus for given p and $V(r)$, E_1 must be *much less than* the critical value shown. The fact that we have an upper limit to E_1 rather than a lower limit as before, is because p and $V(r)$ have been specified and in order for the interaction to last long enough (Δt) for quantum uncertainties in the energy transfer (ΔE_2) to be negligible, the particle must move slowly enough that $\Delta E_2 \Delta t \gg \hbar$.

Typical values of the energy limit $[I(p)]^2/E_1^*$ for the Vineyard Born–Mayer potential in the case of Cu–Cu with $p = 1 \cdot 4\,\text{Å}$ is $\sim 10^7\,\text{eV}$. Since normal recoils in radiation damage are considerably less energetic than this, and p is always less than $1 \cdot 4\,\text{Å}$ ($= \frac{1}{2}D$) the use of the impulse approximation appears to be justified. In the case of very light elements where higher energies and weaker potentials might occur, there could be cases where quantum mechanics should be used.

In collisions which are closer to head-on the '*Hard-Sphere*' approximation is often applicable. In fig. 38, when $p = 0$ and $\phi = \pi$, let the particles come instantaneously to rest when

$$r_1 = R_1 \quad \text{and} \quad r_2 = R_2 \quad (R_1 + R_2 = \rho_0).$$

ρ_0 is then the distance of closest possible approach for the energies concerned. The potential energy $V(\rho_0)$ will equal the asymptotic kinetic energies in the G co-ordinates, given by (2.5), i.e.

$$V(\rho_0) = M_2 E_1/(M_1 + M_2). \tag{4.32}$$

This equation defines ρ_0.

Now
$$R_1 = \rho_0 M_2/(M_1 + M_2),$$
$$R_2 = \rho_0 M_1/(M_1 + M_2)$$

and so far as the head-on collision is concerned one could replace the colliding particles by elastic hard spheres of radii R_1 and R_2, and masses M_1 and M_2, obtaining the same asymptotic momenta. For particles which have $V(r)$ increasing very rapidly with r, in the vicinity of ρ, it is a good approximation to use these same hard spheres to construct the asymptotes for $p \neq 0$. This approximation gets worse for small values of ϕ and to some extent it is complimentary to the impulse approximation.

From fig. 40, if ρ_0 is known,

$$p = \rho_0 \cos(\phi/2).$$

Then, differentiating and using (4.26),

$$d\sigma = \pi\rho_0^2 \sin(\phi/2)\cos(\phi/2)\,d\phi.$$

Differentiating (4.8)

$$dE_2/\Lambda E_1 = \sin(\phi/2)\cos(\phi/2)\,d\phi,$$

hence
$$d\sigma = \pi\rho_0^2\,dE_2/\Lambda E_1. \tag{4.33}$$

To calculate σ requires integration over a part of the E_2 range where glancing collisions must occur, but approximately:

$$\sigma \simeq \pi\rho_0^2. \tag{4.34}$$

Thus
$$P(E_2)\,dE_2 \simeq dE_2/\Lambda E_1 \tag{4.35}$$

and it follows that the mean recoil energy

$$\bar{E}_2 = \tfrac{1}{2}\Lambda E_1. \tag{4.35a}$$

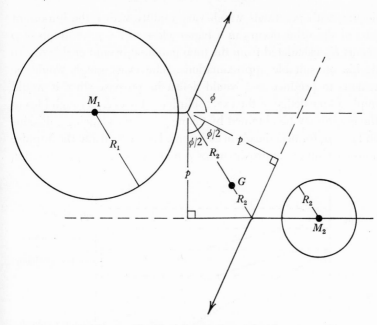

Fig. 40. The Hard Sphere Approximation.

4.3.2. *Classification of ions.* In radiation damage there are three important classes of incident ion. The first is the energetic light ion, such as the proton, deuteron or alpha-particle, with $E_1 > 10^6$ eV. The second is the highly energetic heavy ion with $M_2 \sim 10^2$ and $E_1 \sim 10^8$, which emanates from nuclear fission and which will be referred to as the *fission fragment*. The third category is heavy ions with $E_1 < 10^6$ eV, of importance in the sputtering of solid surfaces, and recoil atoms produced in the primary collision with the incident radiation. In typical cases such recoils occur with $< 10^6$ eV kinetic energy. In Chapter 4 we shall be considering the multiplication of collision cascades induced by recoils, but here we shall lay the foundations for this by dealing with the nature of the individual interatomic collision.

In principle one could solve (4.24) numerically and compute the functions $d\sigma/dE_2$ and $P(E_2)$ for any total potential. This would be cumbersome and would lack the generality of an analytical solution. In deciding which approximations to use for a particular collision a useful guide is ρ, the distance of closest approach. Because one is dealing with potentials which vary rapidly with r the important part of the orbit occurs in a region close to $r = \rho$. A graph of ρ versus E_2, calculated from the total potential, would enable one to decide on suitable approximations. The exact graph would be tedious to produce and would defeat its purpose since it would imply a knowledge of the exact solution. However, a rough idea of the graph may be obtained by taking $\rho \simeq p$ for glancing collisions and $\rho = \rho_0$ for near-head-on collisions. Hence by using the impulse and hard sphere approximations ρ may be sketched as a function

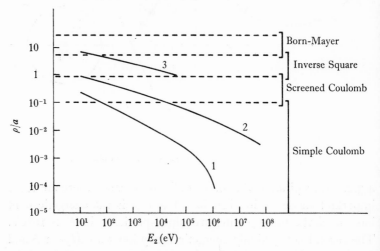

Fig. 41. The distance of closest approach ρ, as a function of E_2. Curve (1) 20 MeV protons in Cu; curve (2) 70 MeV Xe$^+$ ions in Cu; curve (3) 50 keV Cu$^+$ recoils in Cu.

of E_2. This has been done in figure 41 for three representative ions in Cu: 20 MeV protons, 70 MeV fission fragments, 50 keV Cu recoils.

In the first case $\rho \ll a$ for all collisions of interest (i.e. for $E_2 > \check{E}_2$) and one concludes that the Simple Coulomb potential is adequate. In the next section this finding will be confirmed and some simpler criteria established for determining its range of validity.

In the second case, when $E_2 \rightarrow \check{E}_2$ and collisions are close to head on, $\rho \ll a$ once again. For glancing collisions though, when

$$E_2 \rightarrow \check{E}_2, \ \rho \simeq a$$

and the Screened Coulomb term can be taken as an approximation to the total potential. Alternatively the Brinkman potential could be used.

In the third case, $a < \rho < 5a$ and one must take account of both Born–Mayer and Screened Coulomb terms. Either of the Inverse-Square or Brinkman potentials would be suitable here.

4.3.3. *Energetic light ions: simple Coulomb collisions.* It is well known that a solution of the orbits in a Simple Coulomb potential leads to the Rutherford scattering law $d\sigma \propto \mathrm{cosec}^4(\phi/2)\,d\omega$, $d\omega$ being the element of solid angle at ϕ. It is not proposed to reproduce a proof of this here, as it appears in all the standard text-books on Nuclear Physics and may be done by a straightforward extension of the §4.3.1. However, it is instructive to apply the impulse approximation to such collisions and from this small-angle case to infer the general solution.

Differentiating $V(r)$ from (4.15) inserting in (4.29), integrating, and using (4.30) leads to

$$\phi = \frac{1}{p}\frac{M_1+M_2}{M_2}\frac{Z_1Z_2\epsilon^2}{E_1}. \tag{4.36}$$

Using this to obtain p and dp in terms of ϕ and $d\phi$ by differentiation, substituting in (4.26), and putting $\epsilon^2 = 2a_0E_R$ gives

$$d\sigma = \frac{8\pi a_0^2 Z_1^2 Z_2^2 E_R^2 (M_1+M_2)^2}{M_2^2 E_1^2}\frac{d\phi}{\phi^3}.$$

Then with $d\omega = 2\pi\phi\,d\phi$ we have the small-angle approximation to Rutherford's law.

$$d\sigma \simeq \frac{a_0^2 Z_1^2 Z_2^2 E_R^2 (M_1+M_2)^2}{4M_2^2 E_1^2}\frac{d\omega}{(\phi/2)^4}, \quad \phi \ll 1.$$

Replacing $(\phi/2)$ by $\sin(\phi/2)$ gives us the general solution and from this we may obtain the exact ϕ, $d\phi$ expression for $d\sigma$, using the exact relation $d\omega = 2\pi\sin\phi\,d\phi$,

$$d\sigma = \frac{\pi a_0^2 Z_1^2 Z_2^2 E_R^2 (M_1+M_2)^2}{M_2^2 E_1^2}\frac{\cos(\phi/2)\,d\phi}{\sin^3(\phi/2)}. \tag{4.37}$$

Differentiating (4.8) allows this to be expressed in terms of E_2 and dE_2,

$$d\sigma = \frac{4\pi a_0^2 M_1 Z_1^2 Z_2^2 E_R^2}{M_2 E_1}\frac{dE_2}{E_2^2}. \tag{4.38}$$

If this is integrated from $E_2 = 0$ to $E_2 = \Lambda E_1$ an infinite total cross-section is found. This is characteristic of power-law potentials, but in radiation damage one is only interested in recoil energies above $\check{E}_2$. The cross-section for this restricted class of collisions is found by taking limits to the integration at $\check{E}_2$ and ΛE_1, then

$$\sigma_p = \frac{4\pi a_0^2 M_1 Z_1^2 Z_2^2 E_R(1 - \check{E}_2/\Lambda E_1)}{M_2 E_1 \check{E}_2}. \qquad (4.39)$$

From (4.28), (4.37) and (4.38) one finds the probability of recoil with E_2 to $E_2 + dE_2$, above $\check{E}_2$, is

$$P(E_2)\,dE_2 = \frac{\check{E}_2}{1 - \check{E}_2/\Lambda E_1}\frac{dE_2}{E_2^2} \quad (\check{E}_2 \leqslant E_2 \leqslant \Lambda E_1). \qquad (4.40)$$

A comparison of (4.35) with (4.40) shows that the hard sphere approximation would be entirely inappropriate for the simple Coulomb collision. This is because $(1/r)$ does not vary rapidly enough with r.

From (4.40) the mean energy of the primary recoil is easily calculated as:

$$\bar{E}_2 = \frac{\check{E}_2 \log(\Lambda E_1/\check{E}_2)}{1 - \check{E}_2/\Lambda E_1}. \qquad (4.40a)$$

In most cases the denominator can be taken as unity since ΛE_1, the maximum energy transfer, is so much greater than $\check{E}_2$, the smallest value of interest. The variation of $\bar{E}_2$ with E_1 is very slow since it depends on the logarithm, and in most cases $\bar{E}_2$ is only a few times $\check{E}_2$. This illustrates once again the high probability of low energy recoils in such collisions.

We must next decide exactly under what conditions the above results can be applied. One requires that during the encounter the major part of the scattering occurs in the region where $r \ll a$. For collisions that are close to being head-on this condition is that $\rho_0 \ll a$ or that $E_1 \gg E_a$, where E_a is the value of E_1 that would give $\rho_0 = a$ in a Screened Coulomb potential given by

$$E_a = 2E_R(Z_1 Z_2)^{\frac{7}{6}}(M_1 + M_2)/M_2 e \qquad (4.41)$$

using (4.21), (4.6), (4.16) and putting $\epsilon^2 = 2a_0 E_R$.

For glancing collisions one can only neglect those for which $p \gtrsim a$ if they do not contribute recoils of sufficient energy to be of interest; i.e. if $E_2 \ll \breve{E}_2$ for $p = a$. For a simple Coulomb collision with $p = a$ we have from equations (4.36) and (4.8)

$$E_2 = \frac{e^2 \Lambda E_a^2}{4E_1}.$$

Then provided this E_2 is less than $\breve{E}_2$, i.e. $E_1 \gg E_b$ with

$$E_b = \frac{e^2 \Lambda E_a^2}{4\breve{E}_2} \tag{4.42}$$

one can neglect encounters for which $p \gtrsim a$ and use the simple Coulomb potential. Values of E_a and E_b are given in table XI and it will be clear that $E_a < E_b$ in all cases shown. This fact allows us to use the criterion $E_1 \gg E_b$ exclusively as an extreme test of the validity of a simple Coulomb model.

TABLE XI. *The energy limits E_a and E_b, calculated for various particles and target atoms, taking $\breve{E}_2 = 10$ eV for all cases*

Incident particle	Target atom	E_a (eV)	E_b (eV)
D$^+$	C	$1 \cdot 5 \times 10^2$	2×10^3
	Al	4×10^2	5×10^3
	Cu	1×10^3	2×10^4
	Au	4×10^3	1×10^5
C	C	2×10^3	8×10^5
Al	Al	1×10^4	2×10^7
Cu	Cu	7×10^4	1×10^9
Au	Au	7×10^5	1×10^{11}
Xe	U	5×10^5	3×10^{10}

To summarize, if $E_1 \gg E_a$, the simple Coulomb potential may be used for near-head-on collisions; if $E_1 \gg E_b$ it can be used for all collisions of interest in radiation damage. Charged particles such as protons, deuterons and alphas, with energies in excess of 1 MeV, are clearly in this category. Consideration of the heavier ions will be made in the next sections, but it is clear that fission fragments are in the region $E_a < E_1 < E_b$ and recoils generally have $E_1 \lesssim E_a$.

If we take the case of 2 MeV protons in Al as an example, $\Lambda E_1 = 0 \cdot 28$ MeV and if $\breve{E}_2$ is 25 eV,

$$\bar{E}_2 = 250 \text{ eV} \quad \text{and} \quad \sigma_p = 0 \cdot 4 \times 10^{-21} \text{ cm}^2.$$

Knowing that the mean free path between primary collisions is given by $(n\sigma_p)^{-1}$ this is seen to be 4×10^{-2} cm. This is ten times the length of the proton track and thus there is, on average, a primary event with a mean energy of 250 eV on every tenth track.

4.3.4. *Fission fragments.* Fission fragments are products of a nuclear reaction in which a heavy nucleus such as ^{235}U splits asymetrically into two. These fly apart sharing between them 160 MeV in the case of ^{235}U, and carrying with them electron clouds that are almost complete. Fission events are not identical in the nuclear species that result, but the most probable situation is to find a light fragment with $M_1 = 96$ and $E_1 = 95$ MeV, and a heavy fragment with $M_1 = 137$, and $E_1 = 55$ MeV (see fig. 157).

In order to visualize events near the fission track let us first take a Simple Coulomb potential $Z_1 Z_2 e^2/r$ as a rough approximation knowing that its use can only be justified for recoil energies E_2 approaching ΛE_1 where $\rho \ll a$. Equations (4.38), (4.40) and (4.40a) will then give $d\sigma$, $P(E_2)$ and $\bar{E}_2$. The cross-section, when compared with light ions, is increased in the ratio of $Z_1^2 M_1/E_1$. Comparing with the example of 2 MeV protons in Al, given in the previous section, we see that the fission fragment cross-section is increased by a factor of the order 10^4. The mean free path is reduced by the same factor implying that recoils will occur every 10 Å along the fission track. Equation (4.20) implies a probability of recoil with E_2 in dE_2 that varies like $1/E_2^2$. This will only be true near ΛE_1, at lower energies we must expect the screening effect to make the probability less sensitive to energy, perhaps making $P(E_2)$ vary like $1/E_2$. The mean recoil energy predicted by (4.40a) is about 400 eV, but this will be an underestimate due to the expected behaviour of $P(E_2)$ which makes high energy recoils more likely. Hence $\bar{E}_2$ will be of the order of 10^3 eV.

Now the length of a fission track is known to be about 10^4 Å and since there is one 10^3 eV recoil per 10 Å the total energy lost in atomic collisions is of the order of 1 MeV, about 1 % of the total energy. The majority is lost in electron excitation at the rate of 10^4 eV per Å. This tremendous energy density and the overlapping of adjacent recoil cascades gives fission damage some unique features, such as the formation of tracks visible in the electron

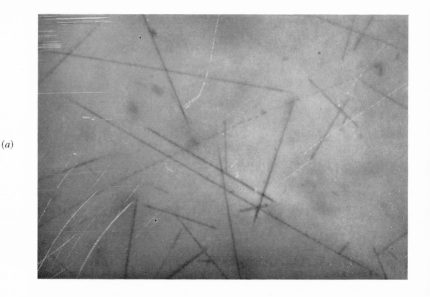

(a)

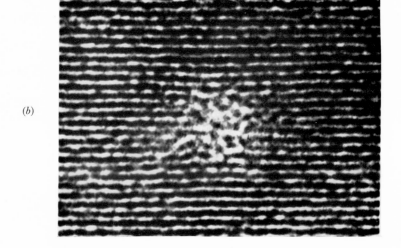

(b)

Plate XII (a) Fission fragment tracks in mica, observed in the electron microscope (by courtesy of E. C. H. Silk and R. S. Barnes). (b) Electron micrograph looking along a track at much higher magnification than (a) showing the disruption of atomic planes in a platinum phthalocyanine crystal caused by a fission fragment. The interplanar spacing is about 30 Å (by courtesy of Bowden & Chadderton, 1961).

(*Facing p.* 117)

microscope apparently associated with a continuous disruption of the atomic planes. These effects can be seen in plates XII a and b.

Let us now make a more formal approach to the problem using a better potential as before.

For such ions as these, fig. 41 shows that as $E_2 \rightarrow \hat{E}_2$, and scattering through wide angles occurs, $\rho \sim 10^{-3}a$. Hence one estimates the cross-section for a wide-angle collision to be of the order $10^{-6}\pi a^2$. On the other hand, the cross-section for transferring energies of the order 10^3 eV is of the order πa^2 and one expects that wide-angle collisions might be neglected as a significant cause of damage. This is borne out by observations of fission fragment tracks in solids (plate XII) since one can detect kinks in only about one track in ten (Silk & Barnes, 1959). We shall therefore use the impulse approximation exclusively and although this will give the wrong answers for small impact parameters $\sim 10^{-3}a$, the recoils generated in such collisions are so rare as to be neglected for most purposes.

An appropriate potential must take account of both Screened Coulomb and Closed Shell repulsion. Brinkman's formulation (4.19) includes both terms and if this is inserted in the impulse approximation, (4.29) and (4.31) give:

$$E_2 = \frac{M_1}{M_2} \frac{A^2}{E_1} [G(\alpha, p/b) - (1-\alpha) G(1+\alpha, p/b)]^2, \qquad (4.43)$$

where α, A and b are given in (4.19) and (4.20) and

$$G(\alpha, p/b) = \frac{p}{b} \int_{\frac{p}{a}}^{\infty} \frac{e^{-x}\, dx}{(x^2 - p^2/a^2)^{\frac{1}{2}} (1 - e^{-\alpha x})^2}$$

$$= \frac{p}{b} \sum_{n=0}^{\infty} (n+1) K_0 \left\{ \frac{p}{b}(1 + n\alpha) \right\}$$

$K_0(y)$ being a Bessel function of the third kind. It will be seen from (4.19) that α is the ratio of Born–Mayer and Screened Coulomb terms at $r = a$; hence $\alpha < 1$ in general. Knowing E_2 as a function of p from (4.43) one may invert to obtain p^2 as a function of E_2, then by differentiation $\dfrac{d\sigma}{dE_2} = 2\pi p \dfrac{dp}{dE_2}$ is found. Hence σ and $P(E_2)\, dE_2$ by the procedure outlined above. Because of the complexity of (4.43) these operations can only be carried out numerically for each case.

The short range of fission fragments ($\sim 10^{-3}$ cm) often results in the entire track being in the solid. In order to take account of the variation in E_1 one must first calculate dN, the number of recoils in dE_2 at E_2, produced by a fragment travelling a distance dx and losing an energy dE_1 in the process.

$$dN = n \, d\sigma \, dx$$

$$= n \frac{d\sigma}{dE_2} \left(-\frac{dE_1}{dx} \right)^{-1} dE_1 \, dE_2,$$

n being the density of atoms.

Using experimental values (Bethe & Ashkin, 1953) of $-dE_1/dx$, the stopping power for fragments, one may then integrate this number over E_1 along the entire track obtaining $N(E_2) \, dE_2$, the number of recoils in dE_2 at E_2 produced by the fission fragment in slowing down to rest, is

$$N(E_2) \, dE_2 = n \int_0^{E_1} \frac{d\sigma}{dE_2} \left(\frac{dE_1}{dx} \right)^{-1} dE_1 \, dE_2. \tag{4.44}$$

Brinkman (1962) has carried out such a calculation for light and heavy fragments from ^{235}U fission slowing down in uranium. The results are reproduced in fig. 42 and confirms that at high energies E_2, where $\rho < a$, $N(E_2)$ varies approximately as E_2^{-2} as is the Simple Coulomb case (4.40).

It will be shown in the next chapter that the number of displaced atoms produced by a recoil is roughly proportional to its energy. From fig. 42 it is clear that $N(E_2)$ decreases more rapidly than E_2^{-1} and hence the majority of displaced atoms are produced by the low-energy recoils. Therefore by using the impulse approximation and neglecting the high energy recoils due to wide-angle collision, the final estimate of damage will not be seriously in error.

4.3.5. *Primary recoils and heavy ions.* The primary recoil has up until now been labelled with suffix 2 and there is some risk of confusion since, in this section, we shall be considering this as the incident particle and using suffix 1. It is hoped that by making this clear at the outset, no difficulty will arise.

Fig. 41 shows that when $E_1 \lesssim E_a$ one cannot afford to neglect head-on encounters and on the other hand that glancing collisions with impact parameters up to ~ 10a must be dealt with. Roughly speaking, the range of interest is a $a < \rho < 10a$. The treatment applied to fission fragments in the previous section is suitable for the glancing collisions but for near-head-on collisions another approach is necessary.

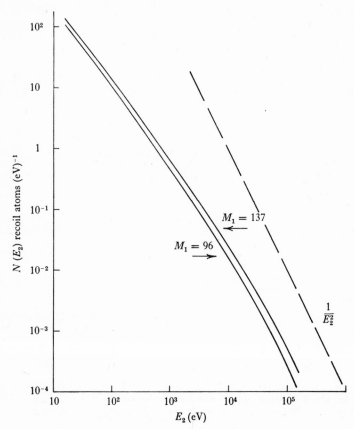

Fig. 42. The energy spectrum of recoils $N(E_2)dE_2$ produced by fission fragments slowing down to rest in U (Brinkman, 1962). Two cases are shown: $M_1 = 96$, $E_1 = 95$ MeV and $M_1 = 137$, $E_1 = 55$ MeV.

An appropriate potential for $a/5 \lesssim \rho \lesssim 5a$ is the Inverse Square approximation of (4.21). Exact solution of the orbits is then possible (J. H. O. Varley, 1957) and (4.24) gives

$$\phi/\pi = 1 - \frac{1}{\sqrt{(1 + a^2 E_a/p^2 E_1)}}. \tag{4.45}$$

Hence using (4.8) E_2 is found:

$$E_2 = \Lambda E_1 \cos^2\{\pi/2\sqrt{(1 + a^2 E_a/p^2 E_1)}\}. \qquad (4.46)$$

Expressing p^2 in terms of E_2 and differentiating, one obtains, with (4.26):

$$\frac{d\sigma}{dE_2} = \frac{4 E_a\, a^2\alpha}{\Lambda E_1^2 (1 - 4\alpha^2)^2 \sqrt{[x(1-x)]}} \qquad (4.47)$$

with $\qquad x = E_2/\Lambda E_1 \quad$ and $\quad \pi\alpha = \cos^{-1}\sqrt{x}.$

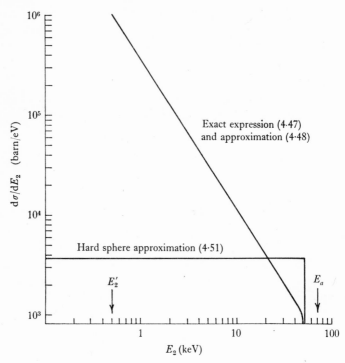

Fig. 43. The differential cross-section for collision between Cu recoil atoms at 50 keV and Cu atoms. Expressions (4.47), (4.48) and (4.51) are compared.

For small x this has the form

$$\frac{d\sigma}{dE_2} \simeq \frac{\pi^2 a^2 E_a \Lambda^{\frac{1}{2}}}{8 E_1^{\frac{1}{2}} E_2^{\frac{3}{2}}}. \qquad (4.48)$$

A typical calculation is shown in fig. 43 for 50 keV Cu recoils in Cu. It will be noticed that (4.48) is a very good approximation to (4.47) for $x < 1$ and one may usefully take $d\sigma/dE_2$ in this form with a cut-off to zero at $E_2 = \Lambda E_1$.

The total cross-section and mean recoil energy are easily calculated from (4.48) as

$$\bar{E}_2 = \sqrt{(\Lambda E_1 \check{E}_2)}, \tag{4.48a}$$

$$\sigma_p = \frac{\pi^2 a^2 E_a \sqrt{\Lambda}}{4 \sqrt{(E_1 \check{E}_2)}}. \tag{4.48b}$$

The range of heavy ions and primary recoils will be dealt with in §§5.3 and 5.4 of the next chapter. We shall see that for a 50 keV heavy ion the range is of the order $\sim 10^3$ Å. From (4.48) we can expect the mean recoil energy to be of order 1 keV with a spacing of about 20 Å between such energetic recoils.

Using (4.25) the following expression for ρ is obtained:

$$\rho^2 = p^2 + a^2 E_a / E_1. \tag{4.49}$$

The Inverse Square approximation is not good for $\rho > 5a$. When E_1 is not too small ($> E_a/100$) $\rho = 5a$ implies a glancing collision for which $\rho \simeq p$ and one may calculate from (4.46) an approximate value $(E_2)_{\rho=5a}$ above which the expressions (4.47) and (4.48) for $d\sigma/dE_2$ are valid

$$(E_2)_{\rho=5a} \simeq \Lambda E_1 \cos^2 \left[\pi/2 \sqrt{(1 + E_a/25E_1)} \right]. \tag{4.50}$$

A graph from which this lower energy limit may be calculated for any ΛE_1 is shown in fig. 44. In the case of Cu–Cu collisions it is found to be of the order 10^2 eV for $E_1 > 500$ eV. The inverse square approximation thus enables us to cover most cases of interest with $E_1 \lesssim E_a$ by using (4.48) in the range $(E_2)_{\rho=5a} < E_2 < \Lambda E_1$.

The remainder of the function $d\sigma/dE_2$, for $E_2 < E_2 < (E_2)_{\rho=5a}$ must be found from the impulse approximation with a Born–Mayer potential.

It has often been assumed that these collisions could be treated by the hard-sphere approximation. The extent to which this is justified may be judged from the third curve in fig. 43 which shows the results expected from a hard-sphere treatment based on the Inverse Square potential.

$$\frac{d\sigma}{dE_2} = \frac{\pi a^2 E_a}{\Lambda E_1^2}. \tag{4.51}$$

The disagreement is particularly marked at low energies. Further reference to this discrepancy will be made in chapter 5 when considering the multiplication of collision cascades.

When dealing with ions or recoils of light elements one frequently finds $E_1 \gg E_a$. In such cases one may treat them in a manner analogous to fission fragments, neglecting wide-angle collisions and using the Impulse approximation with Brinkman's potential. The resulting function $d\sigma/dE_2$ favours low-energy recoils much more than is found in the $E_1 \lesssim E_a$ case above.

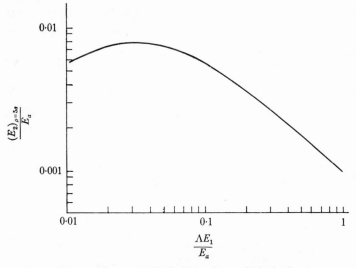

Fig. 44. The energy limit $(E_2)_{5a}$ above which the Inverse Square expressions (4.47) and (4.28) may be used.

4.4.   Relativistic collisions: fast electrons

For electrons having sufficient energy to produce damage we have already seen earlier in § 4.2 that relativistic quantum mechanics must be used. The problem is to calculate the scattering of electrons by the nuclear charge. Since the energies of interest are in excess of 1 MeV the encounters are sufficiently close for screening to be ignored and a simple Coulomb potential used.

The relations between asymptotic energies and scattering angle found in § 4.3.3 require modification for velocities close to that of light. In relativistic form the momentum and kinetic energy of an electron with rest mass m_0 and velocity u_1 are

$$P_1 = mu_1 \tag{4.52}$$

with $\qquad m = m_0/\sqrt{(1 - \beta^2)}$ and $\beta = u_1/c$ $\hspace{2cm}$ (4.53)

$$E_1 = (m - m_0)\,c^2. \hspace{2cm} (4.54)$$

In table XII are listed values of β and $\sqrt{(1 - \beta^2)}$ as a function of E_1.

TABLE XII. *Relativistic parameters:* $x = E_1/m_0c^2$, $\beta = u_1/c$ *and* $1/\sqrt{(1 - \beta)^2}$

x	β	$\dfrac{1}{\sqrt{(1-\beta)^2}}$	E_1 (MeV)
0·1	0·4165	1·1	0·051
0·2	0·5527	1·2	0·102
0·3	0·6390	1·3	0·153
0·4	0·6999	1·4	0·204
0·5	0·7453	1·5	0·255
0·6	0·7806	1·6	0·307
0·7	0·8086	1·7	0·358
0·8	0·8314	1·8	0·409
0·9	0·8502	1·9	0·460
1·0	0·8660	2·0	0·511
1·4	0·9090	2·4	0·715
1·8	0·9342	2·8	0·920
2·2	0·9499	3·2	1·124
2·6	0·9606	3·6	1·329
3·0	0·9682	4·0	1·533
3·5	0·9750	4·5	1·789
4·0	0·9798	5·0	2·044
4·5	0·9833	5·5	2·300
5·0	0·9861	6·0	2·555
5·5	0·9881	6·5	2·811
6·0	0·9898	7·0	3·066

The mass of the electron is sufficiently small relative to the nuclear mass for the L and G systems to coincide. Then referring to fig. 45, the electron approaches and recedes from M_2 with the same velocity u_1 and the momentum transfer in a collision with scattering angle θ_1 is $2mu_1 \sin(\theta/2)$. The corresponding energy transfer given by $(\Delta P)^2/2M_2$ is then

$$E_2 = 2m^2u_1^2 \sin^2(\theta_1/2)/M_2.$$

Using the relations above to introduce E_1 this becomes

$$E_2 = 2E_1(E_1 + 2m_0 c^2) \sin^2(\theta_1/2)/M_2 c^2. \hspace{1cm} (4.55)$$

From this, the maximum possible recoil energy $\hat{E}_2$, found by putting $\theta_1 = \pi$, is
$$\hat{E}_2 = 2E_1(E_1 + 2m_0 c^2)/M_2 c^2. \hspace{1cm} (4.56)$$

An approximate solution of the Dirac equation for light elements by McKinley & Feshbach (1948) gives the following differential cross-section:

$$d\sigma = \frac{4\pi a_0^2 Z_2^2 E_R^2}{m_0^2 c^4} \frac{1-\beta^2}{\beta^4} [1 - \beta^2 \sin^2(\theta_1/2)$$

$$+ \pi\alpha\beta\sin(\theta_1/2)(1 - \sin(\theta_1/2))] \cos(\theta_1/2) \operatorname{cosec}^3(\theta_1/2) \, d\theta_1,$$

$$(4.57)$$

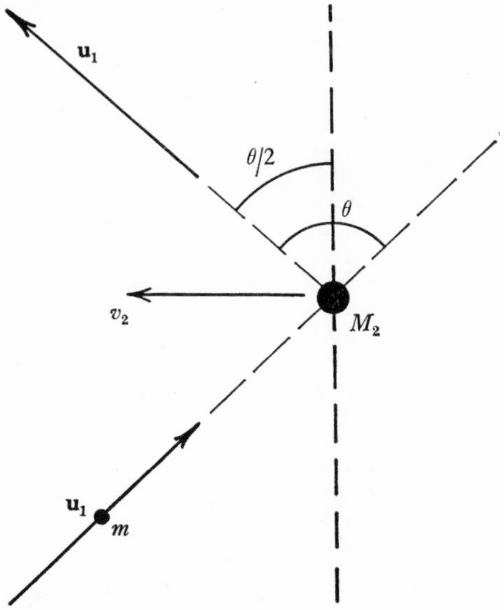

Fig. 45. The scattering of an electron by the nucleus. Because $M_2 \gg m$ the G and L systems coincide and $\mathbf{u}_1$ is little affected by the collision.

where $\alpha = Z_2/137$. This is accurate to terms in β^2 and $\alpha\beta$. For small β this clearly approaches the Rutherford scattering law given in (4.37). Using the relations (4.55), (4.56) above to put this in terms of E_2 and $\hat{E}_2$:

$$d\sigma = \frac{4\pi a_0^2 Z_2^2 E_R^2}{m_0^2 c^4} \frac{1-\beta^2}{\beta^4} \left[1 - \beta^2 \frac{E_2}{\hat{E}_2} + \pi\alpha/\beta \left\{ \left(\frac{E_2}{\hat{E}_2}\right)^{\frac{1}{2}} - \frac{E_2}{\hat{E}_2} \right\} \right] \frac{\hat{E}_2 \, dE_2}{E_2^2}.$$

$$(4.58)$$

Which again has its equivalent for small β in equation (4.38). The total cross-section σ_p for primary collisions with $\hat{E}_2 \geqslant E_2 \geqslant \check{E}_2$ is found by integrating from $\check{E}_2$ to $\hat{E}_2$:

$$\sigma_p = \frac{4\pi a_0^2 Z_2^2 E_R^2}{m_0^2 c^4} \frac{1-\beta^2}{\beta^4}\left(\frac{\hat{E}_2}{\check{E}_2}-1\right) - \beta^2\log\frac{\hat{E}_2}{\check{E}_2} + \alpha\beta 2\left(\frac{\hat{E}_2}{\check{E}_2}\right)^{\frac{1}{2}} - 1 - \log\frac{\hat{E}_2}{\check{E}_2}.$$
(4.59)

For electrons with energies just above the damage threshold $\check{E}_1$ and hence $\hat{E}_2/\check{E}_2$ slightly greater than unity:

$$\sigma_p \simeq \frac{4\pi a_0^2 Z_2^2 E_R^2}{m_0^2 c^4}\left(\frac{1-\beta^2}{\beta^4}\right)^2\left[\frac{\hat{E}_2}{\check{E}_2}-1\right].$$
(4.60)

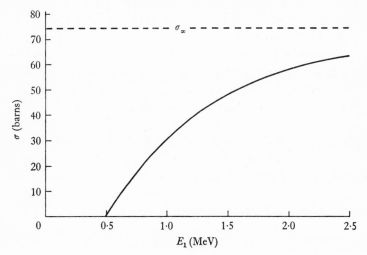

Fig. 46. The damage cross-section for electrons bombarding copper (taking $\check{E}_2 = 25$ eV).

For high enough energies, σ_p approaches an asymptotic value:

$$\sigma_p \to \frac{8\pi a_0^2 Z_2^2 E_R^2}{\check{E}_2 M_2 c^2} \quad \text{when} \quad E_1 \gg m_0 c^2.$$
(4.61)

This is often useful for making order-of-magnitude calculations. A graph showing the behaviour of σ_p for the case of Cu taking $\check{E}_2 = 25$ eV is plotted in fig. 46.

The McKinley and Feshbach formulae are very convenient for rough calculations, but they seriously underestimate σ_p for heavy elements. For example, in Cu ($Z = 29$) at 1 MeV (4.59) underestimates σ_p by only 10 % ,but in Au ($Z = 79$) it gives less than 25 % of the correct value at 1·7 MeV. Oen (1965) has carried out numerical solutions using the method of Mott (see Mott & Massey, 1949, chapter IV) and has produced graphs and tables of σ_p for representative elements and several values of $\breve{E}_2$. The reader is referred to his publication for this data (Oen, 1965).

4.5. Elastic collisions with fast neutrons

Neutrons with energies above 100 eV are generally able to produce damage by direct collision. Because they are the principle cause of damage in the non-fissile materials of a nuclear reactor they are of great practical importance. The forces between the neutrons and a nucleus cannot be expressed in simple terms, but some experimental data on neutron scattering has been accumulated. This shows that for neutron energies below about 2 MeV the majority of nuclei scatter neutrons isotropically in G co-ordinates and the collisions apparently resemble those between elastic hard spheres. Since the non-fissile materials of a reactor are chosen to have small absorption cross-sections we are mainly concerned with elastic scattering cross-sections and these are of the order of barns (1 barn = 10^{-24} cm^2) and may depend on E_1, often showing resonances at high energies (see fig. 47). For isotropic scattering:

$$\frac{d\sigma}{d\omega} = \text{constant}$$

and since

$$d\omega = 2\pi \sin \phi \, d\phi,$$

$$\frac{d\sigma}{d\phi} \propto \sin \phi.$$

Hence in a polar plot $d\sigma/d\phi$ should appear as a circle. The extent to which this occurs in the case of ^{12}C nuclei may be judged from fig. 48 which shows $d\sigma/d\phi$ for three neutron energies. Considerable departures from the circle are observed at 6 MeV, but at 2 and 1 MeV the agreement is fairly good. Carbon is one of the most extensively studied nuclei and a compilation of data has been made by Kalos & Goldstein (1956). For others the measurements near 1 MeV by Walt & Barschall (1954) indicate a hard-sphere collision but at higher energies little data exist. Very often, when resonances

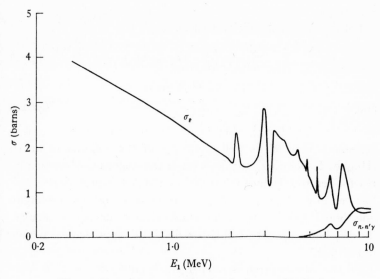

Fig. 47. Neutron cross-sections for elastic σ_e and inelastic $\sigma_{n,n'\alpha}$ collisions with ^{12}C nuclei.

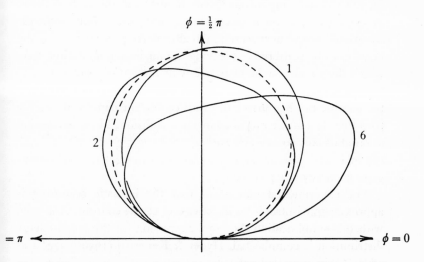

Fig. 48. The angular properties of neutron scattering by ^{12}C nuclei at 1, 2 and 6 MeV shown in a polar plot of $d\sigma/d\phi$ versus ϕ (after Kalos & Goldstein, 1956).

occur in the total cross-section versus E_1, scattering becomes anisotropic.

For the sake of simplicity we shall take

$$\frac{d\sigma}{dE_2} = \frac{\sigma_e}{\Lambda E_1} \quad (E_2 \leqslant \Lambda E_1) \tag{4.62}$$

$$= 0 \quad (E_2 > \Lambda E_1)$$

and $\qquad\qquad \bar{E}_2 = \tfrac{1}{2}\Lambda E_1 \tag{4.62a}$

as in (4.33) and (4.34), where σ_e is the observed elastic scattering cross-section which is well tabulated for the majority of nuclei (Hughes & Schwarz, 1958). Because the majority of damage by reactor neutrons is due to those below 2 MeV this procedure is well justified in most cases of interest. For neutrons of higher energies one should strictly use experimental values of $d\sigma/d\phi$ to calculate $d\sigma/dE_2$, but as the data is seldom available this approach is generally impossible.

Because the neutron is uncharged, it enters the nuclear force field and excitation of the nucleus is possible. This may result in the capture of the neutron to form a nucleus of mass $(M_2 + 1)$. Frequently this new nucleus is unstable and emits a charged particle capable of producing further recoils by the mechanisms considered in § 4.2 and 4.3. Such cases will be dealt with under 'Nuclear Reactions' in the next section. An alternative possibility is for the neutron to be scattered but leaving the nucleus in an excited state which decays almost immediately with the emission of a gamma ray; i.e. an $(n, n'\gamma)$ reaction. The recoiling nucleus is unaltered and the gamma ray is unlikely to cause significant damage itself. We treat this class of reaction as an inelastic collision. The cross-section σ_i is not usually comparable with σ_e until energies above 2 MeV are reached. The case of carbon is illustrated in fig. 47, in which the gamma energy is 4·4 MeV.

The experimental data show that the inelastic scattering is approximately isotropic, as in the case of elastic collisions, but even with this simplification the exact expression for the recoil energy distribution is complicated. Hyder & Kenward (1959) suggest as an approximation that one takes

$$\frac{d\sigma_i}{dE_2} \simeq \frac{\sigma_i}{\bar{E}_2'}, \tag{4.63}$$

where $\hat{E}'_2$ is an *effective* maximum recoil energy given by

$$\hat{E}'_2 = \Lambda E_1 - 2E_\gamma/(1 + M_2)$$

or $\qquad \hat{E}'_2 \simeq \Lambda E_1(1 - E_\gamma/2E_1) \quad \text{for} \quad M_2 \gg 1 \qquad (4.64)$

with E_γ the energy of the gamma photon. For damage by reactor neutrons the inelastic collisions generally play a minor role as they occur at high energies and this appears to be an adequate approximation.

In the energy range $\hat{E}'_2 < E_2 < \Lambda E_1$ the differential cross-section will be given by (4.62). For $\hat{E}'_2 > E_2 > 0$ one must add together the inelastic and elastic contributions obtaining for the net differential cross-section:

$$\frac{\mathrm{d}\sigma}{\mathrm{d}E_2} = \frac{\mathrm{d}\sigma_\mathrm{e}}{\mathrm{d}E_2} + \frac{\mathrm{d}\sigma_\mathrm{i}}{\mathrm{d}E_2}$$

$$= [\sigma_\mathrm{e} + \sigma_\mathrm{i}/(1 - E_\gamma/2E_1)]/\Lambda E_1$$

and for $E_\gamma \ll 2E_1$ we may write approximately:

$$\frac{\mathrm{d}\sigma}{\mathrm{d}E_2} \frac{\sigma_\mathrm{e}}{\Lambda E_1}\left[1 + \frac{\sigma_\mathrm{i}E_\gamma}{2\sigma_\mathrm{e}E_1}\right] \quad (0 < E_2 < \hat{E}'_2). \qquad (4.65)$$

In most cases of neutron irradiation one is confronted, not with monoenergetic neutrons, but with an energy spectrum having $\mathrm{d}\Phi$ neutrons in the interval $\mathrm{d}E_1$ at E_1. This must be taken into account when calculating $P(E_2)\,\mathrm{d}E_2$.

Reliable methods of computing the neutron spectra in reactors have been developed and an extensive compilation of spectra in typical reactors is available (Wright, 1962; Thompson & Wright, 1965). Experimental information about spectra in the region 1 keV to 10 MeV is scanty, but such as does exist confirms the predictions reasonably well. In fig. 49 two spectra are shown as graphs of $E_1(\mathrm{d}\Phi/\mathrm{d}E_1)$ against E_1. Both are for a graphite moderated, natural uranium fuelled, reactor; one is at a position inside a hollow uranium cylinder, the other for a point in the moderator, equidistance from four fuel rods. These are positions frequently used for carrying out irradiations. As one might expect, in the first case the spectrum has a high energy peak corresponding to neutrons coming directly from the uranium fission reaction. In the second, some slowing down has occurred in the moderator and below 2 MeV the spectrum $\mathrm{d}\Phi/\mathrm{d}E_1$ behaves approximately like E_1^{-1}.

9

The number of recoils per atom produced in dE_2 at E_2 by neutrons in dE_1 at E_1 is zero unless $E_1 > \check{E}_2/\Lambda$, but if this condition is satisfied one has

$$d\sigma\, d\Phi \quad \text{or} \quad \frac{d\sigma}{dE_2}\frac{d\Phi}{dE_1}\, dE_1\, dE_2.$$

Integrating for all neutrons with $E_1 > \check{E}_2/\Lambda$ the total in dE_2 at E_2 per atom is

$$\int_{\check{E}_2/\Lambda}^{\infty} \frac{d\sigma}{dE_2}\frac{d\Phi}{dE_1}\, dE_1\, dE_2.$$

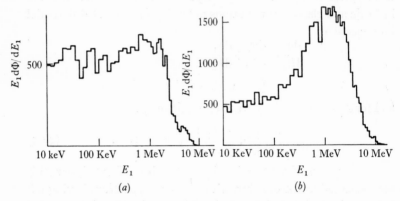

Fig. 49. Neutron spectra in a graphite moderated reactor, showing $E_1(d\Phi/dE_1)$ versus E_1. (a) At a point in the moderator equidistant from four fuel rods; (b) inside a hollow uranium cylinder. (From Thompson & Wright, 1965.)

Then the probability of a recoil having energy in dE_2 at E_2 is

$$P(E_2)\, dE_2 = \frac{\displaystyle\int_{\check{E}_2/\Lambda}^{\infty} \frac{d\sigma}{dE_2}\frac{d\Phi}{dE_1}\, dE_1}{\displaystyle\int_{\check{E}_2}^{\infty_1}\int_{\check{E}_2/\Lambda}^{\infty} \frac{d\sigma}{dE_2}\frac{d\Phi}{dE_1}\, dE_1\, dE_2}\, dE_2. \qquad (4.66)$$

The integrals may be evaluated numerically using the expressions (4.62) or (4.65) for $d\sigma/dE_2$, taking experimental values of σ_e, σ_i, E_γ and computed values of $d\Phi/dE_1$. This has been done for ^{12}C in the two spectra of fig. 49 and graphs of $P(E_2)$ versus E_2 are shown in fig. 50. It will be obvious that near to the uranium the increased proportion of fast neutrons leads to a recoil spectrum with more energetic recoils.

A number of simplifying assumptions enable (4.66) to be evaluated analytically and although the results are not exact they give one a clear impression of how the recoil varies with neutron spectrum. In a reactor the neutron energy spectrum $d\Phi/dE_1$ can be approximated by a flux Φ_{fiss} of fission neutrons with the same energy E_{fiss} plus a total moderated flux Φ_{mod} distributed over the interesting energy interval $\check{E}_2/\Lambda$ to E_{fiss} according to a $1/E_1$ law. For example,

$$\frac{d\Phi}{dE_1} = \Phi_{\text{fiss}}\,\delta(E_{\text{fiss}}) + \frac{1}{E_1}\frac{\Phi_{\text{mod}}}{\log(\Lambda E_f/\check{E}_2)},$$

where $\delta(E_{\text{fiss}})$ is the Dirac delta function. It is easily verified by integration that the second term contributes a total flux of Φ_{mod} when written in this way. Using this expression in (4.66) together with the expression for $d\sigma/dE_2$ given by assuming hard sphere elastic neutron scattering (4.62)

$$P(E_2) = \frac{1}{\Lambda E_{\text{fiss}}}\frac{\dfrac{\Phi_{\text{mod}}}{L\Phi_{\text{fiss}}}\left(\dfrac{\Lambda E_{\text{fiss}}}{E_2} - 1\right) + 1}{\dfrac{\Phi_{\text{mod}}}{L\Phi_{\text{fiss}}}(L-1) + 1} \qquad (4.66\,a)$$

for
$$E_2 < E_{\text{fiss}}$$

with
$$L = \log\frac{\Lambda E_{\text{fiss}}}{\check{E}_2} \quad (\sim 10 \text{ in typical cases}).$$

Three important cases can be distinguished:

(1) $\Phi_{\text{mod}} = 0$, monoenergetic neutrons, which might apply in the case of a fast reactor.

$$P(E_2) = \frac{1}{\Lambda E_{\text{fiss}}} \qquad (4.66\,b)$$

and
$$E_2 = \tfrac{1}{2}\Lambda E_{\text{fiss}}.$$

Here we have all recoil energies equally probable up to the maximum possible: ΛE_{fiss}.

(2) $\Phi_{\text{fiss}} = 0$, which could describe the situation at a position in the moderator of a thermal reactor far removed from any fuel elements

$$P(E_2) = \frac{1}{\Lambda E_{\text{fiss}}}\frac{\dfrac{\Lambda E_{\text{fiss}}}{E_2} - 1}{L - 1} \quad (E_2 < E_{\text{fiss}}) \qquad (4.66\,c)$$

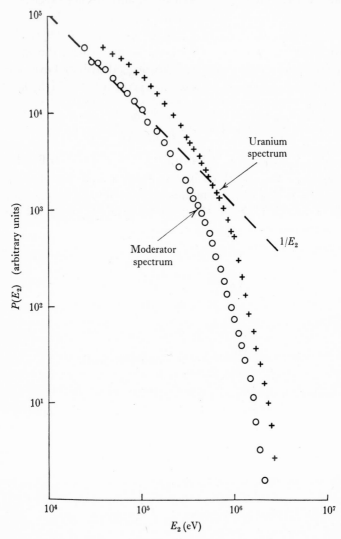

Fig. 50(a). $P(E_2)\,dE_2$ the probability of recoil into dE_2 at E_2 for ^{12}C irradiated with neutrons having the spectra shown in fig. 49 (by courtesy of Dr S. B. Wright).

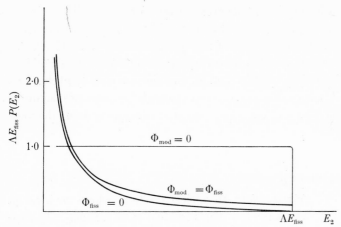

Fig. 50(*b*). Approximation to $P(E_2)$ for the three cases discussed in the text: the core of a fast reactor where $\Phi_{\text{mod}} = 0$, the fuel of a thermal reactor where $\Phi_{\text{mod}} \sim \Phi_{\text{fiss}}$ and the moderator of a fast reactor where $\Phi_{\text{fiss}} = 0$.

and
$$\bar{E}_2 = \frac{\Lambda E_{\text{fiss}}}{2L} \quad (\sim \tfrac{1}{20}\Lambda E_{\text{fiss}}).$$

In this case $P(E_2)$ behaves like $1/E_2$ at energies well below ΛE_{fiss}, but gradually cuts off to zero at ΛE_{fiss}, decreasing more rapidly than $1/E_2$. Thus the lower recoil energies are much more probable and the $P(E_2)$ is very similar to that for fission fragments.

(3) $\Phi_{\text{fiss}} = \Phi_{\text{mod}}$, a typical situation near the fuel of a thermal reactor, then:

$$P(E_2) = \frac{1}{LE_2} \quad (E_2 < \Lambda E_{\text{fiss}}), \qquad (4.66d)$$

$$\bar{E}_2 = \frac{\Lambda E_{\text{fiss}}}{L} \quad (\sim \tfrac{1}{10}\Lambda E_{\text{fiss}}).$$

This is intermediate between cases (1) and (2) as may be seen in fig. 50*b* where the three recoil distributions are compared. Obviously, any reactor situation may be approximated in this way and comparison with the more exact results of figure 50*a* shows the procedure to be fairly good. The main shortcoming is the failure to include a distribution in E_1 for the fission neutrons as these are by no means mono-energetic.

In visualizing the track of a neutron we must remember that it has no interaction with electrons, consequently it loses energy only by nuclear collisions. The cross-section for such collisions in the non-fissile materials is of the order of 10^{-24} cm², hence the mean path between collisions is of the order of several cm. The average collision in a fast reactor (case 1) spectrum transfers several hundred keV, since $E_{\text{fiss}} \sim 2$ MeV and all energies are equally likely. In a thermal reactor the typical recoil energy is an order of magnitude lower, being a few tens of keV, and low energy recoils are the most probable.

4.6.   Indirect processes

4.6.1. *Nuclear reactions.* In nuclear reactions where the incident particle has an energy less than 100 MeV or so, it is generally supposed that a compound nucleus is formed as an intermediate stage. This, being unstable, decays in a very brief time to produce the final reaction products, i.e.

$$(\text{Particle}, M_1) + (\text{Nucleus}, M_2) \rightleftharpoons (\text{Compound Nucleus}, M_1 + M_2)$$

$$\rightleftharpoons (\text{Particle}, M_3) + (\text{Nucleus}, M_4) + (\text{Energy}, Q)$$

The sum of masses on opposite sides of the equation balance to within 1 a.m.u. and the small discrepancy, when multiplied by c^2, gives the reaction energy Q. If Q is positive the forward reaction is exothermic whilst the reverse reaction is endothermic. In principle nuclear reactions are always reversible and this has been verified in many cases, but if more than two bodies are involved at any stage, experimental verification is virtually impossible.

If the target nucleus, M_2, is at rest in the L system and the particle M_1 has E_1 then the energy balance in G co-ordinates in the reaction as:

$$\frac{M_2 E_1}{M_1 + M_2} + Q = E_3' + E_4',$$

where E_3' and E_4' refer to the kinetic energies in G co-ordinates, of the product. In order that momentum shall be conserved:

$$M_3 E_3' = M_4 E_4',$$

and combining these with the approximate relation

$$M_1 + M_2 \simeq M_3 + M_4$$

one has
$$E'_3 = \frac{M_4 Q}{M_1 + M_2}\left\{1 + \frac{M_2}{M_1 + M_2}\frac{E_1}{Q}\right\}, \qquad (4.67)$$

$$E'_4 = \frac{M_3 Q}{M_1 + M_2}\left\{1 + \frac{M_2}{M_1 + M_2}\frac{E_1}{Q}\right\}. \qquad (4.68)$$

To obtain E_3 and E_4, the kinetic energies in L co-ordinates one must know the directions in which the products move relative to the direction of incidence. This requires a detailed knowledge of the angular properties of the reaction, for which data is often available. In many cases of interest a simplification is possible for if $Q \gg E_1$, as in the case of uranium fission or thermal neutron induced reactions, $E_3 = E'_3$, $E_4 = E'_4$ and $E_1/Q \ll 1$.

Then
$$E_3 = \frac{M_4 Q}{M_1 + M_2}, \qquad (4.69)$$

$$E_4 = \frac{M_3 Q}{M_1 + M_2}. \qquad (4.70)$$

If the product particle is a photon of energy E_γ, the momentum relation becomes
$$E_\gamma^2/2c^2 = M_4 E_4$$
which with the energy balance leads to

$$E_4 = \frac{Q^2}{2M_4 c^2}. \qquad (4.71)$$

4.6.2. *Nuclear reactions with light charged-particles.* We have already seen that light charged-particles having energies of the order $10\,\text{MeV}$ have nuclear coulomb collisions with cross-sections of the order 10^2 barn. The cross-sections for nuclear reaction are generally in the range 1 to $100\,\text{millibarn}$ for such particles, which reflects the short range of nuclear forces compared with Coulomb interactions. Only when particle energies reach a few hundred MeV do reaction cross-sections compete favourably with direct collisions as a means of producing damage; because the latter cross-sections must decrease as $1/E_1$ according to (4.39).

The reactions occurring at such energies are very violent and resemble fission of ^{235}U in that nuclei split into several heavy fragments and a shower of light particles. The kinetic energies are of the order $10\,\text{MeV}$ and we have already seen how to deal with recoils produced by such particles through direct collision. Smoluchowski (1956) has utilized protons in the range 100–400 MeV to induce damage by such reactions in tungsten, and similar experiments have been performed by Simon, Denney & Downing (1963) in silicon, but otherwise there have been few attempts to study radiation damage at high energies.

4.6.3. *Nuclear reactions with neutrons.* Because the neutron lacks a charge, its interactions with nuclei come almost exclusively from nuclear forces. Consequently, reaction cross-sections may be comparable with elastic collision cross-sections, or even exceed them, becoming an extremely important generator of damage. This is especially true for thermal neutrons ($E_1 \sim$ 0·025 eV) which could not possibly produce damage by direct collision. Cross-sections for thermal neutron reactions may be exceedingly large ($\sim 10^5$ barn) which is surprising at first sight in view of previous remarks concerning the short range of nuclear forces. However, the thermal neutron moves so slowly that its wavelength is of the order of Å and it is able to interact with the nucleus for relatively long periods $\sim \lambda/u_1$ on the nuclear time-scale ($\lambda/u_1 = h/E_1 \sim 10^{-13}$ sec at $E_1 = $ 0·025 eV).

Unfortunately it is very difficult to make generalizations, as the behaviour of different nuclei, even though isotopes of the same element, is often widely different by many orders of magnitude. We shall therefore consider some as specific examples. A few cases are listed in table XIII. These are chosen because their large values of reaction cross-section makes them of practical importance. The fission reaction in ^{235}U has already been dealt with in § 4.2.4 and will not be considered further.

Next, consider the ^{6}Li, (n, α) ^{3}H reaction. The products are an alpha particle and a triton which fly apart with energies well in excess of E_b. Consequently their collisions occur in a Simple Coulomb field and $d\sigma/dE_2$ is given by (4.38). In specimens whose size exceeds the range of such particles ($\sim 10^{-3}$ cm) the majority will come to rest in the sample. Thus one cannot deal with a single value of E_1 when calculating $N(E_2) \, dE_2$, the number of recoils in dE_2 at E_2. One uses the method described in § 4.2.4 whereby $d\sigma/dE_2$ is integrated along the particle track leading to (4.44) for $N(E_2) \, dE_2$.

TABLE XIIIA. *Some thermal neutron reactions involving charged particles*

Calculations assume that the product nucleus is in the ground state and that $E_3 + E_4 = Q$.

Reaction	Q (eV)	barn	E_3 (eV)	E_4 (eV)
^{6}Li (n, α) ^{3}H	$4 \cdot 7 \times 10^6$	950	$2 \cdot 0 \times 10^6$	$2 \cdot 7 \times 10^6$
^{10}B (n, α) ^{7}Li	$2 \cdot 66 \times 10^6$	3990	$1 \cdot 69 \times 10^6$	$0 \cdot 97 \times 10^6$
^{25}Mg (n, α) ^{22}Ne	$4 \cdot 0 \times 10^5$	0·27	$3 \cdot 4 \times 10^5$	$6 \cdot 2 \times 10^4$
^{235}U $(n, \text{Fission})$	2×10^8	582	$0 \cdot 95 \times 10^8$	$1 \cdot 37 \times 10^8$
			$(M_3 = 137)$	$(M_4 = 96)$

TABLE XIIIB. *Mean recoil energies from (n, γ) reactions*

Values of cross-section and recoil energy are averaged to take account of the natural element after Coltman, Klabunde, McDonald & Redman (1962).

	barn	E_4 (eV)
Al	0·23	879
Cu	3·8	388
Ag	63	132
Au	98	81
Pt	8·8	57
Cd	2500	133
Ni	4·8	567

For (n, α) reactions in heavier elements the product nucleus may have energy below E_b and the methods of §§4.2.4 or 4.2.5 are required to obtain $d\sigma/dE_2$.

In the (n, γ) reactions, low energy recoils result from the small momentum associated with a gamma photon. However, one often finds $E_4 > \breve{E}_2$ and in cases where the reaction cross-section is large this form of damage is important. The reaction energy Q is not usually disposed of in a single gamma decay but rather in a series of transitions each involving the emission of a gamma ray. Since the nucleus reaches its ground state in a time that is short compared to the flight time of the recoil, the calculation of the mean recoil energy is not simple. Coltman *et al.* (1962) have published values of $\bar{E}_4$ in various (n, γ) reactions, some of which are reproduced in table XIII.

4.6.4. *Gamma irradiation.* The production of recoils by gamma irradiation is not of great practical importance. In most nuclear reactors for instance, the fast neutron flux is about equal in magnitude to the gamma flux but because neutrons generate recoils with much more energy they cause about 10^3 times more damage in an element of maximum atomic weight. In spite of this, gamma sources have frequently been used in basic studies, and we shall consider here some mechanisms of recoil production, other than photonuclear reactions which belong in the preceding sections.

In priniciple it is possible for a high-energy photon, or gamma ray, to transfer momentum directly to a nucleus by the nuclear Compton process. However, the cross-section for such events is insignificant when photon energies are less than 100 MeV. The principle means by which they produce damage are indirect processes in which fast electrons are produced; the photoelectric effect, the Compton effect and pair production. A detailed discussion of these mechanisms will be found in W. Heitler's book *The Quantum Theory of Radiation* (1944). The way in which the electrons cause damage by direct Coulomb collisions has been considered in §4.3.

In the photoelectric process the entire photon energy is absorbed by the atom as a whole and this results in the ejection of an electron from one of the inner shells, generally the K shell. The kinetic energy of the electron is sharply defined according to the relation:

$$E_1 = E_\gamma - E_K,$$

where E_γ is the gamma photon energy and E_K the ionization energy of a K electron. In most cases of interest in radiation damage $E_\gamma \gg E_K$ and

$$E_1 \simeq E_\gamma. \tag{4.72}$$

The total cross-section per atom, σ_{pe}, for the photoelectric process rises sharply to a maximum at the 'K absorption edge' when $E_\gamma = E_K$. Thereafter it decreases, roughly as $E_\gamma^{-\frac{7}{2}}$. No single theoretical expression for σ_{pe} covers the energy range of interest, but the graph of σ_{pe} versus E_γ for lead is shown in fig. 51. From this one may calculate σ_{pe} for any other element using the well-established fact that $\sigma_{pe} \propto Z_2^5$.

In the Compton process a gamma ray photon is scattered by an individual electron, giving it some energy in the process, the photon proceeding with reduced energy in a different direction. The differential cross-section per atom for producing an electron in dE_1 at E_1 as a function of E_1, is strongly peaked near $E_1 = E_\gamma$. For many purposes it is enough to take $E_1 = E_\gamma$. Since the interaction is with individual electrons the cross-section per atom depends on Z_2, the

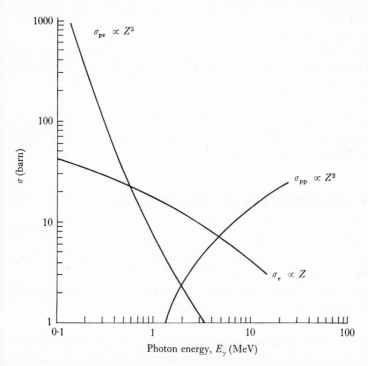

Fig. 51. Atomic cross-sections for the photo-electric effect,
the Compton effect and pair production in Pb.

number of electrons in the atom. The total cross-section per atom, σ_c, is shown as a function of E_γ for lead in fig. 51 and from this curve one may obtain σ_c for any other element by scaling in the ratio of atomic numbers.

Before pair production can occur E_γ must exceed $2m_0 c^2$ (= 1·02 MeV), the rest energy of the electron-positron pair. This process occurs in the vicinity of a nucleus which recoils with a negligibly small energy. The energies E_1^+ and E_1^- of the positron and electron related to E_γ by

$$E_1^+ + E_1^- = E_\gamma - 2m_0 c^2. \qquad (4.73)$$

The available energy may be distributed in any proportion between the pair, but there is a slight tendency for E^+ to exceed E^- due to the unsymmetrical nature of the Coulomb interactions near the catalysing nucleus. The cross-section per atom depends on Z_2^2 and is shown as a function of energy in fig. 51.

Qualitatively it will be clear from fig. 51 that for light elements the Compton effect will be the most important when $E_\gamma < 10\,\mathrm{MeV}$. Although σ_{pe} may exceed σ_c at low energies the electrons so produced do not have enough energy to cause damage.

In heavy elements with $E_\gamma < 5\,\mathrm{MeV}$ the photoelectric process becomes important since $\sigma_{\mathrm{pe}} \propto Z^5$. Above 5 MeV the main process is pair production.

4.7. Summary of primary recoil distributions

We are now in a position to summarize the primary recoil distributions, both in space and in energy, that characterize particular types of radiation. The most important of these are listed in table XIV.

It will be seen that many of the energy distributions are quite similar, behaving like $1/E_2$ or $1/E_2^2$ but there are wide differences between the recoil spacings and the mean recoil energies. Thus, although reactor neutrons and the 50 keV heavy ion produce a very similar energy distribution and mean energy, the spacing between primary recoils is 10^7 times greater for the neutron. Comparison of the fission fragment and the reactor neutron spectrum shows that although $P(E_2)$ is somewhat similar the mean recoil energy is an order of magnitude greater for the neutron and the spacing is 10^8 times greater. This question of spacing will prove to be very important when we come to investigate the consequences of the overlapping of neighbouring collision cascades. The fission tracks of plate XII are one example.

In table XIV two particles stand out quite distinctly from the rest. The 2 MeV neutron has the highest mean recoil energy of all, and the recoils are distributed uniformly in energy. The 1 MeV electron is barely able to transmit the minimum energy necessary for displacement to occur, consequently the mean energy is extremely small. The recoil spacing is large and we shall see in Chapter 6 that these properties can be utilized in experiments to produce isolated defects throughout a specimen.

Having considered the generation of primary recoils in some detail we now proceed to the collision cascades by which they produce defects.

TABLE XIV. *Properties of recoil atoms from the main types of radiation for $M_2 = 50$*

Particle	Spacing between primary recoils	Track length of particle	Mean recoil energy	Recoil energy distribution $P(E_2)\, dE_2$
1 MeV proton	10^{-3}	10^{-3} cm	200 eV	$\propto \dfrac{dE_2}{E_2^2}$, low energies strongly favoured
100 MeV fission fragment	10^{-7} cm	10^{-4} cm	1000 eV	$\propto \dfrac{dE_2}{E_2^2}$ at high energy, $\propto \dfrac{dE_2}{E_2}$ at low energy
50 keV heavy ion	10^{-6} cm	10^{-5} cm	7 keV	roughly $\propto \dfrac{dE_2}{E_2^{\frac{3}{2}}}$
1 MeV electron	0·1 cm	0·1 cm	50 eV	Peaked near mean energy
2 MeV neutron	5 cm	100 cm	160 keV	$\dfrac{dE_2}{\Delta E_1}$ all energies equally probable
Neutron (thermal reactor spectrum)	5 cm	100 cm	10 keV	Roughly $\propto \dfrac{dE_2}{E_2}$ at low energy, faster decrease at high energy

CHAPTER 5

THE COLLISION CASCADE

5.1. The simple theory

In this section we shall consider the production of displaced atoms by the primary recoil. The final aims are to determine the *damage function* $\nu(E_2)$ which gives the average number of displaced atoms due to a recoil at E_2, and to estimate the spatial distribution of the defects produced.

From the previous chapter we know how to obtain $d\sigma$, the differential cross-section for the incident radiation to produce primary recoil in dE_2 at E_2. Hence, if the flux of bombarding particles is $\Phi\,\mathrm{cm}^{-2}.\mathrm{sec}^{-1}$, and the irradiation lasts for t sec, the atomic concentration of displaced atoms will be

$$C_{\mathrm{d}} = t\Phi \int_{\check{E}_2}^{\hat{E}_2} \nu(E_2) \frac{d\sigma}{dE_2}\, dE_2. \tag{5.1}$$

A preliminary calculation of the *damage function* $\nu(E_2)$ based on a much oversimplified model, first used by Kinchin & Pease (1955) will be useful to introduce the subject. The cascade is represented diagrammatically in fig. 52 and the following initial assumptions will be made:

(*a*) Atoms in collision behave like hard spheres.

(*b*) All collisions are elastic and no energy is dissipated in electron excitation.

(*c*) The cascade proceeds as a series of two-body collisions.

(*d*) These collisions are independent of each other and any spatial correlations implied by the periodicity of the crystal structure will be ignored.

(*e*) In a collision such as that illustrated in fig. 53, when an atom with E emerges with E' and generates a new recoil with E'' it is assumed that no energy passes to the lattice and

$$E = E' + E''. \tag{5.2}$$

(*f*) A stationary atom which receives less than a critical energy E_{d} is not displaced. Similarly if an incident atom emerges from collision with $E' < E_d$, it does not contribute further to the cascade. These statements form our definition of the *mean displacement energy* E_d. For the simple theory we shall take it that the probability of displacement is zero below E_d, rising sharply to 1 at $E = E_d$.

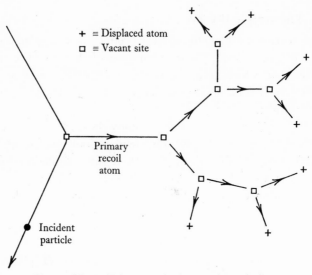

+ ≡ Displaced atom
□ ≡ Vacant site

Primary
recoil
atom

Incident
particle

Fig. 52. The collision cascade shown schematically.

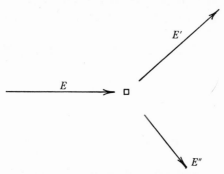

Fig. 53. A single branch of the cascade.

It follows from (*f*) that an atom with $E_d < E < 2E_d$ cannot produce any increase in the number of displaced atoms and $\nu(E)$ is a step function:

$$\left.\begin{array}{l} \nu(E) = 0, \quad \text{if } E < E_d, \\ \nu(E) = 1, \quad \text{if } E_d < E < 2E_d. \end{array}\right\} \tag{5.3}$$

In a collision with $E \gg E_d$ the probability of finding the scattered atom with energy in dE' at E' and the recoil atom with energy in dE'' at E'' is $d\sigma/\sigma$ where σ and $d\sigma$ are the total and differential cross-sections for the atomic collision. Because of assumption (a) we can use the hard-sphere results of §2.3 to give the probability:

$$\frac{d\sigma}{\sigma} = \frac{dE''}{E} \quad \text{or} \quad = \frac{dE'}{E}.$$

Now the average number of displaced atoms produced by the scattered atom will be found by integrating the product of $\nu(E')$ and this probability over the energy range from E_d to E:

$$\int_{E_d}^{E} \nu(E') \frac{dE'}{E}.$$

Similarly the average number produced by the recoil with E'' is

$$\int_{E_d}^{E} \nu(E'') \frac{dE''}{E}.$$

The sum of these must equal the average number produced by an atom with E: $\nu(E)$. This fact gives us the integral equation

$$\nu(E) = \frac{2}{E} \int_{E_d}^{E} \nu(x)\,dx. \tag{5.4}$$

It is easily verified that this is satisfied approximately by

$$\nu(E) = kE \quad \text{for} \quad E^2 \gg E_d^2.$$

Where k is a constant whose value may be found by putting $\nu(2E_d) = 1$, whence $k = 1/2E_d$.

$$\therefore \quad \nu(E) = E/2E_d \quad \text{for} \quad E \gg E_d. \tag{5.5}$$

Relations (4.3) and (4.5) give the required function $\nu(E_2)$.

A simple, but less rigorous, argument shows (4.5) to be reasonable. One expects any atom in the cascade to continue the multiplication process until its energy falls below $2E_d$, and the number of atoms with $2E_d$ should be about $E_2/2E_d$.

An alternative model has been used by Snyder & Neufeld (1955) who make the assumption that E_d is lost at each collision. For example,

$$E = E' + E'' + E_d \tag{5.6}$$

and that both atoms move off after collision, no matter how small their energy. Comparing this with the Kinchin & Pease model one expects the introduction of the energy loss to decrease $\nu(E)$, but the inclusion of some atoms leaving the collision site with less than E_d increases it. In fact the two effects tend to compensate and the expression for $\nu(E)$ is little different, for large E_2/E_d, from the expression (5.5)

$$\nu(E) = 0.56(1 + E/E_d) \quad \text{for} \quad E > 4E_d. \tag{5.7}$$

It must be emphasized that $\nu(E_2)$ is an *average* number and deviations from the mean are to be expected. If one were unlucky the primary could dissipate all its energy in sub-threshold collisions, making $\nu = 1$. Alternatively it could be extremely lucky and use all its energy in displacement making $\nu = E/E_d$. As one might expect the average lies midway between the two extremes at $E/2E_d$.

The weakest point of these models is the assumption (a) of hard-sphere collisions. However, it has frequently been shown that by setting up an integral equation analogous to (5.4) using more realistic differential cross-sections, the proportionality of (5.5) is preserved though there is a slight effect on the magnitude (e.g. Brown & Goedecke, 1960; Lehmann, 1961; Robinson, 1965; Sanders, 1967). For example, using an inverse power potential (r^{-l}) Sanders used the Kinchin & Pease model to obtain:

$$\nu(E) = l(2^{1/(l+1)} - 1)\frac{E}{2E_d}. \tag{5.5a}$$

For the inverse square potential $(l = 2)$ this is approximately

$$\nu(E) \simeq 0.52\frac{E}{2E_d}. \tag{5.5b}$$

One must have some reservation about this result because it applies the inverse square potential to all collisions in the cascade and for some of these E is below the limit $(E)_{\rho=5a}$ defined in §4.3.5.

Physically, the effect of a realistic scattering is to make a larger number of collisions generate E'' in the sub-threshold region below E_d where they are lost to the multiplication process.

We shall see that other corrections to (5.5) can be equally important, but this form of damage function is suitable for rough calculations.

An important quantity in an irradiated solid is the density $\rho(E)\,dE$ of recoil atoms with energies in the interval dE at E. Suppose primary recoils with energy E_2 are produced at a rate $\dot{q}$ per second per unit volume. The average time τ that these spend before their first collision is

$$\tau = \frac{\lambda_2}{v_2} = \frac{1}{n\sigma v_2},$$

where $\lambda_2 = 1/n\sigma_2$ is their mean free path, and we note that

$$\sigma_2 = \int_{E_d}^{E_2} \frac{d\sigma}{dE''}\, dE''.$$

At any instant the number of primaries in unit volume is $\dot{q}\tau$, and this is then the density of primary recoils ρ_2, i.e.

$$\rho_2 = \frac{\dot{q}}{n\sigma_2 v_2}.$$

By multiplication in the collision cascades, let each primary produce $\nu(E_2, E)$ recoils slowing down through the lower energy level E. When $E = 2E_d$ this function is clearly the same thing as the damage function, hence we know its form from (5.5) by substituting E for $2E_d$. Each recoil takes a time dt to slow down through an interval dE at E and we may put

$$dt = \frac{dE}{\dfrac{dE}{dt}} = \frac{dE}{v\dfrac{dE}{dx}}.$$

The rate of energy loss dE/dx is given by

$$dE = \int nE''\, d\sigma\, dx \quad \text{or} \quad \frac{dE}{dx} = n\int_{E_d}^{E} E\frac{d\sigma}{dE''}\, dE''.$$

The density of recoils in dE at E is given by taking the density arriving at E per second: $\dot{q}\nu(E_2, E)$, and multiplying by the time that each spends in dE: dt. For example,

$$\rho(E)\,dE = \frac{\dot{q}\nu(E_2, E)\,dE}{v\dfrac{dE}{dx}}.$$

If we now take the ratio of this density to ρ_2 the density of primaries we shall obtain $N(E_2, E)\,dE$, the number of recoils in dE at E

for each primary at E_2, then

$$N(E_2, E)\,dE = \frac{n\sigma_2 v_2 \nu(E_2, E)\,dE}{v\dfrac{dE}{dx}}$$

or $\qquad N(E_2, E)\,dE = \nu(E_2, E)\sqrt{\dfrac{E_2}{E}}\dfrac{\displaystyle\int_{E_d}^{E}\dfrac{d\sigma}{dE''}\,dE''}{\displaystyle\int_{E_d}^{E}E''\dfrac{d\sigma}{dE''}\,dE''}\,dE.$

If we take the inverse square potential, (4.48) and (5.5b) give $d\sigma/dE''$ and $\nu(E_2, E)$, then:

$$N(E_2, E)\,dE \simeq \frac{E_2\,dE}{2E_d^{\frac{1}{2}}E^{\frac{3}{2}}}.$$

A more general expression for a $1/r^l$ potential was obtained by Sanders (1967) using a slightly different method:

$$N(E_2, E)\,dE \simeq \frac{(1 - 1/l)E_2\,dE}{E_d^{(1-1/l)}E^{(1+1/l)}}.$$

This is clearly in agreement with expression derived here when $l = 2$.

Equation $\rho(E)$ can easily be generalized to the case of a primary recoil distribution such that $\dot{q}(E_2)\,dE_2$ primaries are generated per unit volume per second with energies in the range dE_2 at E_2

$$\rho(E)\,dE = \frac{\displaystyle\int_{E}^{\Lambda E_1}\dot{q}(E_2)\,\nu(E_2, E)\,dE_2}{v\dfrac{dE}{dx}}\,dE.$$

Then using the fact that

$$\dot{q}(E_2) = n\Phi\frac{d\sigma_1}{dE_2},$$

with Φ the flux of bombarding particles and $d\sigma_1$ the cross-section for these to create the primary recoil in dE_2 at E_2, and putting in the expression for dE/dx from above:

$$\rho(E)\,dE = \frac{\Phi\displaystyle\int_{E}^{\Lambda E_1}\dfrac{d\sigma_1}{dE_2}\nu(E_2, E)\,dE_2}{v\displaystyle\int_{E_d}^{E}E''\dfrac{d\sigma}{dE''}\,dE''}\,dE.$$

It will be shown that this quantity is of great importance in understanding the sputtering experiments to be described in § 5.5.5.

The simple proportionality, expressed in (5.5), (5.5 a) and (5.5 b) between the number of displacements and the recoil energy allows (5.1) to be rewritten in a very simple form. We note that by definition

$$\bar{E}_2 = \frac{1}{\sigma_p} \int E_2 \frac{d\sigma}{dE_2} dE_2.$$

Then putting (5.5) into (5.1) and comparing with the above definition

$$C_d = t\Phi \sigma_p \frac{\bar{E}_2}{2E_d} \quad \text{for} \quad \Lambda E_1 \gg E_d. \tag{5.1 a}$$

Thus we can calculate C_d by using σ_p the total cross-section for producing a primary and treating each primary recoil as though it had an energy $\bar{E}_2$. Corrections to the simple theory, to be detailed in later sections, unfortunately spoil this simple situation, but as a rough working approximation it can be extremely useful. Expressions for $\bar{E}_2$ and σ_p have been derived for various types of radiation in chapter 4, using those for which $\bar{E}_2 \gg E_d$ a list of formulae for C_d has been collected below:

(1) Energetic light ions

$$C_d = t\Phi \frac{2\pi a_0^2 M_1 Z_1^2 Z_2^2 E_R^2}{M_2 E_1 E_d} \log \frac{\Lambda E_1}{E_d}. \tag{5.1 b}$$

(2) Fission fragments. No simple expression can be given, but equation (5.1 b) represents a lower limit to C_d since this underestimates the relative importance of higher energy recoils. The average recoil will produce about 20 displacements and there are $\sim 10^4$ recoils making of the order of 10^5 displaced atoms per fission track.

(3) Heavy ions with $E \lesssim E_a$

$$C_d = t\Phi \frac{\pi^2 a^2 \Lambda E_a}{8E_d}. \tag{5.1 c}$$

This has a particularly interesting form, for it is independent of bombarding energy and implies a constant linear density of displaced atoms along the ion's track. This comes about by the decreasing probability of collision with E_1 being exactly compensated by the increase in number of displaced atoms per collision.

(4) Fast neutrons (mono-energetic)

$$C_d = t\Phi\, \sigma_{\text{total}} \frac{\Lambda E_1}{4E_d}. \tag{5.1d}$$

(5) Moderated neutrons (total flux Φ_{mod}, $\dfrac{d\Phi}{dE_1} \propto \dfrac{1}{E_1}$ up to a maximum energy E_{fiss})

$$C_d = t\Phi\, \sigma_{\text{total}} \frac{1}{\log\dfrac{\Lambda E_{\text{fiss}}}{E_d}} \frac{\Lambda E_{\text{fiss}}}{4E_d}. \tag{5.1e}$$

The effect of moderation into a $(1/E_1)$ spectrum is to reduce the damage per neutron by the logarithmic factor which is of the order of $\frac{1}{10}$.

5.2. Electron excitation and energy loss from moving atoms

Some energy may be lost to the cascade by excitation of the electrons in the solid. Consider some possible mechanisms, taking first the excitation of inner shells by an interatomic collision. Morgan & Everhart (1962) have made a close study of inelastic collisions between inert gas atoms with energies up to 10^5 eV. Inelasticity, due to electron excitation, first occurs when the distance of closest approach permits the L shells to overlap. A second stage involving the absorption of rather more energy occurs when the K shells overlap. Since the excitation energy is less than 10^3 eV the percentage effect on the kinetic energies after collision is rather small as these are approaching 10^5 eV. Furthermore, since close enough penetration can only occur in near-head-on collision, this type of inelastic collision is rather infrequent.

A much more likely event is the excitation of valence electrons.

Let us take the simple case of an atom moving at velocity v past an electron in an energy level E_1 below ionization. During the collision the Coulomb field is momentarily perturbed. The duration of the perturbation is of the order a_0/v and a Fourier analysis of the perturbing field would show a strong term with frequency $\omega \sim v/a_0$. One may therefore anticipate that any excitation requiring the absorption of a quantum $\hbar\omega$ will be highly favoured. In effect, there is a resonant coupling between the motion of the atom and a natural vibration mode of the electron.

The energy above which ionization, and hence strong excitation, occurs is given in order of magnitude by

$$E_{\text{ex}} = \tfrac{1}{2}M_2 v^2,$$

where
$$\hbar v/a_0 = \hbar\omega = E_1,$$

hence
$$E_{\text{ex}} \sim \frac{M_2 E_1^2}{4M_0 E_{\text{R}}}. \tag{5.8}$$

Energy loss by this mechanism is enhanced when $\hbar\omega$ exceeds the first ionization energy of the moving atom, when this itself may become ionized and, as it then carries an electric charge, is able to apply a stronger perturbing field in any future collision.

In a solid, the band model requires that the valence electrons, for which E_{ex} must be least, be distributed, and treated apart from the localized shell electrons. A moving atom in a solid interacts with these valence electrons. In an insulator the expression for E_{ex} is easily modified by putting the energy gap between valence and conduction band for E_1. In a metal the valence band is partly filled and electrons near the Fermi surface may be excited by arbitrarily small amounts. But the majority of electrons are below the Fermi surface and can only be excited by quanta large enough to take them above the Fermi surface (see below). To a rough approximation we take E_1 as one half the Fermi energy E_{F}. In practice, E_1 turns out to be several eV in all solids and one finds

$$E_{\text{ex}}(\text{keV}) \sim M_2 \,(\text{a.m.u.}). \tag{5.8a}$$

We wish to know $(\mathrm{d}E/\mathrm{d}x)_{\text{ex}}$ the rate of energy loss in electron excitation along the path of a moving atom in its own solid. Two cases can be treated simply: $E \gg E_{\text{ex}}$ and $E < E_{\text{ex}}$.

(1) $E < E_{\text{ex}}$. Here we treat the case of metal in which the valence electrons can be treated as a Fermi–Dirac gas. The energy E is too small to excite electrons in the atomic cores and these can be neglected, leaving only the electron gas to be considered.

The density of electrons is given in order of magnitude by n the density of atoms, and these will follow the well known Fermi–Dirac velocity distribution with a maximum velocity of v_{F}.

If v is the velocity of the atom the average change in velocity of a free electron colliding with the atom will be of the order v. But since $v < v_F$ only the fraction v/v_F of the electrons that lie near the Fermi surface can be excited, thus the effective density of electrons is only

$$n_e \sim n \frac{v}{v_F}.$$

The energy change of an electron near the Fermi surface that changes its velocity by v is

$$\bar{E}_2 \sim m v_F v.$$

The electron atom collision cross-section σ is of the order πa_0^2 and then since the flux of effective electrons is $\sim n_e v_F$, the number of electron collisions per unit time is $\sim \pi a_0^2 n_e v_F$. Each carries away an energy $\bar{E}_2$, hence the rate of energy loss is

$$\frac{dE}{dt} \sim \pi a_0^2 n_e v_F \bar{E}_2.$$

Then using the relations above and the fact that

$$\frac{dE}{dx} = \frac{dE}{dt} \Big/ \frac{dx}{dt}$$

$$= \frac{1}{v} \frac{dE}{dt}$$

we have

$$\left(\frac{dE}{dx}\right)_{ex} \sim \pi a_0^2 n m v_F v$$

or

$$\left(\frac{dE}{dx}\right)_{ex} \sim 2\pi a_0^2 n \sqrt{\left(\frac{m}{M} E_F E\right)}.$$

This treatment is based on that of Fermi & Teller (1947).

More detailed work by Lindhard goes into the Z dependence of σ and other refinements, leading to the expression:

$$\left(\frac{dE}{dx}\right)_{ex} \sim 8\pi a_0^2 n Z^{\frac{7}{6}} \sqrt{\left(\frac{m_0}{M} E_R E\right)} \tag{5.9}$$

for recoil atoms in their own solid with $E < E_{ex}$ (Lindhard & Thomsen, 1962).

Data has been accumulated for several ions in a variety of solids at energies below 1 MeV by Ormrod & Duckworth (1963) and Porat & Ramavataram (1959, 1961). Their results for C^+ ions in graphite are shown in fig. 54. Qualitatively the agreement is good in all cases, with an energy dependence roughly as $\sqrt{E}$, but (5.9) predicts an energy loss two or three times too great. There is a clear need for a better theory. In the meantime for the collision cascades the best procedure is to use experimental data where possible, extrapolating to low energies with $\sqrt{E}$. Where data does not exist either one can interpolate between known ion–solid combinations, or risk using (5.9) with an empirical factor of 0·4 inserted.

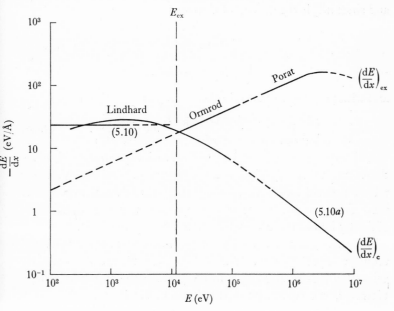

Fig. 54. The energy loss from carbon recoil atoms in graphite. $(dE/dx)_{ex}$ is the energy lost per unit path length to electron excitation. Experimental data are extrapolated with the theoretical $\sqrt{E}$ expression. $(dE/dx)_c$ is the component due to elastic collisions with stationary atoms calculated from the (5.15) and the theory of Lindhard.

(2) $E \gg E_{ex}$. This case is unlikely to be met in the atomic recoils of radiation damage, but is of interest because it often applies to the bombarding particles (e.g. 1 MeV protons or 100 MeV fission fragments).

Take the case of a light charged particle, with M_1, Z_1 and E_1, which will be completely ionized because the excitation criterion will apply to all its electrons. The electrons in the solid interact with the particle by a Coulomb potential, hence we may use the *impulse approximation* of §4.3.3 if we replace M_2 by m_0, the electron mass Z_2 by 1, and $\check{E}_2$ by an effective ionization energy $(E_i)_{\text{eff}}$.

Hence we calculate σ the cross-section for collision and $\bar{E}_2$ the mean energy transfer per primary collision from (4.39) and (4.40 a). Then if λ is the mean free path between collisions

$$\left(\frac{\mathrm{d}E}{\mathrm{d}x}\right)_{\text{ex}} = \frac{\bar{E}_2}{\lambda}$$

and since nZ_2 is the density of electrons

$$\lambda = \frac{1}{nZ_2\sigma},$$

hence

$$\left(\frac{\mathrm{d}E}{\mathrm{d}x}\right)_{\text{ex}} = nZ_2\sigma\bar{E}_2$$

and with (4.39) and (4.40 a) in their new form

$$\left(\frac{\mathrm{d}E}{\mathrm{d}x}\right)_{\text{ex}} = \frac{4\pi a_0^2 nZ_1^2 Z_2 E_{\text{R}}^2 M_1}{m_0 E_1} \log \frac{4m_0 E_1}{M(E_1)_{\text{eff}}}. \tag{5.9 a}$$

This expression is in good agreement with experimental observations when values of $(E_1)_{\text{eff}}$ are taken between 10 and 100 eV. These relatively small values are a consequence of (5.8) which implies that ionization probability depends on $1/E_1^2$ and that the contribution of lightly bound electrons to $(E_1)_{\text{eff}}$ is strongly weighted.

Equation (5.9 a) is not quite correct because it uses the classical impulse approximation. If transition probabilities are correctly calculated it is reduced by a factor $\frac{1}{2}$ (see Bethe & Ashkin, 1953). However, the treatment is still incomplete since we have only accounted for energy loss in the excitation of single electrons. If the collective excitation of electrons (i.e. plasmon generation) is introduced, a similar equation with the plasmon energy E_p ($\sim$ 10 eV) replacing $(E_1)_{\text{eff}}$ is obtained. The two contributions are of comparable magnitude and since $E_p \sim (E_1)_{\text{eff}}$, and both appear in the logarithm we can take (5.9 a) as a good approximation (Lindhard & Winther, 1964).

The case of energetic heavy ions like fission fragments is difficult, for their energy does not sufficiently exceed E_{ex} that they are completely ionized and the effective value of Z_1 in (5.9 a) is uncertain. Bohr (1948) suggests a velocity dependent $(Z_1)_{\text{eff}}$ which makes (5.9 a) almost independent of energy. This is close to the observed excitation loss from fission fragments (Bethe & Ashkin, 1953).

5.3. Energy loss from moving atoms in the cascade

Returning to the collision cascade where moving atoms have $E \lesssim E_{ex}$ in general, we have to compare $(dE/dx)_{ex}$ with the elastic collision loss $(dE/dx)_c$. In calculating the latter quantity we usefully can identify two energy regions:

$$E \lesssim E_a \quad \text{and} \quad E \gg E_a.$$

(1) $E \lesssim E_a$. Here we can use the cross-section and mean recoil energy deduced for the Inverse Square potential in § 4.3.4, equations (4.48). Then again using the fact that

$$\frac{dE}{dx} = \frac{\bar{E}_2}{\lambda} = n\sigma\bar{E}_2$$

we have

$$\left(\frac{dE}{dx}\right)_c = \frac{\pi^2}{4} a^2 n E_a. \tag{5.10}$$

This is often referred to as Nielsen's formula (Nielsen, 1956) and its important feature is the lack of dependence on E, leading to interesting consequences in § 5.4.1.

(2) $E \gg E_a$. Using the simple Coulomb potential the cross-section and mean recoil energy were deduced for this case in § 4.3.3; (4.39) and (4.40a). These lead directly to

$$\left(\frac{dE}{dx}\right)_c = \frac{4\pi a_0^2 n Z^4 E_R^2}{E} \log \frac{E}{\check{E}_2}. \tag{5.10a}$$

The minimum recoil energy $\check{E}_2$ can be taken as the value obtained when impact parameter $p = a$, since collisions beyond this will be prevented by electron screening, then

$$\check{E}_2 = 4Z^4 E_R^2 a_0^2/Ea^2.$$

In radiation damage we are often concerned with recoil energies below E_a and (5.10) should apply to the moving atoms in the cascade.

The range measurements of J. A. Davies et al. (1960a, b, 1961) agree with this expression to about 50 %. It can be improved on slightly by using a true Thomas–Fermi potential and Lindhard with his colleagues have computed $(dE/dx)_c$ in this way (see for instance Lindhard & Thomsen, 1962). Expression (5.10) will generally be good enough for our purpose.

In fig. 54 we plot $(dE/dx)_c$ for C recoils in graphite using (5.10) and Lindhard's Thomas–Fermi result, the latter being used in order to show that (5.10), which would have predicted a constant value of 25 eV/Å, is a fair approximation for energies up to at least E_a (2 keV in this case). It will be observed in fig. 54 that at high energies when $E \gg E_{ex}$ the electronic losses predominate by several orders of magnitude but at low energies $\ll E_{ex}$ the situation is reversed. The cross-over comes somewhere near E_{ex} in general. The trend through the periodic table is illustrated by table XV which compares the maximum recoil energy $\hat{E}_2$ expected from 1 MeV neutron bombardment, with E_{ex} and E_a.

TABLE XV. *Comparison of the maximum recoil energy in* 1 MeV *neutron irradiation with* E_a *and* E_{ex}

Element	C	Al	Cu	Au
$\hat{E}_2$ (keV)	375	150	63	20
E_a (keV)	2	10	70	700
E_{ex} (keV)	12	27	65	197

For elements heavier than Cu it is usually the case that $\hat{E}_2 < E_{ex}$ and the electronic losses may be disregarded. For lighter elements the larger values of Λ favour high energy recoil and often $\hat{E}_2 > E_{ex}$. One must then take excitation losses into account in the cascade.

Fortunately, because of departures from the hard-sphere model, the primary recoil creates secondaries with average energies far below $\frac{1}{2}\hat{E}_2$. These will almost always be in the range where excitation can be neglected and the amount of energy lost to the cascade is simply that lost by the primary. Then to obtain $\nu(E_2)$ to a fair approximation we calculate the energy E_c, dissipated in elastic collisions by the primary:

$$E_c = \int_0^{\hat{E}_2} \frac{(dE/dx)_c \, dE}{(dE/dx)_{ex} + (dE/dx)_c} \tag{5.11}$$

using expressions (5.10) and (5.10a) as appropriate, empirical data for $(dE/dx)_{ex}$ if available, or expression (5.9) modified as suggested above. The modified damage function is the original (5.5) with E_c for E_2:

$$\nu(E_2) = E_c/2E_d, \tag{5.12}$$

A calculation for graphite using Lindhard's $(dE/dx)_c$ gives the curve shown in fig. 55. It will be seen that for recoils with energy below E_{ex} the simple theory of § 5.1 gives an adequate description, but that above E_{ex} the losses in electron excitation are important. Above about $10E_{ex}$ the net energy available for displacing atoms, E_c, increases rather slowly with E_2.

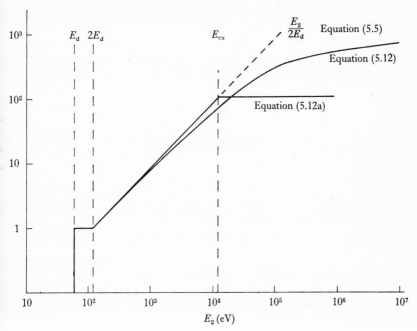

Fig. 55. $\nu(E_2)$, the number of displaced atoms per primary recoil with E_2. Calculated for carbon with expressions (5.16) and (5.17) with the experimental data of fig. 37.

A rough approximation was introduced by Kinchin & Pease (1955) who equated E_{ex} and E_c, implying that *all* energy above E_{ex} is lost in electron excitation and displacement accounts for all the energy loss below E_{ex}, thus if

$$\left.\begin{array}{l} E_c \simeq E_{ex}, \\ \nu(E_2) \simeq E_{ex}/2E_d. \end{array}\right\} \qquad (5.12a)$$

This approximation is shown to be very crude by figs. 54 and 55, nevertheless it is frequently used in making rough calculations and can be improved a little by increasing E_c to about $2E_{ex}$.

5.4. The cascade at high energy

5.4.1. *The path of the primary recoil.* In the early stages of a collision cascade energy is dissipated to a number of secondaries, which on the average have energies very much lower than the primary. In order to estimate the distribution of damage within the cascade one must know the path of the primary recoil. Several important quantities must first be defined. In fig. 56 the path is shown schematically with the vector $\mathbf{x}_n$ as the segment of path

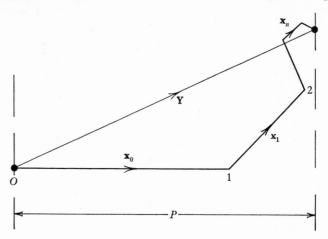

Fig. 56. Path of the primary recoil, defining x_n and Y.

between the nth and $(n+1)$th collisions. The *total path length* X is given by:

$$X = \sum_n x_n. \tag{5.13}$$

The vector joining the primary recoil site and the point at which the atom finally comes to rest, the *vector range* $\mathbf{Y}$, is given by

$$\mathbf{Y} = \sum_n \mathbf{x}_n. \tag{5.14}$$

For comparison with experiment, the distance travelled in the direction of the first segment $\mathbf{x}_0$ is of importance. This we shall call the *penetration* P and it is found by summing the resolved components of the segments along the direction of $\mathbf{x}_0$. For example,

$$P = \sum_n \mathbf{x}_n \cdot \mathbf{x}_0 / x_0. \tag{5.15}$$

Provided the primary's energy is below E_{ex} there is no need to take electronic losses into account. Most cases of radiation damage in elements of medium or large atomic weight fall into this category but in light elements the picture to be presented below may require modification. In the previous section we saw how to deal with electronic losses. In general, when these are important $E \gg E_a$ and the segments between wide angle collision are long, giving an almost straight path.

A very simple picture due to Holmes & Leibfried (1960) may be based on the hard-sphere collision model when $E_1 < E_a$. Equation (4.51) shows that the total hard-sphere cross-section σ is proportional to $1/E_1$ and (4.35) shows that, on the average, the primary loses half its energy at each collision. The mean free path between collisions is $1/n\sigma$, hence it follows that the average length of the first segment $\bar{x}_0$ is just twice the average length of the second segment $\bar{x}_1$, etc. In general

$$2^n \bar{x}_n = \bar{x}_0.$$

Under these circumstances, and with the wide angle scattering that occurs in hard-sphere collisions, the primary cannot travel very much further from the origin than the length of the first segment, i.e.

$$\overline{Y} \sim \bar{x}_0 \quad \text{but} \quad \bar{x}_0 = \frac{1}{n\sigma},$$

hence with (4.51)

$$\overline{Y} \sim \frac{E_2^2}{\pi a^2 n E_a^2}. \tag{5.16}$$

It will be obvious that the assumption that $\overline{\mathbf{Y}}$ has the same direction as $\bar{\mathbf{x}}_0$ would be very crude but as regards the *magnitude* $\overline{Y}$, (5.16) is fairly well justified. To the same order of approximation (5.16) may be taken as giving $\overline{P}$ since the assumption that $\overline{P} \sim x_0$ is equally justified by the arguments used above.

A good estimate of the mean total path $\overline{X}$ may be obtained following Nielsen (1956) and using the inverse-square potential (4.21) to obtain (5.10). For example,

$$\left(\frac{dE}{dx}\right)_c = \tfrac{1}{4}\pi^2 a^2 n E_a \quad \text{for} \quad E_2 \lesssim E_a. \tag{5.17}$$

Hence

$$\overline{X} = \frac{4E_2}{\pi^2 a^2 n E_a}. \tag{5.18}$$

A closer approximation to $\overline{P}$ and $\overline{Y}$ is probably obtained by taking them midway between this expression for $\overline{X}$ and expression (5.16).

Further sophistication is possible by using more realistic potentials. Lindhard & Scharff (1961) have calculated $\bar{X}$ for a Thomas–Fermi potential, but as this requires the use of numerical methods their results cannot be quoted in analytic form. The path length they find is not much different from (5.18) but there is a slight decrease in (dE/dx) at low energies, as shown in fig. 37.

It will be noticed that

$$\bar{X} = \frac{4}{\pi}(E_a/E_2)x_0$$

from (5.16) and (5.18). It will provide a useful picture of the cascade and the meaning of E_a to remember that the total distance travelled by a recoil at E_a is about the same as the mean free path for a wide-angle collision.

Furthermore we observe that since (5.10) gives a constant energy loss along the track and since the number of displaced atoms is proportional to this energy loss, the number of displaced atoms per unit length of track must be approximately constant.

Calculations of this type can be checked by experiments in which the range of energetic atoms are determined in solids. One of the first experiments in this class was performed by Schmitt & Sharp (1958) (for a review see van Lint & Nichols, 1966 or Holmes, 1964). They subjected Cu, amongst other metals, to intense gamma irradiation capable of inducing (γ, n) reactions. In general such neutrons are emitted with several MeV which cause the atom to recoil with energies in the desired range between 10 and 100 keV. The recoil nucleus is generally radioactive, for instance ^{64}Cu decays by β, γ emission with a half life of 12 h. Inert collector foils were placed near a specimen during irradiation and any recoil atoms that emerged from the specimen landed upon them. The ^{64}Cu activity of the collectors was compared with that of the specimen and from these measurements the mean range could be calculated. Fairly good agreement was found with (5.18), which gives some confidence in this simple approach.

5.4.2. *Extended cascade calculations.* We now come to the spatial distribution of point defects generated by the collision cascade. One may picture the primary recoil starting out at high energy, at first leaving a trail of well separated interstitial-vacancy pairs. As the energy falls, the distance between collisions becomes less until x_n approaches the interatomic spacing, its lower limit, and the region around the track is violently disturbed. Each of the energetic secondary recoils will imitate this behaviour in miniature.

The simple model above can be extended to show how the distribution of defects behaves in the high energy part of the cascade. The low energy region cannot be treated in this way, because the Inverse Square potential is a poor approximation, and also the crystal lattice must be taken into account, first because x_n approaches the lattice spacing, and secondly because the recoil energies become comparable with E_d. We shall calculate x_d the mean distance travelled by an atom with energy E between collisions transferring more than E_d. This will give some measure of the spacing between defect pairs. The corresponding cross-section for transferring greater than E_d is given by

$$\sigma_d = \int_{E_d}^{E} \frac{d\sigma}{dE} dE$$

and using (4.48), which gives $d\sigma/dE_2$ for the Inverse Square potential, with the relation $x_d = 1/n\sigma_d$ we have

$$x_d = \frac{4\sqrt{(E_d E)}}{\pi^2 a^2 n E_a}. \tag{5.19}$$

Comparing this expression with (5.18), shows x_d to be $\bar{X}$ divided by $\sqrt{(E/E_d)}$, and therefore in most cases the primary recoil will come to rest after fewer than 10 super-threshold collisions.

In a classic paper Brinkman (1954) suggested that when x_d approaches the interatomic spacing D, the track enters a violently disturbed region where every atom is flung outwards and which he called a *displacement spike*. The crystal lattice must be introduced into any detailed discussion of this region and this will be left until § 5.5. There is a characteristic energy E_D where the recoil track should enter a displacement spike, found by equating x_d from (5.19) to D obtaining:

$$E_D = \frac{\pi^4 a^4 n^{1/3} E_a^2}{16 E_d}. \tag{5.20}$$

It is instructive to see the numerical values of x_d and $\bar{X}$ in some typical cases, such as 10 keV recoils in Al, Cu, or Au. These have been calculated in table XVI together with E_D for the three metals Al, Cu and Au.

TABLE XVI. *Comparison of metals, assuming*
$E_d = 25$ eV *and* $D = 3$ Å

	Al	Cu	Au
Atomic number Z	13	29	79
$\overline{X}$(Å) for 10 keV	130	21	3
x_d (Å) for 10 keV	7	3	3
E_D (keV)	1	46	2700
$\overline{E}_2$ (keV) for $\frac{1}{2}$ MeV neutron	37	16	5
$\overline{X}$(Å) for $\overline{E}_2$ $\frac{1}{2}$ MeV neutron	480	34	3
x_d (Å) for $\overline{E}_2$ $\frac{1}{2}$ MeV neutron	12	3	3

The first noteworthy fact is that in heavy elements $\overline{X}$ is only a few times x_d, indicating that the primary soon loses its energy to the secondary recoils. Next one sees that in light elements, the displacement collisions are well separated, but in elements with $Z > 20$ the 10 keV recoil produces closely spaced displacements. This is emphasized by the rapid increase of E_D with Z.

To illustrate a typical irradiation, the case of $\frac{1}{2}$ MeV neutrons is taken and the behaviour of the average recoil calculated. Because heavy elements receive least energy in collision, the difference between light and heavy is accentuated further. In light elements many of the displacements occur outside the displacement spike, but for $Z > 20$ almost all occur within such regions.

The site of the displacement collision will be roughly the position of the vacancy produced. We must defer consideration of the interstitial until § 5.5, but the vacancy distribution can be discussed, for the high energy region, at this stage. Clearly they will lie near the centre of the cascade and in cases such as the Cu or Au in table XVI some will be formed as a group in close proximity. It is easy to imagine that such a group could collapse to form a small cluster, or at least form a nucleus for later condensation of vacancies to turn into a cluster. This clustering tendency will clearly be greatest in heavy elements.

When large electronic computers became available in the late nineteen fifties, several calculations were made in which the movement of the atoms involved in a collision cascade were computed in detail. These had the great advantages that any potential $V(r)$ could be used and the crystal lattice could be introduced. Such calculations are exceedingly difficult to handle by analytical methods in the statistical model of a cascade. Since classical scattering prevails, the collision orbits can be followed in detail and the main difficulty is accounting for the binding of atoms to the lattice. For the high energy part of the cascade a fairly crude assumption will serve, since $E_d \ll E_2$, and the calculation should be an accurate guide to cascade behaviour.

Two similar attempts have been made to follow the complete cascade, one by Yoshida (1961), the other by Beeler (1964 a, c) and with Besco (1963 a, b, 1964). We shall describe the latter in which the individual collisions were treated by calculating the angle of scattering and energy transfer, from $V(r)$ and the impact parameter. Binding to the lattice was allowed for by subtracting E_d from the energy of recoil, which is the same as Snyder and Neufeld's assumption in §5.1. As we have pointed out above, this proceedure should give a fair picture of the cascade at high energy, but we must treat any other results with care. In fact this proceedure prevents focused collision sequences from forming, and these will be shown in §5.5 to be an essential feature of the cascade at low energies.

Beeler and Besco have used several potentials including the Born–Mayer and the Screened Coulomb, but the essential features of the cascade are not affected by altering the potential within reasonable limits. The crystals studied were BeO (wurtzite) and Fe (b.c.c.). In fig. 57 their results for a 10 keV cascade in Fe are shown as a projection on the {100} plane. Table XVI shows that E_D is of the order 3 keV so that this recoil should just be entering the violent region. It is evident that our simple picture was largely correct with the primary track terminating rather soon, handing its energy on to secondaries, tertiaries, etc. One can see the significance of E_D in the high density of displaced atoms where the recoils have moved with about 3 keV. The spacing between vacancies is of the order of D in this region and there is plenty of evidence for small clusters. In a statistical survey of several hundred cascades with energies between 5 and 20 keV Beeler and Besco drew the following conclusions about vacancy clusters. When $E_2 > 15$ keV 50 % cascades had at least one cluster of more than 10 vacancies. The largest cluster observed contained 21 vacancies. The total number of displaced atoms $\nu(E_2)$ was also studied and found in rough agreement with (5.5). We shall return to this point in § 5.4.4.

One can also see the tendency for interstitials to form in an outer zone with their centroid ahead of the vacancy centroid, but we must be cautious of the magnitude of distances between interstitials and

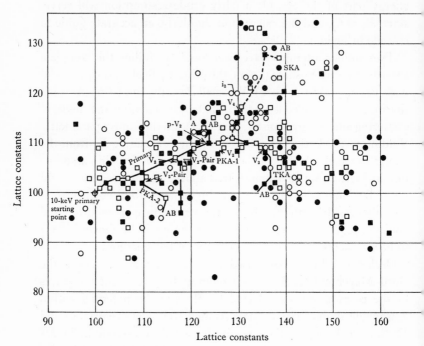

Fig. 57. Projection of a 10 keV Fe atom collision cascade in Fe onto the xy-plane as given by computer calculations. Note that the original 10 keV Fe atom (primary) was slowed down principally by collisions with two primary knock-on atoms (PKA-2 and PKA-1). □■ Vacancy; ○● interstitial atom; PKA primary knock-on atom; SKA secondary knock-on atom; TKA tertiary knock-on atom; ⊕ starting point for 10 keV Fe; V_2 divacancy; V_4 tetravacancy; p−V_2 pseudo-divacancy; i_2 di-interstitial. Open symbols indicate even z coordinates; filled symbols odd z coordinates. (From Beeler, 1963.)

vacancies since this involves the lower energy part of the cascade and focusing mechanisms which are excluded from this calculation.

Another approach with the computer was tried with great success by Robinson, Oen and Holmes at Oak Ridge. By restricting their interest to the movement of only the primary recoil, they were able to make a more detailed calculation for similar expenditure of computer time. The scattering by lattice atoms was computed exactly for a variety of potentials. The crystal lattice was f.c.c. and potentials were chosen to simulate Cu, some calculations were also made in a random array of atoms.

They started their atom off with a given energy E_2 and direction and followed it to rest, where its position was noted. The shot was repeated many times with a random variety of starting directions to produce a statistically significant distribution of ranges, path lengths or penetrations, suitable for comparison with experiments. This technique of averaging computed solutions with random starting conditions is referred to as the *Monte Carlo* method (see Holmes, 1964; Robinson & Oen, 1963 a; Holmes, Leibfried, Oen & Robinson, 1962).

The more sophisticated calculations called for more detail than the range experiments of Schmitt and Scharp could provide. Fortunately a technique had been developed by J. A. Davies and his colleagues at Chalk River and this came upon the scene at exactly the right moment (Davies *et al.* 1960 a, b and 1961). They simulated the behaviour of recoil atoms in Al and W by bombarding specimens with beams of either ^{24}Na, ^{85}Kr or ^{133}Xe at energies in the range 1 to 80 keV. All these isotopes were radioactive and easy to detect by beta counting. The sensitivity of detection was so great that the density of injected atoms could be kept well below the level where the chance of an incoming ion disturbing an atom already present was negligible.

After injecting the ions, the surface layers of the specimen were removed in stages, by first anodizing in an electrolytic cell, then dissolving the resulting film of oxide in a reagent that did not attack the substrate metal. The thickness removed could be deduced from calibration against weight loss or radiochemically. This method of sectioning can be extremely uniform and permits layers as thin as 10 Å to be removed in a reproducible way. After each stage of removal the radio-activity was measured. Taking the ratio of this to the initial activity gave $F(P)$, the fraction of atoms with penetration greater than P. The results were generally shown as a graph of $F(P)$ *versus* P. Alternatively the derivative of this curve showed the distribution of injected atoms; i.e. the fraction having penetration in dP at P. Examples are shown in figs. 58 and 59.

Using polycrystalline targets it was hoped to obtain data for direct comparison with Monte Carlo calculations in a random array of atoms. Figure 58 shows such a comparison for the case of the

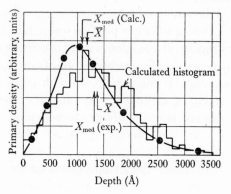

Fig. 58. Experimental and calculated density distribution of 60 keV ^{24}Na in Al. (From Holmes, 1964.)

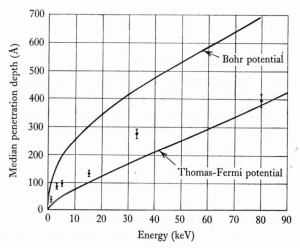

Fig. 59. Median penetration depths of ^{133}Xe ions in aluminium. (From Holmes, 1964.)

distribution of ^{24}Na in Al at 60 keV. The histogram was obtained by computer, assuming a Screened Coulomb type of potential, the curve by experiment. Out to 3500 Å the agreement is remarkably good and gives one some confidence in the calculation. Indeed comparisons of this type have been used to test interatomic potentials, as shown in fig. 59 where the median range of ^{133}Xe in Al is compared with calculations using, first, the screened Coulomb potential of (4.6) with Bohr's expression for the screening radius and, second, the exact Thomas–Fermi potential. The Thomas–Fermi potential seems to give the better fit, but one should remember that electron excitation has been neglected, even though for Xe in Al the effect should be small since $E \ll E_{ex}$.

5.4.3. *The channelling effect.* A surprising result of the experiments with polycrystalline targets was a very penetrating tail on the distribution, extending beyond 5000 Å. This was not found in calculations with a random array, but the effect had been anticipated from calculations with atoms arranged on a lattice. Here it had been found that about 1 trajectory in 10^4 was extremely long and straight, indeed the computer had been unable to follow such events to their termination (Robinson & Oen 1963 a). On closer examination these were found to lie along the open channels that exist in any lattice between closely packed rows of atoms. The trajectory was apparently retained in the channel by glancing collisions with these rows of atoms. In any lattice the most favourable channels are bounded by the most closely packed rows in the lattice. Because the $\langle 110 \rangle$ rows are closest packed, one should expect the $\langle 110 \rangle$ channels to be most highly favoured, and the Monte Carlo calculations confirm this.

The importance of channelled trajectories is two-fold. Firstly they allow a recoil to slow down gradually, giving its energy to many thousands of secondaries, none of which receives enough to be displaced. Such a recoil is lost so far as the damage function is concerned. Secondly, they permit the recoil to travel far away from the vacated lattice site.

By changing to single crystal targets, Davies' experiments were able to confirm this prediction in a striking way. Figure 60 shows how the penetration of 40 keV ^{85}Kr in Al is drastically altered by varying the direction of ion incidence relative to the crystal axes. This may be compared with a graph calculated by Robinson & Oen (1963 b) for the case of 5 keV Cu in Cu (see fig. 61). The general agreement is extremely good in both the order of preference of directions, and in the shape of the curves.

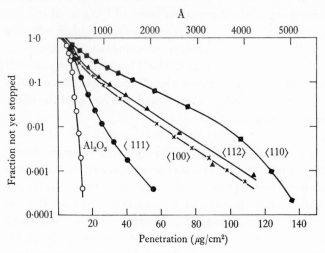

Fig. 60. Penetration of 40 keV ^{85}Kr ions in the principal crystallographic directions of aluminium and in amorphous Al_2O_3. (From Piercy *et al.* 1964.)

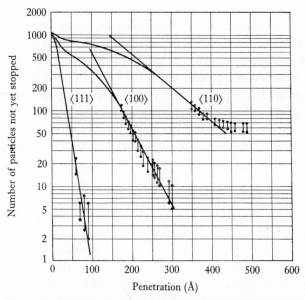

Fig. 61. 5 keV Cu atoms slowing down in Cu, Born–Mayer potential, static lattice; initial directions as given on curves. (From Piercy *et al.* 1963.)

The reality of channelling was also demonstrated by firing a beam of 75 keV protons at a gold crystal 3500 Å in thickness; just sufficient to prevent a majority of ions from penetrating under normal conditions. As the crystal was rotated the transmitted current of ions increased roughly tenfold whenever incidence was along a $\langle 110 \rangle$, direction (Nelson & Thompson, 1963). Furthermore it was found that after a prolonged ion bombardment the effect was diminished and eventually disappeared. Presumably the channels were being blocked by accumulating defects in the lattice, and this effect proves that channelling occurs on the atomic scale and is not due, for instance, to oriented macroscopic holes in the crystal. It also suggests an important consequence for radiation damage; for if the damage function $\nu(E_2)$ is affected by channelling, and channelling is reduced by accumulated damage, then $\nu(E_2)$ must be a function of radiation dose. We shall return to this point in later chapters.

Another experiment, described in the same paper by Nelson & Thompson, illustrates the great multiplicity of channels. This is shown in fig. 62 where a beam of ions (H^+, Ne^+, or Xe^+ at 50 keV) struck the $\{110\}$ surface of a thick Cu crystal. The current to a neighbouring electrode was due to phenomena such as ion reflexion, Cu ion emission, and photon emission by the photoelectric effect, all of which depend on the density of energy deposited in surface layers of the crystal. The current diminished sharply whenever the crystal was turned to give incidence parallel to a low-index direction. An attractive interpretation is that whenever channelling occurs the surface energy density, and hence the current, are sharply reduced. In fig. 63 the directions in which some of the minima occur are relative to the crystal axes on a $\{110\}$ stereogram.* The relative magnitudes of the minima are indicated roughly by the size of the shaded rings. Very favourable channels appear along $\langle 110 \rangle$, $\langle 100 \rangle$ and $\langle 121 \rangle$ directions, a number of lesser ones appear along higher index directions and some correspond to incidence parallel simply to low-index planes.

This last result is perfectly reasonable, for *planes* in which atoms are closely packed will have spaces between them in which *inter-*

---

* Readers who are unfamiliar with the stereogram and practical crystallography will find a good introduction in C. S. Barrett's book *The Structure of Metals*, McGraw-Hill, 1945.

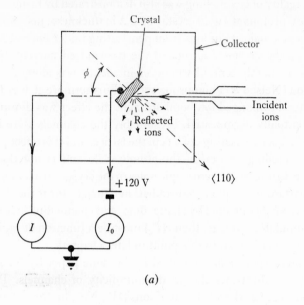

(a)

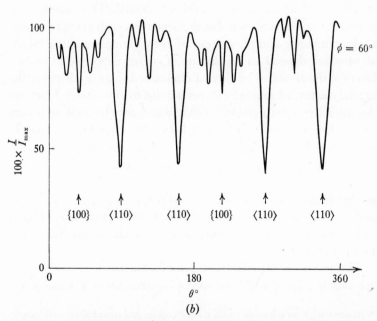

(b)

Fig. 62. (a) The reflexion experiment. (b) Current versus rotation about
⟨110⟩ for 50 keV Ne⁺ ions (Nelson & Thompson, 1963).

planar channelling might occur. Alternatively this can be regarded in a slightly different way. The stereogram demonstrates how low index *directions* occur in families each of which defines a low index plane. For instance the {111} plane contains directions like ⟨110⟩

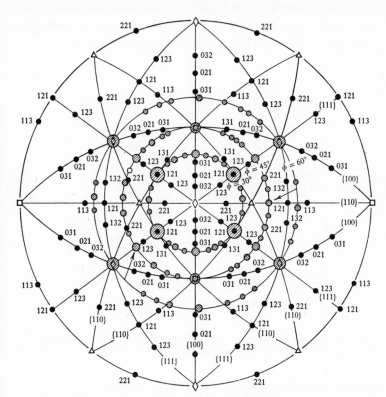

Fig. 63. {110} stereogram showing positions of reflection minima relative to the principal crystal directions and planes for 50 keV protons on Cu. The radius of the rings gives a rough indication of the depth of each minimum. (From Nelson & Thompson, 1963.)

and ⟨112⟩ along which inter-row channels exist. If one takes such a plane and considers channels in it with progressively higher indices, the lateral restraint in the plane diminishes as the indices increase and for practical purposes it may be advantageous to think of interplanar channels with preferred inter-row channels super-imposed.

The detailed channelled trajectories have been investigated by

Robinson and Oen. In fig. 64, the projection of several 1 keV trajectories in Cu onto a plane normal to the $\langle 110 \rangle$ channel axis are shown. The trajectories oscillate back and forth between rows. The tendency to wander from one channel to the next illustrates incipient interplanar channelling. At higher energies the effective size of the rows diminishes with decreasing collision cross-section, and interplanar channelling will clearly become better favoured.

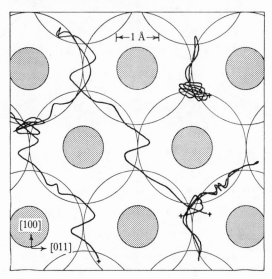

Fig. 64. Projection of some $\{00\bar{1}\}$ channel trajectories onto the $(0\bar{1}\bar{1})$ surface of f.c.c. Cu. 1 keV Cu slowing down according to the truncated Bohr potential. The points at which the primaries entered the crystal are shown by the crosses. Each primary penetrated approximately 300 Å into the crystal in the part of its trajectory shown (250 collisions). (From Robinson & Oen, 1963.)

Though the computer calculations give accurate results, it is both instructive and valuable to follow through a rough analytical treatment based on work by the author (Nelson & Thompson, 1963) and by Lehmann & Leibfried (1963 a). The momentum approximation, (4.29), shows that in a glancing collision with impact parameter p the momentum transfer perpendicular to the trajectory is:

$$\Delta P = \frac{2}{v_2} \int_p^\infty -\frac{\mathrm{d}V(r)}{\mathrm{d}r} \frac{p\,\mathrm{d}r}{\sqrt{(r^2 - p^2)}}.$$

Call this integral $I(p)$ since for a given potential $V(r)$ it depends only on p.

Referring now to fig. 65, consider the channel along the crystal direction $\langle hkl \rangle$, axis Ox, with a periodic distance D^{hkl}. The distances of the N atoms in the interval D^{hkl} from Ox are $p_1, p_2, \ldots p_n \ldots p_N$. In order to simplify the calculation take $\langle hkl \rangle$ as a crystal axis with even symmetry of rotation. Thus each atom to one side of Ox has a counterpart on the opposite side. Initially we shall consider motion in the plane Oxy and neglect the influence of atoms not in this plane.

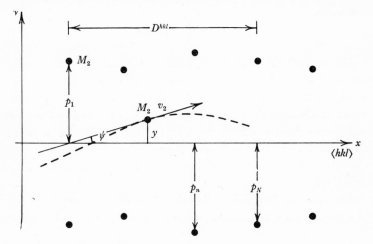

Fig. 65. Channelling between a pair of atomic rows.

In general the recoil will not travel directly along Ox; let y and ψ be the displacement and angular deviation from Ox respectively. Each atom passed by contributes a change in ψ:

$$(\Delta\psi)_n = -(\Delta P)_n / M_2 v_2,$$

where $(\Delta P)_n$ is given by (4.24) with $p = p_n - y$. In the time interval $\Delta t = D^{hkl}/v_2$, during which the recoil travels through one periodic length of the lattice D^{hkl}, the total $\Delta\psi$ is the summation over N deviations like $(\Delta\psi)_n$ and $\Delta\psi/\Delta t = \dot{\psi}$ is given by

$$\dot{\psi} = -\frac{2}{D^{hkl} M_2 v_2} \sum_{n=1}^{N} I(p_n - y).$$

Now, for small angles,

$$\dot{y} \simeq \psi v_2 \quad \text{and} \quad \ddot{y} = \dot{\psi} v_2,$$

since v_2 is approximately constant.

Hence
$$M_2 \ddot{y} = -\frac{2}{D^{hkl}} \sum_{n=1}^{N} I(p_n - y) = -F(y),$$

where $F(y)$ is an effective restoring force directed against the displacement y. Note that this is independent of E_1, and that this is a consequence of using the impulse approximation which implies that Δp varies like $1/v$. The situation may then be expressed in terms of an effective potential well $U(y)$ or *channel potential* defined as

$$U(y) = \int_0^y F(y)\,dy. \tag{5.21}$$

The crystal has now been replaced formally by an extended potential well along the channel axis. The motion of the channelled atom will be similar to that of a ball rolling up an inclined gully, whose cross-sectional profile is given by $U(y)$. The lateral motion will be oscillatory whilst the longitudinal velocity is smoothly attenuated. Provided the kinetic energy of forward motion greatly exceeds that of the lateral motion, we can treat the two motions independently.

The nature of $U(y)$ is best illustrated by taking the simple example of a two-dimensional channel formed between two rows of atoms, each with interatomic spacing D and each set at a distance p from the axis. Then from equations (5.20) and (5.21)

$$U(y) = \frac{2}{D} \int_0^y [I(p-y) + I(p+y)]\,dy.$$

If $V(r)$ is in either Born–Mayer (4.14) or Screened Coulomb (4.16) form, the integrals can be evaluated to give approximate expressions:

Born–Mayer

$$U(y) = \frac{A}{D} \sqrt{(2\pi pb)}\, e^{-p/b} \{ \sqrt{(1 - y/p)}\, e^{+y/b} + \sqrt{(1 + y/p)}\, e^{-y/b} - 2 \}$$

$$\tag{5.22a}$$

Screened Coulomb

$$U(y) = \frac{2Z_2^2 a_0 E_R \sqrt{(2\pi a/p)}}{D}\, e^{-p/a} \left\{ \frac{e^{y/a}}{\sqrt{(1 - y/p)}} + \frac{e^{-y/a}}{\sqrt{(1 + y/p)}} - 2 \right\}.$$

$$\tag{5.22b}$$

Taking a Cu atom between adjacent $\langle 110 \rangle$ rows in a Cu crystal as an example, these two functions are plotted in fig. 66.

Unfortunately the inverse square potential cannot be used here, as the impact parameters near the channel centre are much greater than $5a$, and we have seen in chapter 4 that this potential is too large under these circumstances.

If $V(r)$ is compounded as a sum of Born–Mayer and Screened Coulomb, the resultant $U(y)$ is simply the sum of equations (5.21)

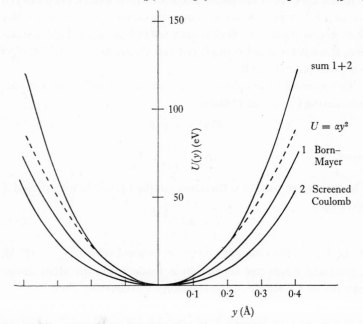

Fig. 66. The channel potential $U(y)$ for a Cu recoil between two $\langle 110 \rangle$ rows.

and (5.22). This is also shown in fig. 66. We have seen earlier that this compound potential is likely to be a realistic approximation. Near to the channel axis the Born–Mayer contributes most to $U(y)$, but the infinity at $r = 0$ for the Screened Coulomb potential ensures that this dominates for wide amplitudes.

To obtain more manageable expressions than (5.22) one can expand the contents of their curly brackets in powers of y, when all odd terms disappear. For small y we can neglect terms of higher order than y^2 and obtain y^2/b^2 for the curly bracket of (5.22a) and y^2/ap in (5.22b).

Then, writing
$$U(y) = \alpha y^2 \tag{5.23}$$
we find

$$\alpha = \begin{cases} \dfrac{A}{Db}\sqrt{\left(\dfrac{2\pi p}{b}\right)}\, e^{-p/b} & \text{for Born–Mayer,} \tag{5.24} \\[2ex] \dfrac{2Z_2^2 a_0 E_R}{Dp}\sqrt{\left(\dfrac{2\pi}{ap}\right)}\, e^{-p/a} & \text{for Screened Coulomb.} \tag{5.25} \end{cases}$$

For the compound potential one simply adds these two together.

In the Cu $\langle 110 \rangle$ example, figure shows how well (5.23) is obeyed out to 0·2 Å, which is quite a large displacement. In higher order channels the parabola will obviously not be such a good approximation, though for small enough y it can always be used and $U \propto y^2$ is a quite general result.

The general problem is now greatly simplified since by differentiating (5.23) one obtains

$$F(y) = -2\alpha y$$

hence
$$\ddot{y} = -\frac{2\alpha}{M}y.$$

The transverse motion is therefore simple harmonic with period T given by

$$T = 2\pi\sqrt{\frac{M}{2\alpha}}. \tag{5.26}$$

Assuming for the moment a constant forward velocity $v = \sqrt{2E/M}$, a complete transverse oscillation is made while the atom moves forward by a distance λ, the trajectory wavelength. Then

$$\lambda = vT$$

or
$$\lambda = 2\pi\sqrt{\left(\frac{E}{\alpha}\right)}. \tag{5.27}$$

The equation of the trajectory is

$$y = y_0 \sin\sqrt{\left(\frac{\alpha}{E}\right)}\,x. \tag{5.28}$$

The transverse amplitude y_0 is found by equating the maximum transverse kinetic energy to the potential energy at y_0. If the trajectory crosses the axis at angle ψ_0 the maximum kinetic energy is $\psi_0^2 E$, then y_0 is given by:

$$U(y_0) = \psi_0^2 E. \tag{5.29}$$

This is the exact form, in the harmonic approximation

$$y_0 = \psi_0 \sqrt{\frac{E}{\alpha}} \qquad (5.30)$$

and

$$y = \psi_0 \sqrt{\left(\frac{E}{\alpha}\right)} \sin \sqrt{\left(\frac{\alpha}{E}\right)} x. \qquad (5.31)$$

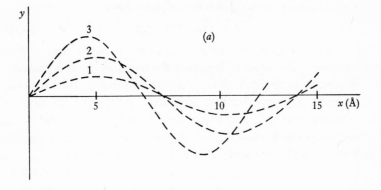

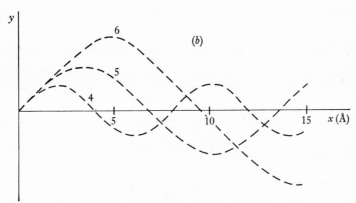

Fig. 67. (a) Channelled trajectories for constant energy.
(b) Channelled trajectories for constant angle.

For constant energy the amplitude changes in proportion to ψ_0 and λ remains constant. This is illustrated by trajectories 1 and 2 in fig. 67 a. If y_0 exceeds the limit of the harmonic approximation, $U(y) > \alpha y^2$ and the period of oscillation becomes smaller; λ is also

12

shorter. Case 3 illustrates this, and the trajectory becomes more zig-zag than sinusoidal. In the limit of a perfect zig-zag:

$$\psi = \pm \psi_0,$$
$$y = \psi_0 x \quad \text{for} \quad x < \lambda/4 \Big\} \tag{5.32}$$

and
$$\lambda = 4y_0/\psi_0. $$

Using (5.29) to give ψ_0 in terms of y_0 we have

$$\lambda \simeq 4y_0 \sqrt{[E/U(y_0)]} \tag{5.33}$$

for a zig-zag trajectory. Since $U(y_0)$ increases faster than y_0^2 this shows how λ decreases with y_0 once the harmonic limit is passed.

Returning to fig. 67, as E increases for constant ψ_0 the harmonic solution (5.31) shows that both amplitude and wavelength increase in proportion to $\sqrt{E}$, and this is seen in trajectories 4 and 5. In trajectory 6 the harmonic limit is passed and there is a zig-zag tendency with both λ and y_0 increasing less fast than $\sqrt{E}$.

It is clear that some limits must be imposed on the amplitude y_0 since our potential $U(y)$ is only valid for a limited range of y. Obviously, when y approaches p the collisions can no longer be treated by the impulse approximation, then $F(y)$ is dependent on E, and an effective channel potential cannot be defined. For these atoms one may expect wide angle collisions to occur and the trajectory to terminate violently. However, one must also consider the stability of the trajectory when there are fluctuations in the impact parameter $(p - y)$ due to the vibrational motion of atoms on the crystal lattice. Suppose there is a root-mean-squared displacement x_{rms}, then we must demand that

$$(p - y_0) \gg x_{rms}.$$

If this is not satisfied Lindhard (1965) has shown that the amplitude y_0 fluctuates and there is a progressive perturbation of the trajectory outwards to the point where wide-angle collisions terminate it violently as in fig. 68. A maximum amplitude $\hat{y}_0$ for a stable trajectory can be defined by

$$p - \hat{y}_0 = mx_{rms}, \tag{5.34}$$

where m is a constant that will be a few times unity.

The maximum amplitude implies a maximum angle $\hat{\psi}_0$ given by (5.29)

$$\hat{\psi}_0 = \sqrt{(U(\hat{y}_0)/E)} \tag{5.35}$$

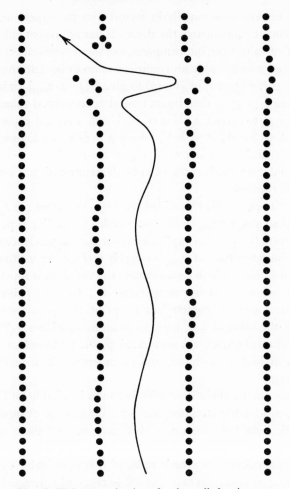

Fig. 68. Violent termination of a channelled trajectory.

and with equation (5.34) this becomes

$$\hat{\psi}_0 = \sqrt{\left(\frac{U(p - mx_{\mathrm{rms}})}{E}\right)}. \qquad (5.35a)$$

Assuming that m is independent of E this equation shows that the angular width of the channelling effect depends on $1/\sqrt{E}$, thus becoming smaller at high energy, and on the root of the channel potential strength. In the case of the Screened Coulomb potential therefore $\hat{\psi}_0$ is proportional to $\sqrt{(Z_1 Z_2)} = Z$, in this case.

This behaviour is roughly in accord with the experiments with heavy ions. Unfortunately the theory is not at present able to give an explicit value to m, but comparison of the theoretical expressions and observations enable an empirical value to be determined. We can use either (5.22 a) or (5.22 b) to give $U(p-mx_{rms})$ for insertion in equation (5.35 a), then insert typical experimental values of $\hat{\psi}_0$, E_1 and x_{rms} to extract a value of m. Using a typical experimental value of $5°$ for $\hat{\psi}_0$ at 10 keV, taking $p = D/2$ and knowing from lattice dynamics that at 300 °K $x_{rms} \simeq 0.09$ Å in Cu one finds that m is about 2 for the Screened Coulomb assumption and 2·8 for the Born–Mayer case.

Thus it appears that provided the trajectory does not approach closer than about $2x_{rms}$ to the row, stable channelling is possible.

Equation (5.35 a) also implies a temperature dependence through x_{rms}. The magnitude of x_{rms} is roughly $0.1D$ at the melting point, for most solids. At high temperatures the variation is roughly with $\sqrt{T}$ but below the Debye temperature Θ_D, the zero-point motion dominates, with its characteristic temperature independence. Very roughly the value of x_{rms} for zero-point motion is $0.1D\sqrt{(\Theta_D/T_m)}$. Thus values of order 0·1 Å are typical. $\hat{\psi}_0$ should therefore decrease slightly at high temperature and be independent of temperature below Θ_D.

At this point a slight digression becomes inevitable for in recent years considerable attention has been devoted to channelling of energetic light ions, such as 4 MeV protons, and there are some important comparisons to be made. Because the cross-section for wide angle collision is greatly reduced it is permissible to use the impulse approximation over the whole range of y, hence $U(y)$ is a valid concept even for particles in the channel wall. There is a class of trajectories with ψ_0 slightly *greater* than $\hat{\psi}_0$ that can travel through the crystal for considerable distances, which is not possible for recoil atoms. The value of m deduced from these high energy experiments is roughly the same as in the present case.

We now consider some further limitations of our treatment of channelled recoil atoms. First, some caution is required when applying classical mechanics to channelling and we have seen in §§ 4.1 and 4.3 that the impulse approximation can only be used when E is much less than $[I(p-y)]^2/E_1^*$. In the case of Cu–Cu collisions in channelling this limit is of the order of 10^7 eV for $y = 0$ and there is little cause for concern in radiation damage problems. However, for lighter elements the classical treatment could be wrong for trajectories near the axis where $I(p-y)$ is small.

Another way of looking at this quantum limitation *for a particular projectile* is to calculate the angle ψ_q at which the first minimum occurs in the intensity of diffracted waves from an aperture of width $2b$ with plane waves of wavelength λ_p incident upon it. This is given by

$$\psi_q \sim \frac{\lambda_p}{2b}.$$

The particle wavelength for the channelled particle is

$$\lambda_p = \frac{h}{\sqrt{(2ME)}},$$

then

$$\psi_q \sim \frac{h}{2b\sqrt{(2ME)}}.$$

This must be compared with the critical angle for channelling, $\hat{\psi}_0$ and from above

$$\frac{\hat{\psi}_0}{\psi_q} \sim 2b\sqrt{(2M\,U(\hat{y}_0)/h)}.$$

It is seen that because both ψ_q and $\hat{\psi}_0$ depend on $E^{-\frac{1}{2}}$, their ratio is independent of energy and if there is a problem in applying classical mechanics to channelling of a particular projectile it will be of the same magnitude at all energies.

If classical mechanics is to be used we require that the ratio above shall be much greater than unity.

For protons in Cu the ratio is about 20; thus the critical angle for channelling is about ten times the angle at which diffraction effects should occur. One is justified in using classical mechanics for most of the protons and certainly for any channelled recoil atom. But there is clearly a need to develop methods of solving the problem of proton channelling quantum mechanically.

If one calculates the ratio $\hat{\psi}_0/\psi_q$ for electrons or positrons it is clear that the smaller mass makes the channelling and diffraction angles very similar and there are well-known solutions to this problem in the dynamical theory of electron diffraction. Unfortunately these use approximations which cannot be carried over into the realm of protons.

Secondly, we must ask what happens as the forward energy is progressively reduced. The trajectory wavelengths follows the decrease, proportional to $\sqrt{E}$. But the treatment is only valid if $\lambda \gg D^{hkl}$. Lehmann & Leibfried (1963 a) have concluded that instability of the trajectory occurs when $\lambda = nD^{hkl}$, where n is a small integer. Essentially there is a resonance between the impulses from the channel walls and the transverse oscillation. When this happens the trajectory terminates by a set of violent collisions with the channel walls as in fig. 68. Taking the Cu $\langle 110 \rangle$ example table XVII shows how λ varies with E.

TABLE XVII. *Trajectory wavelengths for the*
$\langle 110 \rangle$ *channel in* Cu $D = 2 \cdot 56$ Å

E (keV)	1	3	10	30	100	300
λ (Å)	8·5	15	27	47	85	150

Clearly the situation is becoming critical near 1 keV. For each channel one can define a critical *channelling energy* E_{ch} below which channelling is unlikely to proceed. This is given in order of magnitude by equating $\lambda/4$ and D^{hkl}. For example,

$$E_{ch} \sim D^2\alpha/10. \qquad (5.36)$$

This gives $E_{ch} \sim 300$ eV for Cu.

We are now in a position to predict Z dependence in channelling. For a particular crystal there will be a set of E_{ch} values, one for each type of channel. If these are all higher than the primary recoil energy, channelling is unlikely to occur. But low values of E_{ch} will favour channelling. In order of magnitude the interatomic spacings D, do not vary much through the periodic table and we look to α for the Z dependence. Equations (5.24) and (5.25) show that α follows a high positive power of Z, roughly Z^2 for the screened Coulomb and perhaps Z^5 for the Born–Mayer if we use Brinkman's expression (4.18). Clearly, channelling is more important for lighter elements and though present in Cu it is unlikely to be found in the heaviest elements like Au unless at very high energies.

In conclusion we must mention the effect of lattice vibration on channelling. Although the theory is not well developed at present the experiments by Davies & Lutz and Sizmann have shown channelled ranges to be reduced though the random part of the distribution curve is little effected.

As we have seen, if channelled trajectories approach within a few vibration amplitudes of the channel wall we can expect our approximations to break down and the trajectory to be scattered by a close encounter. The effect on trajectories near the axis will be simply to increase $U(y)$ with an increased energy loss from the atom. This explains the experimental observations.

Robinson and Oen introduced some small displacements into their lattice model to simulate lattice vibration. This had a marked effect on the range of channelled trajectories. The implication for radiation damage is important, for if $\nu(E_2)$ is reduced by channelling and channelling is in turn reduced by vibration, $\nu(E_2)$ should increase with temperature.

5.4.4. *The importance of channelling in the collision cascade.* Channelling can influence the collision cascade in four ways. First it will reduce $\nu(E_2)$, the average number of displaced atoms, for whilst an atom is channelled it is less likely to make displacement collisions. Secondly, the long range of channelled trajectories will spread the damage over a larger volume. Thirdly, $\nu(E_2)$ may become dose dependent since long channelled trajectories will be less likely in a damaged lattice. Fourthly, it may be temperature dependent.

An extension of the hard sphere cascade theory in § 5.1 will show the physical situation in a rough way. We saw how the struck atom in fig. 5.2 made an average of

$$\int_{E_d}^{E} \nu(E'') \, dE'' / E$$

displacements, whilst the scattered atom goes on to make

$$\int_{E_d}^{E} \nu(E') \, dE' / E.$$

Now suppose there is a probability c of the atom with energy E becoming channelled and lost to the cascade. In this event the channelled atom is the only one displaced and $\nu(E) = 1$. There is also the probability $(1 - c)$ that it goes on to make

$$2 \int_{E_d}^{E} \nu(x) \, dx / E$$

displacements. Then the average number of displacements is

$$\nu(E) = c + 2(1-c) \int_{E_d}^{E} \nu(x) \, dx/E. \qquad (5.37)$$

If we assume c to be independent of energy, it is easily verified that this has the solution

$$\nu(E) = (E/2E_d)^{1-2c} \qquad (5.38)$$

for $\qquad\qquad E \gg E_d \quad \text{and} \quad c \ll 1.$

There is therefore a departure from the simple proportionality between $\nu(E)$ and E, becoming more marked as E increases. Although the expression (5.38) may be a rather poor approximation in not accounting for any dependence of c on energy, it shows that the larger E is, the more recoils are involved and the greater the overall chance of channelling.

A useful way of expressing departures from the simple cascade theory is to use the *displacement efficiency* function $k(E)$ defined by

$$\nu(E) = k(E)(E/E_d). \qquad (5.39)$$

Then $k(E) \simeq \frac{1}{2}$ at low energies but will decrease progressively as energy increases and channelling exerts a greater influence. With the simple model above

$$k(E) = \frac{1}{2}(2E/E_d)^{2c}. \qquad (5.40)$$

The treatment so far is based on a more general analysis by Oen & Robinson (1963) and Oen, Holmes & Robinson (1963). Further refinement of this hard sphere model by Oen introduces different probabilities c_1 and c_2 for a scattered atom and a struck one. Physically, this seemed desirable since the latter starts from a lattice site and the former does not, which could clearly affect the channelling probability. Equations (5.37) and (5.38) are then modified by replacing $2c$ with $(c_1 + c_2)$.

In simple cascade theory we saw that changing the scattering law from the hard sphere approximation makes little difference to $\nu(E)$. When channelling is taken into account P. Sigmund showed this is no longer true. Using a more realistic differential cross-section in which there is a higher probability of low energy transfer and hence small angular deviation, he found that a given value of c gave smaller $k(E)$ than before. His differential cross-section was chosen for analytical convenience rather than reality, though it is of the same general shape as that in fig. 43.

$$\frac{d\sigma}{dE''} = \frac{\sigma \exp\left(\tfrac{1}{2}E - E''\right)/E_m}{2E_m \sinh E/2E_m}, \quad (5.41)$$

where σ is the total cross-section and E_m a fitting parameter with dimensions of energy and numerically of the same order as the mean energy transfer.

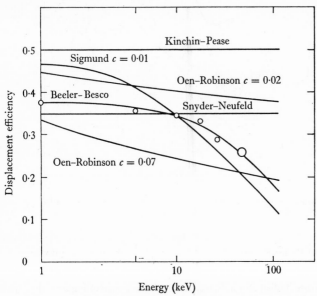

Fig. 69. The displacement efficiency $k(E_2)$ as a function of recoil energy E_2 in Fe. (From Beeler & Besco, 1963.)

Setting up the integral equation for $\nu(E)$ and solving gives

$$\nu(E) = 1 + \frac{1 - \exp\left\{-c(E - 2E_d)/E_m\right\}}{2c \sinh E_d/E_m}. \quad (5.42)$$

For large E this has the asymptotic value

$$\nu(\infty) = E_m/2cE_d \quad (5.43)$$

and this gives a good indication of the influence of c. The detailed behaviour of $k(E)$ is shown in fig. 69 which clearly shows that in this model a given value of c is much more effective in reducing $\nu(E)$ than in the hard-sphere case.

These analytical treatments form a valuable background to the more direct approach by computer simulation, which is not limited by the drastic assumption of an energy independent c. Here, the Beeler and Besco method really comes into its own. Figures 70a and

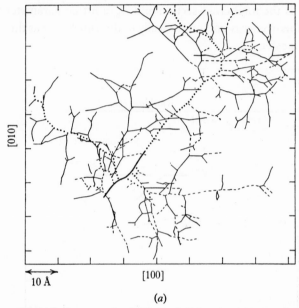

[010]

10 Å   [100]

(a)

Fig. 70. (a) Projection on a (001) plane of all knock-on atom trajectories in a 5 keV displacement spike in Fe. The short, heavy track is that of the initiating, 5 keV PKA; the heavy dotted tracks are those of its secondary knock-ons. Thereafter, alternate dashed and solid tracks describe subsequent higher order knock-ons. (From Beeler, 1964.)

(b) Projected damage patterns [(001) plane] created by collision trajectories in (a). Open squares are vacancies, filled circles are interstitials. Note that interstitials outline the periphery of the damage pattern. Damage above the diagonal line vanished when a slight change in initial conditions allowed the head-on secondary in (a) to channel. (From Beeler, 1964.)

b show first the trajectories of a 5 keV cascade in Fe, then the resulting distribution of interstitials and vacancies. In this particular case we see one of the secondaries in a near-channelled state travel a considerable distance before initiating a new branch of the cascade, some distance from the main body. If the initial conditions are slightly altered this secondary becomes truly channelled and whole

branch cascade disappears. This clearly indicates the nature of the
effect, but to establish the statistical significance about 100 cascades
were run for a given energy, with randomly chosen starting conditions. Of course the Beeler and Besco calculation is much less
suitable for the Monte Carlo method than the Robinson type,
because a single shot requires so much more computer time.

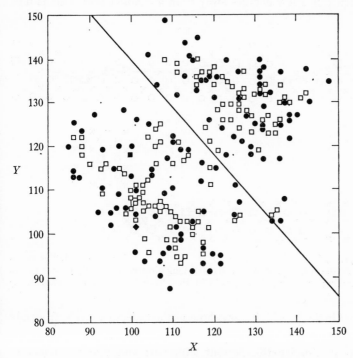

Fig. 70(b). (For legend see facing page.)

Nevertheless, a rough idea of $k(E)$ was obtained for 5, 10, 15 and
20 keV, with a significant decrease above 10 keV as shown in fig. 69.
A definite correlation between low values of $\nu(E)$ and channelling
events was established. The comparison with analytical results is
most encouraging and, as we suspected, shows the realistic scattering law to give the $k(E)$ curve the right shape. However, the
computer was able to show that below 2 keV in Fe channelling was
not observed and the assumption of an energy independent c is not
justified. Thus the analytical $k(E)$ starts to fall too soon. For recoils

at 5 keV in Cu and Fe Robinson and Beeler find that c is of the order 10^{-2}. In the hard sphere model this is not really large enough to give agreement between computed and analytical $k(E)$ functions, as shown in fig. 69.

The computed cascades also showed the importance of an imperfect type of channelling, occurring when the direction of incidence is a few degrees away from a channel axis. This is illu-

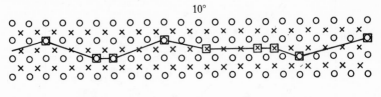

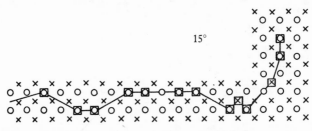

Fig. 71. Projections of 2 keV Fe atom trajectories in a ⟨100⟩ channel. The atoms were started at the centre of the channel along directions parallel to ⟨001⟩ planes and making an angle $\phi = 10°$, $15°$ with the channel axis. (From Beeler, 1963.)

strated in fig. 71 which shows how, as the trajectory jumps from one channel to the next, a few defects are generated.

To summarize the present situation: although the effect of channelling is clearly established and given roughly by equations (5.40) and (5.42), there is no easy way in which $\nu(E)$ can be estimated in any situation, for we rely on the computer to give us c. From § 5.4.3, however, we expect channelling to be most important in lighter elements.

5.5 The cascade at low energy

5.5.1. *Introduction.* Whilst the collision energies are very much greater than the energy binding the atom to its lattice, the results are rather insensitive to assumptions about the nature of binding and the mean displacement energy provides a convenient way of expressing the binding. Below 1 keV however, the situation demands, closer examination.

The method developed by Vineyard and his colleagues at Brookhaven for simulating crystal behaviour by computer have been described in chapter 2. (Gibson *et al.* 1960; Vineyard, 1961, 1962; Vineyard & Erginsoy, 1962; Erginsoy, Vineyard & Englert, 1964; Erginsoy, 1964). There the context was defect configurations, but the model is even better suited to calculation of dynamic events since these are less sensitive to assumptions about cohesive forces. Figure 72a shows a collision cascade started at 400 eV in the {100} plane of Cu. Figure 72b shows the resulting array of defects. It will be recalled that this model predicted a dumbbell configuration for interstitials, and these are seen to lie in an outer zone surrounding an inner core of vacancies. The number of displaced atoms $v(E_2)$ is 11 for $E_2 = 400$, and although this is less than the 16 or so predicted by simple cascade theory, it is not a remarkable difference.

Close examination of fig. 72a shows two phenomena of great importance. First, none of the early recoils themselves end up as interstitial atoms but fall into a space vacated by an atom they have struck. Such *replacement* events greatly outnumber the number of permanent displacements. Secondly each interstitial lies at the end of a sequence of replacement collisions and these collision sequences lie along simple directions in the lattice. For instance at C and D replacement sequences travel down $\langle 100 \rangle$ rows, at B down a $\langle 110 \rangle$ row. One also sees a great number of $\langle 110 \rangle$ sequences which do not involve replacement, but simply transport energy. These are apparently started with less energy.

There appears to be a mechanism which focuses momentum along the simple rows. Where replacement is involved we refer to *focused replacement sequences*, where only energy is transported the name *focused energy packet* is applied. In later sections we shall investigate their properties in considerable detail, since they are the dominant feature of the cascade at low energy, and lead us to a modified damage function $v(E_2)$.

The limitation of the previous cascade calculations is now obvious, for if one subtracts 25 eV from each collision, as they did, the focused collision sequences will stop very rapidly and the i–v

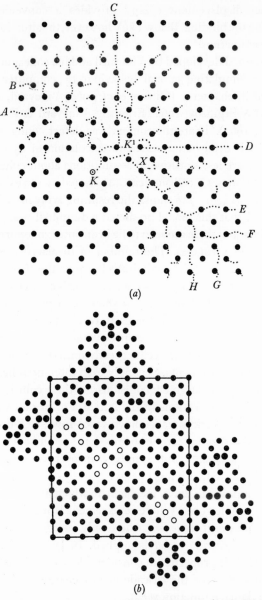

Fig. 72. (a) Shot in (100) plane at 400 eV, 100° away from {011}. Orbits in plane to time 45 are shown. Knock-on started at K, moves to K'. At end of run collision chains $A, B...H$ are still active. (b) Estimated array of 11 vacancies (circles) and 11 interstitials (double dots) that could result from shot in (a). Set used in (a) indicated by rectangle. Indicated vacancy arrangement may not be stable. (From Gibson et al. 1960.)

spacing is badly underestimated. In fact, we shall see that in a focused sequence the energy lost per collision can be as little as 1 eV. Unfortunately the greater sophistication of the Brookhaven method is bought at great expense in computer time and each run requires almost 1 hour to complete on an IBM-7090 computer. This rules out any serious attempt at a Monte Carlo calculation of quantities such as $\nu(E_2)$. In later sections we shall see how a combination of computer and analytical methods can nevertheless build up an accurate picture of the cascade, and give an improved function $\nu(E_2)$.

5.5.2. *Displacements near threshold.* At this point we must define our terms rather carefully. *Replacement energy* E_r^{hkl} will be taken to mean the energy at which a replacement sequence is just produced along a $\langle hkl \rangle$ row of the lattice; this does not imply that the resulting i–v pair is stable. The *displacement energy* E_d^{hkl} is the minimum energy required to produce a stable i–v pair in the $\langle hkl \rangle$ direction. Clearly, E_d^{hkl} may be either greater than or equal to E_r^{hkl}. For a particular direction, therefore, the number of displaced atoms will rise sharply from zero to one at E_d^{hkl}. The damage function $\nu(E_2)$ will be an average over all directions, rising slowly from zero at some *displacement threshold* E_t, which is the smallest E_d^{hkl}, to unity at some higher energy. The *mean displacement energy* E_d is redefined as the energy when $\nu(E_2) = \frac{1}{2}$ averaged over all directions. Note that in the simple cascade models E_t and E_d were synonymous.

One of the most valuable studies with the Brookhaven method has been the investigation of the damage function close to E_d. The procedure was to start an atom in a particular direction with several energies in successive runs. It was then possible to deduce values of E_d^{hkl}.

The orbits for a typical shot, close to the $\langle 100 \rangle$ direction at 40 eV, is shown in fig. 73. At the instant illustrated, a focused replacement sequence has run from A to D, leaving a vacancy at A. When the atoms eventually settle down, a dumb-bell configuration develops at D. Two prominent focused energy packets have run from A to E and B to H.

The threshold energy must obviously depend on the initial direction. Figure 74 summarizes the results of 29 shots in Cu and the threshold is shown by the dashed line. The easiest direction is the $\langle 100 \rangle$ where E_d^{100} about 22 eV, the most difficult is the $\langle 111 \rangle$ with E_d^{111} at about 80 eV. Since $\nu(E_2)$ is an average over all directions one can imagine how it must rise from zero at 22 eV to unity somewhere near 50 eV. We note that the easier directions are making several i–v pairs per shot before the energy has risen enough for the $\langle 111 \rangle$

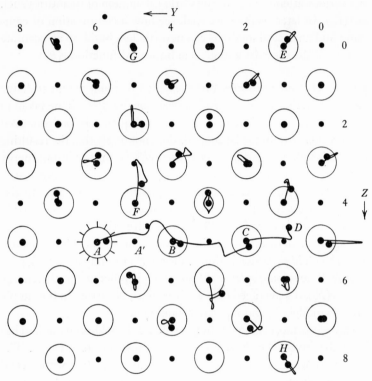

Fig. 73. Atomic orbits produced by shot in (100) plane at 40 eV. Knock-on was at A and was directed above—y axis. Large circles give initial positions of atoms in plane; small dots are initial positions in plane below. Vacancy is created at A, split interstitial at D. Run to time 99. (From Vineyard, 1962.)

shot to be productive, and this is the reason that $\nu(E_2)$ reaches unity some way before 80 eV. It seems quite possible that $\nu(E_2)$ is not a smooth curve but could contain some steplike features.

This calculation has now given some quantities which can be directly compared with experiment. The minimum threshold and E_d have been estimated in irradiation experiments with electrons of about 1 MeV which transfer a maximum energy near E_d. The rate of increase in C_{1v} is found by observing some physical property, such as resistivity, at several fixed electron energies. Then plotting the experimental cross-section for displacement one can estimate the energy E_d for which $\nu(E_2) = \frac{1}{2}$ and also E_t. For Cu the experiments indicate E_d in the region 22 to 24 eV, with $\nu(E_2)$ becoming zero at E_t just below 20 eV. The experimental results for various metals are summarized in table XXIII, Chapter 6, where a full account of the experimental techniques and method of analysis is given.

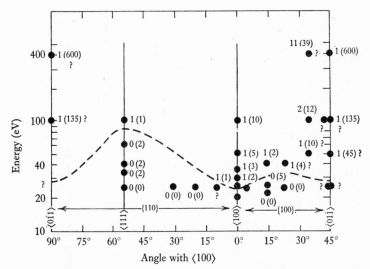

Fig. 74. Diagram showing all dynamic events calculated with Cu potential. A dot is shown for each event and indicates kinetic energy and direction of knock-on atom. First figure attached gives number of stable Frenkel pairs created, figure in parentheses gives number of replacements. Dashed line is estimated threshold for creation of at least one stable Frenkel pair. (From Vineyard, 1962.)

The reason that the Brookhaven prediction is so close to observations is very simple; the potential for the calculation was chosen to give 24 eV as the mean displacement energy! One hopes therefore that the same potential will correctly predict other dynamic events and we shall see later that this is indeed so. The limitation of these calculations is the absence of lattice vibration. One can easily imagine that a favourable configuration near the primary recoil site could make displacement easier. Conversely there may be some cases made more difficult by vibration. The overall effect will be to broaden the range of energy over which $\nu(E_2)$ rises to unity, certainly giving a lower value for E_t. However, one hopes that E_d, where $\nu = \frac{1}{2}$, will not be greatly affected and this is more likely to be a useful fitting parameter than E_t.

The model has also been applied to the b.c.c. case of Fe, again using E_d as a fitting point to fix the potential. The investigation was

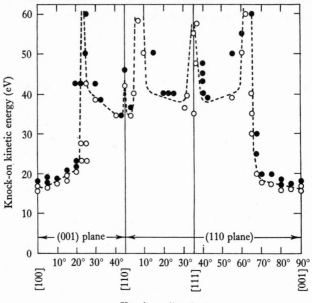

Fig. 75. Directional dependence of the displacement threshold energy. The direction of easiest displacement is the [100], giving a maximum threshold of $\approx$ 17 eV in our model. Note the highly asymmetrical angular region in which [110]-type replacements are possible. ○ No stable defect; ● stable defect; – – – – estimated displacement threshold. (From Erginsoy, 1964.)

carried out in more detail than in the case of Cu and in fig. 75, which shows the result of various shots in different directions, some interesting new features appear.

The easiest direction in which to form a stable defect is the $\langle 100 \rangle$ with a threshold at 17 eV. *Near* to $\langle 110 \rangle$ and $\langle 111 \rangle$ about 40 eV are required, but in the exact directions the threshold is much higher. We shall see later that replacement sequences are formed rather than energy packets when the focusing is not perfect, and this is the origin of the effect. However, it is unlikely that the effect is as sharp as fig. 75 suggests because of lattice vibration, and one should not take this prediction too seriously. The most difficult directions of all are roughly mid-way between the principal axes. A glance at the lattice will show that this is because the primary recoil has then to share its energy in two simultaneous collisions with a pair of neighbours. Neither of these can be effective unless the primary energy is very high. Presumably the Cu calculations would have shown similar effects if they had been more detailed.

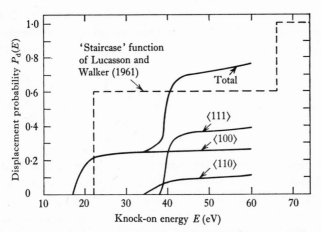

Fig. 76. The integrated displacement probability $P_d(E)$ for a knock-on of energy E and random direction. Note the big changes in value and slope due to the directional dependence of the threshold. The 'staircase' function (– – – –) of Lucasson & Walker (1962) which gave the best fit to the experimental data on electron-irradiated α-iron is shown for comparison. (From Erginsoy, 1964.)

By averaging over all directions the function $\nu(E_2)$ was estimated and this is shown in fig. 76. A two-step curve is predicted, the first being due to the region around $\langle 100 \rangle$, the second being a combination of the $\langle 111 \rangle$ and $\langle 110 \rangle$.

The anisotropic behaviour of E_d^{hkl} can be seen in experiments with Fe single crystals irradiated in a well-defined beam of electrons (Lomer & Pepper, 1967) and these agree with the Brookhaven predictions that E_d^{100} is smaller than E_d^{111}. Some experiments have been done with Si for which no detailed predictions are available (George & Gunnersen, 1964). These show that the lowest displacement energy is along the $\langle 111 \rangle$ direction, as one might expect in the diamond lattice where the recoil along $\langle 111 \rangle$ finds an open space.

5.5.3. *Simple focusing.* The simplest type of focused collision sequence propagates along the closest-packed rows of atoms, such as the ⟨110⟩ in the f.c.c. lattice of fig. 72. An approximate treatment based on the hard sphere collision model of §4.3.1 will illustrate the physical principles.

We remember that the hard sphere radius R increases as the collision energy falls in the later stages of a cascade, then eventually the atomic radius R becomes comparable in magnitude to the interatomic spacing, thus forbidding the channelling of atoms between the lattice planes. Focused collision sequences arise because the lattice is then able to impose even more rigid conditions on the

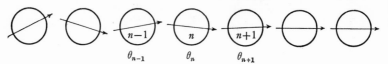

Fig. 77. A focused collision sequence.

possible modes of momentum transfer, for when penetration of the channels in the lattice becomes impossible, a struck atom must receive its momentum from one of its near neighbours. Where closely packed rows of atoms exist in the crystal a directional correlation is imposed on successive collisions and one may expect the situation illustrated in fig. 77 where momentum propagates along a row in a focused sequence of collisions.

Consider the collision illustrated in fig. 78 where two atoms are represented by elastic hard spheres of radius R, separated initially by a distance D. Suppose the first one is given momentum along the direction $A_1 P$ making θ_1 with the line of centres $A_1 A_2$. It moves along $A_1 P$ until collision occurs when its centre reaches P, then the second atom moves off along PA_2 at an angle $-\theta_2$ to $A_1 A_2$. For small angles, $A_1 P \simeq D - 2R$ and

$$(D-2R)\,\theta_1 = -2R\theta_2,$$

hence if we define the *focusing parameter f* by

$$\frac{\theta_2}{\theta_1} = (-f)$$

then

$$f = \frac{D}{2R} - 1.$$

If $D > 4R$ $f > 1$ and $|\theta_1| < |\theta_2|$,

$D < 4R$ $f < 1$ and $|\theta_1| > |\theta_2|$.

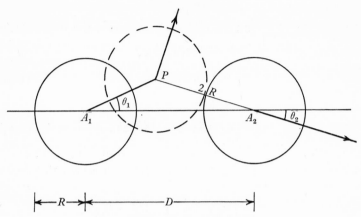

Fig. 78. The simple focusing process.

Suppose one has an equally spaced row of atoms, as illustrated in fig. 77, and a sequence of collisions passes from atom to atom. The relation between θ_n and θ_{n-1} is

$$\theta_n = (-f)\theta_{n-1}$$

$$= (-f)^2\theta_{n-2}$$

$$= (-f)^3\theta_{n-3}$$

.... etc.

until finally $\theta_n = (-f)^n\theta_0.$

If $D > 4R$ and $f > 1$, θ_n diverges and eventually atoms will cease to collide in the line. But if $D < 4R$ and $f < 1$, θ_n converges to zero with $(-f)^n$ and, because the masses are equal, momentum transfer occurs with 100 % efficiency. One then has a *focused collision sequence*. This effect was first analysed by Silsbee (1957).

Considering next the rows of atoms in a crystal, f will be least for the closely packed rows which have the smallest value of D^{hkl}. There will therefore be a tendency for momentum to be focused along lines of closest packing. A critical *focusing energy* E_f^{hkl} exists, below which R is great enough for focusing to occur along the

$\langle hkl \rangle$ direction. In the hard sphere model, when the kinetic energy is E, R is given from the interatomic repulsion potential $V(r)$ by:

$$V(2R) = \tfrac{1}{2}E$$

according to equation (4.32). If $V(r)$ is given in the Born–Mayer form (4.14) then

$$\tfrac{1}{2}E = A \exp(-2R/b)$$

or

$$R = \tfrac{1}{2}b \log(2A/E). \tag{5.44}$$

When $E = E_{\mathrm{f}}^{hkl}$, $R = D^{hkl}/4$ and it follows that

$$E_{\mathrm{f}}^{hkl} = 2A \exp(-D^{hkl}/2b). \tag{5.45}$$

For finite small angles there will be a critical value of θ_1, called θ_{f}, when $\theta_1 = \theta_2$. From the triangles in fig. 78 it is clear that under these circumstances

$$\cos\theta_{\mathrm{f}} = D/4R \quad \text{or} \quad 1 - \tfrac{1}{2}\theta_{\mathrm{f}}^2 \simeq D/4R.$$

Now for $\theta_1 \leqslant \theta_{\mathrm{f}}$ a focused collision sequence is started, hence the probability $P_{\mathrm{f}}(E)$ of generating one at energy E for random starting directions is

$$P_{\mathrm{f}}(E) = \frac{\pi\theta_{\mathrm{f}}^2}{4\pi}$$

by considering the solid angles

or

$$P_{\mathrm{f}}(E) = \frac{1}{2}\left(1 - \frac{D}{4R}\right)$$

from above, using relations (5.45) and (5.46) to give R in terms of E and E_{f} we obtain

$$P_{\mathrm{f}}(E) = \frac{b \log E_{\mathrm{f}}/E}{D + 2b \log E_{\mathrm{f}}/E} \tag{5.46}$$

In most cases of interest the second term in the denominator can be neglected and

$$P_{\mathrm{f}}(E) \simeq \frac{b}{D} \log \frac{E_{\mathrm{f}}}{E}. \tag{5.46a}$$

Vineyard's potential for Cu–Cu repulsion, used in the Brookhaven computations and given in table X, predicts a value of 67 eV for E_{f}^{110} in equation 5.45. To obtain rough estimates for the close packed directions of other crystals one may use Brinkman's relations for A and b (4.19), (4.20). Some values are shown in table XVIII from which it will be seen that E_{f}^{hkl} should vary from ~ 10 eV for light elements to $\sim 10^3$ eV for heavy elements. Clearly, *one expects focused collision sequences to exert their greatest influence on cascades in heavy elements*. The extent to which Brinkman's potential agrees with Vineyard's for Cu may be judged from the values of 40 eV and 67 eV that they respectively predict in (5.45).

TABLE XVIII. *Calculated values of the simple focusing energy E_f^{110} eV in f.c.c. metals*

Potentials referred to are given in table X and § 4.2.

Potential	Method of calculation		
	Hard sphere	Lehmann & Leibfried	Brookhaven
Cu (Gibson *et al.* 1960)	67	36	40
(Brinkman 1962)	40	—	—
Ag (Brinkman 1962)	87	—	—
Au (Brinkman 1962)	620	—	—
(Thompson 1968)	300	170	—

Although the hard sphere model presents us with a clear picture of momentum focusing, and its use may easily be justified for the near-head-on collisions involved, it is not entirely adequate. If one allows for the 'soft' nature of interatomic collision we have already seen, in chapter 4, that the true scattering angle for a given impact parameter is less than the hard sphere approximation would predict. This leads to an overestimate of the focusing by using the hard sphere model. In addition, with a realistic potential, atoms ahead of the main energy packet start moving forward before its arrival and this leads to a shortening of D^{hkl}. The effect is to reduce f^{hkl} and so enhance the focusing, in opposition to the first effect. Of course, the mathematical problem can be solved to a high accuracy by the Brookhaven computer method, but this lacks the convenience of an analytical solution when one comes to test the effect of potential, etc. Leibfried & Lehmann (1961) have developed such a solution using the computer results as a direct check on their approximations. The reader is referred to their paper for details of the calculation, but their conclusion for Cu with the Gibson–Vineyard potential in table X is that the modified scattering alone would reduce E_f^{110} from

67 to 22 eV. If the effective reduction in D^{hkl} is also taken into account the final value is 36 eV. Their expression for f^{hkl} is

$$f^{hkl} = \frac{D^{hkl}}{2R} - 1 + 0.347\frac{D^{hkl}}{R}\left[\frac{b}{R}\right] - \frac{0.48 + 0.17 D^{hkl}/R}{1 - 2R/D^{hkl}}\left[\frac{b}{R}\right]^2$$

(5.47)

Equation (5.45) for R enables one to express this in terms of E. The two expressions (5.44) and (5.47) are compared in fig. 79 by plotting

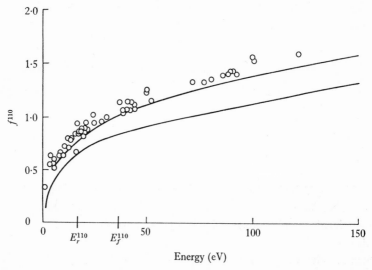

Fig. 79. The focusing parameter f^{110} for Cu as a function of energy. The Brookhaven simulation (open circles) is compared with the hard sphere expression (lower curve) and the improved result of Leibfried & Lehmann (1961) (upper curve).

f^{hkl} as a function of E, for a potential with $b = D^{hkl}/13$, which is the value in the Vineyard potential and would be reasonable for Cu along $\langle 110\rangle$. The extent to which either treatment agrees with the computed results is also seen in fig. 79 where the points represent the individual collisions in a number of $\langle 110\rangle$ Brookhaven runs. Clearly (5.47) is very good.

Perhaps the most important difference between the hard sphere model and the soft collision lies in the possibility of *focused replacement*, where each atom in a collision sequence is replaced by its predecessor. We have already seen that many displaced atoms are generated by such events. A reasonable criterion for replacement is that the nth atom must come to rest beyond the mid point between the nth and $(n+1)$th sites. Referring to fig. 78, the hard sphere model requires that $A_2 P < D/2$ or that $D > 4R$ for replacement. Then in this model one cannot have focusing and replacement together and focused replacement sequences are impossible.

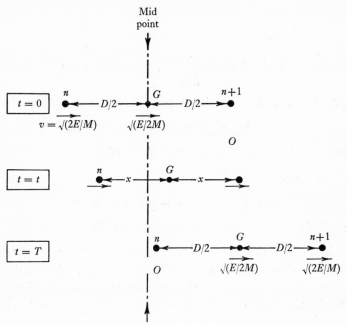

Fig. 80. The replacement process in simple focusing.

In a soft collision the $(n+1)$th atom starts to move forward before the nth comes to rest, and one must examine the situation further to decide whether or not focused replacement can occur. In fig. 80 a head-on collision between n and $(n+1)$ is shown in G co-ordinates. The distance between the atoms is $2x$, and the time T that elapses whilst x changes from $D/2$ to R, and back again to $D/2$, is given by

$$T = 2 \int_{D/2}^{R} \mathrm{d}x/\dot{x}.$$

Now the energy equation for the G system is

$$\tfrac{1}{2}E - M\dot{x}^2 = V(2x)$$

neglecting $V(D)$, which is reasonable for the exponentially varying potentials that we are concerned with, then

$$\dot{x} = \sqrt{[(1 - 2V/E)E/2M]}$$

writing $V(2x) = V$ for brevity. With a Born–Mayer potential (4.14) one has

$$V = A \exp(-2x/b),$$

hence

$$x = \tfrac{1}{2}b \log(A/V).$$

Transforming the integral in x to one in V one obtains

$$T = b\sqrt{(2M/E)} \int_{V(D)}^{V(2R)} dV/V \sqrt{(1 - 2V/E)}.$$

Then since $V(2R) = \tfrac{1}{2}E$ this can be evaluated as

$$T = 2b\sqrt{(2M/E)} \tanh^{-1}\sqrt{(1 - 2V(D)/E)}.$$

With

$$\sqrt{(1 - 2V(D)/E)} \simeq 1 - V(D)/E,$$

since $E \gg V(D)$, and using

$$\tanh y \simeq 1 - 2\exp(-2y) \quad \text{for} \quad y \gg 1$$

one obtains

$$T = b\sqrt{(2M/E)} \log(2E/V(D)).$$

G moves with uniform velocity $\sqrt{(E/2M)}$ in the L system, and this is equal and opposite to the velocity the first atom has in the G system as it passes $x = D/2$ on the rebound. Thus after a time interval T the first atom is at rest in L coordinates and its position then will determine whether or not replacement will occur.

The situation at time T is sketched in fig. 80. G has moved forward by $T\sqrt{(E/2M)}$ and the nth atom comes to rest $D/2$ behind this point. For its final position to be beyond the mid-point and hence for replacement to occur

$$T\sqrt{E/2M} > D/2 \quad \text{or} \quad E > E_{\mathrm{f}}^{hkl}$$

with

$$E_{\mathrm{r}}^{hkl} = \tfrac{1}{2}A \exp(-D^{hkl}/2b). \tag{5.48}$$

It will be noticed that E_{r}^{hkl} is just $1/4$ the focusing energy E_{f}^{hkl} given by the hard sphere model in (5.50).

Then, with an initial energy in the interval $E_{\mathrm{f}}^{hkl} > E > E_{\mathrm{r}}^{hkl}$ one can generate a *focused replacement sequence* in which each atom replaces its neighbour in the focusing line. We shall refer to E_{r}^{hkl} as the *replacement energy*. The importance of the replacement sequence is that it moves every atom forward by one place, leaving a vacant site at one end and forming an interstitial at its terminus and, as we saw earlier, all i–v pairs are produced by them in the Brookhaven calculations on Cu and Fe. In a crystal of an ordered alloy or a compound such a process will cause disordering, in addition to the damage due to the interstitial and vacancy.

Sequences starting with $E < E_r^{hkl}$ will only carry energy, and are referred to as *focused energy packets*. They do not cause damage in an ideal crystal but they are capable of ejecting atoms from a surface, although it is not possible to distinguish them from focused replacement sequences in a sputtering experiment. In circumstances to be discussed below they can be defocused to form i–v pairs.

The analytical model has so far been restricted to small angles. In consequence the effective value of E_r^{hkl} is over-estimated for we can see from fig. 75 that in the $\langle 111 \rangle$ rows of Fe the replacement energy drops from 57 eV at $\theta_0 = 0$ to 39 eV when $\theta_0 = 5°$. Thus the value from (5.48) is too large. The analytical approach is so far unable to decide the finer points such as this, nor the energy at which the replacement sequence drops its interstitial, nor the behaviour of E_r^{hkl} as a function of θ_n.

We turn, therefore, to the Brookhaven method, though we must bear in mind that even this is not perfect as lattice vibration is neglected and we shall see below that this has a very important influence. By noting the sucessive angles θ_n in a large number of $\langle 110 \rangle$ collision sequences in Cu f^{110} may be calculated as a function of E and fig. 79 compares values obtained in this way with the analytical curve from (6.25). It will be seen that although agreement is good, the computed points lie above the curve and show that focusing is less strong for finite angles, as one might expect.

The behaviour of $\langle 110 \rangle$ replacement sequences indicates that they can only be started above E_r^{110}, roughly given by equation (5.48) for $\theta_0 = 0$, and that once started they continue as replacement sequences to energies far below E_r^{110}. The reason is that atoms in the focusing line start to move forward before the arrival of the main impulse and that after this has passed, the relaxation in neighbouring rows favours replacement. Thus the interstitial will be separated from the vacancy by almost the total range of the collision sequence.

Amongst metals only Cu and Fe have so far been treated by the Brookhaven technique and without these computations it is unlikely that such successful analytical treatments could have been developed. Historically, they preceded the more refined analytical approaches which were greatly assisted by having an exact calculation, based on the same model, against which to check the approximations. As one of the first direct simulations of a physical system by digital computer, they represent a major revolution in the methods of theoretical physics.

We next consider the attenuation of simple focused collision sequences for it is clear that they cannot travel indefinitely. There are four important ways in which they can lose energy:

(1) Energy loss directly to the rings of atoms surrounding the focusing axis; this effect may be enhanced by thermal vibration of the lattice.

(2) Scattering of energy out of the line by misalignment of atoms caused by lattice vibration.

(3) Scattering by structural defects in the crystal; for instance dislocation lines.

(4) In crystals containing atoms of differing mass, energy transfer from one atom to the next may be inefficient (i.e. $\Lambda < 1$).

The Brookhaven method cannot at present (1967) simulate the presence of lattice vibration as computer capacity is not sufficient. Hence in mechanisms 1 and 2 we must use analytical methods.

Consider mechanism 1 with reference to fig. 81. B is one of the four atoms which lie in a $\{110\}$ plane and form a ring around the $\langle 110 \rangle$ line. When A_n reaches C with $CA_{n+1} = 2R$ and $CB = c$ the hard sphere collision occurs but the potential energy gained by A_n in moving to C is lost to the sequence. Eventually this energy appears as thermal energy in the lattice when the four B atoms and A_n relax and then oscillate about equilibrium positions. Using the cosine law in triangle BCA_{n+1} and putting in the value of R from (5.44)

$$c = D^{110} \sqrt{\left(\frac{3}{4} + \left\{ \frac{a}{D^{110}} \right\}^2 \log \left\{ \frac{E_f^{110}}{E} \right\} \right)}.$$

Taking all four B atoms into account and subtracting the potential energy at A_n from that at C gives ΔE, the energy loss

$$\left\{ \frac{\Delta E}{E_f^{110}} \right\}_0 = 4[V(c) - V(D^{110})]/2V(\tfrac{1}{2}D^{110}).$$

Since $E_f^{100} = 2V(\tfrac{1}{2}D^{110})$ in the hard sphere approximation (see (5.45)), and putting $V(r)$ into the Born–Mayer form (4.14)

$$\left\{\frac{\Delta E}{E_f^{110}}\right\}_0 = 2\left[\exp\left\{-\frac{D^{110}}{2b}\left(\frac{2c}{D^{110}}-1\right)\right\} - \exp\left\{-\frac{D}{2b}\right\}\right]. \quad (5.49)$$

This expression is of order 10^{-2} and if this were the only mechanism of energy loss one might expect a range of ~ 100 collisions for a sequence starting with the maximum energy of E_f^{110}. The

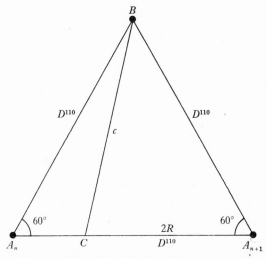

Fig. 81. Mechanism of energy loss in a static lattice.

dependence on E is shown in fig. 82, from which $\{\Delta E/E_f\}_0$ will be seen to decrease at low energies. This behaviour is due to the increasing radius R making the atom from A_n come to rest at a lower point on the potential hill between A_n and A_{n+1}.

In a real crystal the atoms are in a continual state of vibration and expression (5.49) requires some modification. Even at $0\,^\circ$K the amplitudes are approximately $D^{110}/20$ due to zero-point motion. At non-zero temperatures amplitudes increase up to about $D^{110}/10$ at the melting point. The displacement of atoms from their equilibrium sites in this way influences the propagation of collision sequences.

Because the time interval between successive collisions near E_f is of the order 10^{-14} sec. ($\sim \tfrac{1}{2}D^{110}\sqrt{[M/2E_f]}$), whereas the period of atomic vibration is of the order 10^{-13} sec, the lattice appears to the collision sequence as though it were frozen in a static configuration. Two effects are important. First there is the movement of the B atoms in the rings causing fluctuations in the above value of ΔE. On average this causes an increase, and Sanders & Fluit (1964) find that expression (5.49) must be replaced by

$$\left\{\frac{\Delta E}{E_f}\right\}_1 = \left\{\frac{\Delta E}{E_f}\right\}_0 \exp{\frac{(\mathbf{r}_n - \mathbf{r}_{n+1})^2}{2b^2}}. \tag{5.50}$$

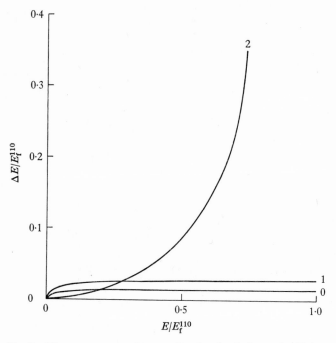

Fig. 82. The components of energy loss for simple focused collision sequences in Cu: (0) (5.49); (1) (5.50) and (5.51); (2) (5.59).

The vectors $\mathbf{r}_n$ and $\mathbf{r}_{n+1}$, are the vector displacements of two neighbouring atoms. Their mean squared difference was calculated for Cu by Sanders, using the method of Mazur & Montroll (1960), to be

$$\overline{(\mathbf{r}_n - \mathbf{r}_{n+1})^2} = 1\cdot 77 \times 10^{-4}T. \tag{5.51}$$

As one might expect, for harmonic motion, the square of the amplitude is proportional to energy and hence temperature. This is valid for temperatures above the Debye temperature Θ_D, which is $315\,°K$ for Cu. At $315\,°K$ the exponential factor in expression (5.50) is then $1\cdot9$, showing that the effect of thermal vibration in Cu near room temperature is almost to double the loss to rings in a static lattice (see fig. 82).

The method of Mazur & Montroll gives a rather accurate result but the calculation does not yield an analytic expression. To obtain a rough expression, suppose we make the simplifying assumption that the moduli r_n and r_{n+1} are equal, then

$$\overline{(\mathbf{r}_n - \mathbf{r}_{n+1})^2} = \overline{2\mathbf{r}_n^2} = 6\overline{x_n^2}$$

and

$$\left\{\frac{\Delta E}{E_f^{110}}\right\}_1 \simeq \left\{\frac{\Delta E}{E_f^{110}}\right\}_0 \exp{(3\overline{x_n^2}/b^2)}, \qquad (5.52)$$

where $\overline{x_n^2}$ is the mean squared displacement parallel to a given axis and is given by the well known Debye–Waller expression, familiar in the context of X-ray diffraction:

$$\overline{x_n^2} = \frac{3\hbar^2}{Mk\Theta_D}\left[\frac{T^2}{\Theta_D^2}\int_0^{\Theta_D/T} \frac{x\,dx}{\exp{(x)} - 1} + \frac{1}{4}\right]. \qquad (5.53)$$

The integral sometimes referred to as Debye's function is well tabulated (see for example Klug & Alexander, 1954). For temperatures above Θ_D this expression gives a roughly linear dependence of $\overline{x_n^2}$ on T, as for a harmonic oscillator.

The second effect of lattice vibration arises from the transverse displacements of atoms from the focusing axis (see Nelson, Thompson & Montgomery, 1962). These prevent perfect focusing and hence lower the efficiency of momentum transfer in the individual collisions. In fig. 83 this scattering effect is shown as a projection of the motion on to a plane. For small angles the focusing relation $\theta_{n+1} = -f\theta_n$ holds equally well when θ is the angle in the projection. When the transverse displacements of the two atoms due to lattice vibration, in this plane, are x_n and x_{n+1}, their relative displacement is $(x_n - x_{n+1})$. In a collision where the momentum vector is initially along the axis the angular deviation produced is

$$\phi_n = (x_n - x_{n+1})/2R. \qquad (5.54)$$

Then following the methods used earlier in this section, the relation between θ_0 and θ_1 is

$$\theta_1 = (-f)\theta_0 + \phi_1$$

and

$$\theta_2 = (-f)^2\theta_0 + (-f)\phi_1 + \phi_2$$

and

$$\theta_3 = (-f)^3\theta_0 + (-f)^2\phi_1 + (-f)\phi_2 + \phi_1$$

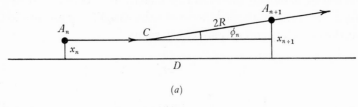

(a)

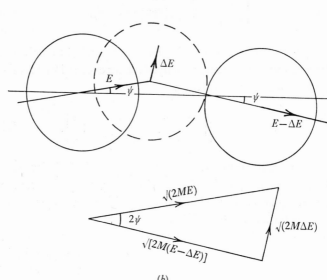

(b)

Fig. 83. (a) The scattering effect of lattice vibration.
(b) Momenta in the scattering process.

and so on, until
$$\theta_N = (-f)^N\theta_0 + \sum_1^N (-f)^{N-n}\phi_n. \tag{5.55}$$

Comparing with the corresponding equation at the beginning of this section one sees that the effect of lattice vibration is to prevent θ_N converging to zero as N becomes large; rather, one has

$$\theta_N \to \phi_N + (-f)\phi_{N-1} + (-f)^2\phi_{N-2} + .. \tag{5.56}$$

Because $f < 1$ in the focusing range $E < E_f$, the effect of a scattering event that occurred n collisions previous to the one considered is reduced by the factor f^n. In other words, although the collision sequence remembers its past history, the memory becomes weaker as the historical event becomes more remote.

The average energy loss per collision is calculated by first squaring each side of (5.56), having put ϕ_n in terms of x_n by (5.54). Averaging over many collision sequences in which the x_n's range over all possible values, the squared equation may be averaged term by term to give $\overline{\theta_N^2}$, which turns out as a function of $\overline{x_n^2}$ and mean products like $\overline{x_n x_{n+1}}$.

To put these last quantities into terms of temperature, etc., a model for the lattice vibration must be assumed. An adequate treatment is obtained by using Debye's well-known theory of lattice vibration, which leads to $\overline{\theta_N^2}$. The mean squared angular deviation in three dimensions $\overline{\psi_N^2}$ is just $2\overline{\theta_N^2}$ given by:

(a) For $T \ll \Theta_D$

$$\overline{\psi_N^2} = \frac{8\overline{x_n^2}}{(D^{110})^2 L(1+L)} \left[1 - \frac{2}{1+L} \left(0 \cdot 139 - 0 \cdot 047 \frac{1-L}{1+L} \right) \right] \quad (5.57)$$

with
$$L = \frac{2b}{D^{110}} \log \frac{E_f^{110}}{E}.$$

(b) For $T > \Theta_D$

$$\overline{\psi_N^2} = \frac{8\overline{x_n^2}}{(D^{110})^2 L(1+L)} \left[1 - \left(\frac{\pi}{6\sqrt{2}} \right)^{\frac{1}{3}} \frac{1}{1-L} \log \frac{2}{1+L} \right]. \quad (5.58)$$

In both cases $\overline{x_n^2}$ is given by the Debye–Waller expression equation (5.53).

In a sequence of collisions, the angle between successive momentum vectors is, on the average, $2\sqrt{(\overline{\psi_N^2})}$, due to the oscillation of the vector from side to side of the axis. This is illustrated in fig. 83 b and from the triangle of momenta it is clear that the magnitude of the momentum loss is $2\sqrt{(\overline{\psi_N^2})}\sqrt{(2M\Delta E)}$ and hence

$$\Delta E = 4\overline{\psi_N^2} E,$$

$$\left\{ \frac{\Delta E}{E_f^{110}} \right\}_2 = 4\overline{\psi_N^2} \frac{E}{E_f^{110}}. \quad (5.59)$$

The two components of energy loss $(\Delta E/E_f^{110})_1$ and $(\Delta E/E_f^{110})_2$ are plotted together in fig. 82, the case illustrated is Cu at $T = \Theta_D$

(315 °K) using (5.51). Below about $E_t^{110}/4$ the first effect, due to penetration of the rings, is greatest. But above this energy the scattering effect completely dominates the total energy loss. This is because f^{110} is close to unity at high energy, and the focusing cannot control the scattering. To see how the situation changes with

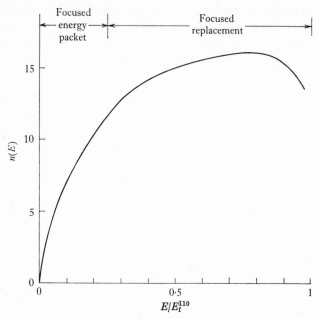

Fig. 84. Number of collisions versus energy for simple focused collision sequences in Cu at 315 °K.

temperature or $\overline{x_n^2}$ (5.52) and (5.59) give a ratio of the two mechanisms. It is then easy to see that the scattering effect increases its dominance if one goes to higher temperatures.

The number of collisions $n(E)$ made by a sequence starting with energy E is calculated as follows. To increase $n(E)$ by one requires an extra ΔE, the total energy loss per collision. Thus to increase it by $\mathrm{d}n(E)$ requires $\Delta E\,\mathrm{d}n(E)$. For example,

$$\mathrm{d}E = \Delta E\,\mathrm{d}n(E),$$

$$\mathrm{d}n(E) = \frac{\mathrm{d}E}{\Delta E},$$

hence
$$n(E) = \int_0^E \frac{\mathrm{d}E}{\Delta E}$$

or
$$n(E) = \int_0^{E/E_f^{110}} \frac{\mathrm{d}(E/E_f^{110})}{\Sigma(\Delta E/E_f^{110})_n}, \qquad (5.60)$$

allowing for the sum of all components of energy loss $(\Delta E/E_f)_n$.

The result of carrying out this calculation for Cu at $315\ ^{\circ}$K, taking our first two mechanisms $(\Delta E/E_f^{110})_1$ and $(\Delta E/E_f^{110})_2$, is shown in fig. 84. The range $n(E)$ increases rapidly with E up to about $E_f/4$.

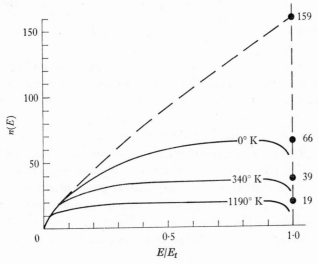

Fig. 85. Number of collisions in a $\langle 110 \rangle$ sequence in Au starting at energy E at various temperatures.

Here, the scattering effect becomes important and prevents much further increase. A decrease occurs at high energies due to defocusing that occurs for finite angles even below E_f (Sanders & Fluit 1964). The maximum possible number of collisions is about 16.

We next examine the range at low temperature where $\overline{3x^2} < b^2$ and we can take $(\Delta E/E_f^{110})_1 \simeq (\Delta E/E_f^{110})_0$ without introducing much error. On this basis a set of $n(E)$ versus E curves has been calculated for Au with $D^{110}/b = 15$, and these show the approximate effect of increasing temperature in fig. 85. It will be clear that even at o $^{\circ}$K the scattering effect of zero point motion, which appears in (5.53) as the term $(+1/4)$ in the square bracket, is considerable.

It is always useful to have a rough but simple expression to describe behaviour such as this and we notice that the total energy loss $\Sigma(\Delta E)_n$ as compounded from fig. 82, behaves *very* roughly like

$$\begin{aligned} \Delta E &\simeq \alpha DE \quad \text{for} \quad E < E_f, \\ &= \infty \qquad\qquad E > E_f, \end{aligned} \right\} \tag{5.61}$$

where α is a constant of proportionality of the order of 10^{-1}, which increases with temperature. In (5.60) this leads to

$$n(E) \simeq \frac{1}{\alpha D} \log \frac{E}{E_0}, \tag{5.62}$$

E_0 being a lower limit to the energy at which the focused collision sequence ceases to be defined.

We must now answer the question: what *is* this lowest energy E_0 for which the focused collision sequence can exist? Suppose we compare the duration of a single collision $\frac{1}{2}D\sqrt{(M/2E)}$ with the natural vibration period of atoms in the lattice $\sim h/k\Theta_D$. If the duration is much shorter than the period we can localize the energy on a small group of atoms and talk about collision sequences. But if the energy is small enough for the two to become comparable the energy will be distributed amongst a larger group of atoms, since they are all coupled together with a communication time of the order of $h/k\Theta_D$. Then the focused collision sequence ceases to exist. The energy E_0 at which this occurs is given by

$$\frac{h}{k\Theta_D} = \frac{D}{2}\sqrt{\left(\frac{M}{2E_0}\right)},$$

i.e.

$$E_0 = \frac{D^2 k^2 \Theta_D^2 M}{8h^2} \tag{5.63}$$

which is of the order of 1 eV in most solids.

The velocity v_0 corresponding to E_0 is simply

$$v_0 = \frac{Dk\Theta_D}{2h}$$

and from the Debye theory of lattice vibration this will be recognized as a typical velocity of lattice waves. Thus provided the velocity of the collision sequence greatly exceeds the sound velocity it retains the characteristics of a supersonic energy packet, but near the sound velocity its energy spreads out into the crystal and one cannot localize the disturbance on to a small group of atoms as the above treatment of focused collision sequences requires.

In mechanism (3) the collision sequence loses energy by passing through a defective region of the lattice. For instance, in the neighbourhood of a dislocation line, the atomic rows are distorted as

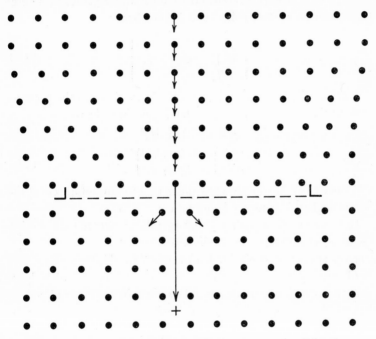

Fig. 86. Interaction of a focused collision sequence with a dislocation.

shown in fig. 86. If a row actually passes through the stacking fault region they are sheared discontinuously (see Chapter 3). A collision sequence passing along one of these rows which end abruptly will have some of its energy refocused into new sequences starting from the opposite side of the fault, but inevitably a large proportion will be lost. If enough energy is available an atom may be displaced permanently.

The chief importance of this process lies in its ability to generate damage in the vicinity of a dislocation line, which has two significant implications. The first is to make the rate of production of displaced atoms dependent on defect structure of the solid. Secondly the damage is produced in the place where its effect on mechanical properties is likely to be greatest. These will be referred to again in later chapters.

Finally, let us consider mechanism (4). If the atoms in a row are not of equal mass, there will be energy lost due to the fact that Λ, given by expression (4.8) is less than unity. In fact, since the energy transferred in a single collision is ΛE one has

$$\Delta E = E(1 - \Lambda)$$

and
$$\left\{ \frac{\Delta E}{E_f^{hkl}} \right\}_4 = \frac{E}{E_f}(1 - \Lambda). \right\} \tag{5.64}$$

For a small difference in mass ΔM, due perhaps to the presence of several isotopes, Λ can be written approximately as

$$\Lambda \simeq 1 - (\Delta M/M)^2.$$

Then
$$\left\{ \frac{\Delta E}{E_f^{hkl}} \right\}_4 \simeq \frac{E}{E_f^{hkl}} \left\{ \frac{\Delta M}{M} \right\}^2 \tag{5.65}$$

and since $(\Delta M/M)$ is only a few times 10^{-2} for an isotope effect this loss is negligible compared with others.

However, if one is dealing with larger differences in mass, such as might be encountered in an alloy, one expects a significant effect. For instance, take the alloy CuAu which would have an average Λ of 0·75 giving $(\Delta E/E_f^{hkl})_4 = 0·25(E/E_f)$, which is rather large. But the above approach was shown to be over-simplified by Vineyard & Erginsoy (1962) in a simulation of collision sequences in the alloy Cu_3Au.

Suppose one has a row of alternate Cu and Au atoms. When an Au atom moves off with E its Cu neighbour will receive 0·75E and move off rapidly because of its small mass to collide with the second Au atom. From this it will rebound and have a second collision with the first Au atom which is still moving forward. The process is repeated several times, with the Cu atom effectively ferrying energy between the two Au atoms. The multiple collision process greatly increases the efficiency of energy transfer. In a typical simulation Vineyard found the following maximum energies on successive atoms in a $\langle 110 \rangle$ sequence: 15, 13, 12, 8, 9 and 6 eV. It is clear that the energy loss is much less drastic than might at first have been expected.

5.5.4. *Assisted focusing.* In the Brookhaven simulation of collision cascades in Cu one can certainly see collision sequences passing along $\langle 100 \rangle$ rows. Signs of this are evident in fig. 73 but fig. 87 shows the effect more clearly. The interatomic spacing D^{100} is $\sqrt{2}\, D^{110}$ and

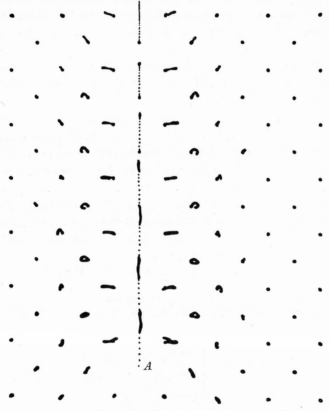

Fig. 87. An assisted focused replacement sequence along a $\langle 100 \rangle$ row in Cu at 50 eV. (From Gibson *et al.* 1960.)

hence *simple* focusing would have a focusing energy of about $E_f^{110}/\exp{(D/2b)}(\sqrt{2}-1)$. This is approximately 4 eV for Cu, yet the sequence in fig. 87 is being focused when travelling with 50 eV when simple focusing could not possibly operate. The clue to the mechanism lies in the atoms of neighbouring rows which are seen to recoil outwards as the sequence passes through. It is the influence of these recoils that produce the *assisted* focusing effect. The low magnitude of the energy for simple focusing along $\langle 100 \rangle$ rows shows that $D \gg 4R$, hence $E_f^{100} \gg E_r^{100}$, and that replacement will accompany assisted focusing. Collision sequences with assisted focusing are therefore *focused replacement sequences*.

A single collision of a $\langle 100 \rangle$ sequence in a f.c.c. crystal is shown in figs. 88 and 90. In going to strike A_2 the atom A_1 suffers four glancing collisions with the atoms B, which form a ring about its path. If its trajectory deviates from $A_1 A_2$ the glancing collisions will transfer more momentum to one side of the ring than to the other, and a focusing will occur. This is somewhat similar to channelling over a single cell of the crystal.

A crude but instructive analysis may be made using the impulse approximation for the A_1–B collisions, and the hard sphere approximation for the A_1–A_2 collision. Take the case illustrated in fig. 88 where motion occurs in a $\{001\}$ plane with A_1 moving off at θ_1 to the $\langle 100 \rangle$ axis. By restricting the impact parameters p to the range of interest in this problem, $D^{110}/\sqrt{3} < p < D^{110}/\sqrt{2}$, and using a Born–Mayer potential with b in the range $D^{110}/10 < b < D^{110}/20$, it is possible to write the impulse approximation approximately as

$$\Delta P = \frac{A D^{110}}{b u_1} \exp\left[-\frac{D^{110}}{4b} \sqrt{(1 + 16 p^2/(D^{110})^2)} \right]$$

(see Thompson & Nelson (1961) for proof).

The four impact parameters for the glancing A_1–B collisions are given by

$$p = \begin{cases} (D^{110}/\sqrt{2})(1 - \theta_1), \\ (D^{110}/\sqrt{2})(1 + \theta_1), \\ D^{110}/\sqrt{2}, \\ D^{110}/\sqrt{2}. \end{cases}$$

Inserting these in turn into the expression for ΔP one calculates the net momentum change of A_1, transverse to its path. Dividing this by $M u_1$ gives the angular deviation ϕ of A_1's trajectory caused by passing through the ring, but since the ring is only half penetrated when the collision with A_2 occurs we reduce this by half, giving:

$$\phi = \theta_1 \frac{A(D^{110})^2}{3 E b^2} \exp\{ -3 D^{110}/4b \}.$$

Focusing occurs if $\theta_2 < \theta_1$ but at the limit when $\theta_2 = \theta_1$, and

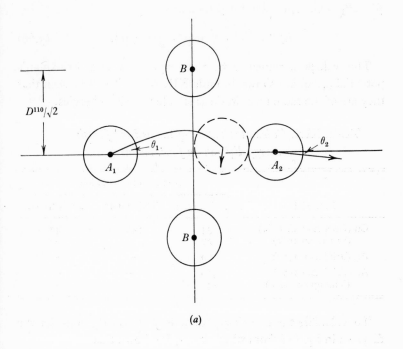

(a)

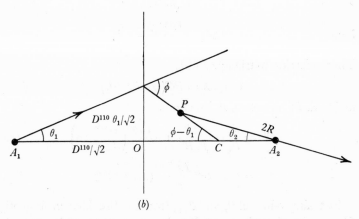

(b)

Fig. 88(a) and (b). ⟨100⟩ assisted focusing in a f.c.c. lattice.

$E = E_f^{100}$, the extended trajectory of A_1 passes through A_2 and $\phi = 2\theta_1$. The expression for ϕ then gives

$$E_f^{100} = \frac{A(D^{110})^2}{6b^2} \exp\{-3D^{110}/4b\}. \qquad (5.66)$$

The values obtained with this expression and Brinkman's potential (4.19) are shown in table XIX where it will be seen that they are of the same magnitude as simple focusing energies.

TABLE XIX. *Calculated values of the assisted focusing energy E_f^{100} eV in f.c.c. metals*

Potential	Method of calculation		
	Equation (5.66)	Weijsenfeld	Brookhaven
Cu (Gibson *et al.* 1960)	34	42	40
(Brinkman 1962)	27	—	—
Ag (Brinkman 1962)	60	—	—
Au (Brinkman 1962)	400	—	—
(Thompson 1968)	140	—	—

To calculate the focusing parameter $f^{100} = \theta_2/\theta_1$, consider the diagram in fig. 88 *b* from which it may be shown that

$$OC = \frac{D^{110}}{\sqrt{2}} \frac{\theta_1}{\phi - \theta_1},$$

hence

$$CA_2 = \frac{D^{110}}{\sqrt{2}} \frac{\phi - 2\theta_1}{\phi - \theta_1}.$$

Then referring to triangle PCA_2

$$CA_2 . \theta_2 \simeq PC(\phi - \theta_1 - \theta_2)$$

and with $PC \simeq 2R - CA_2$ one finds

$$f^{100} = \frac{\theta_2}{\theta_1} = \frac{D^{110}}{\sqrt{2}R} - 1 - \frac{E_f^{100}}{E}\left(\frac{D^{110}}{\sqrt{2}R} - 2\right) \qquad (5.67)$$

$$\simeq 1 \cdot 8 - 0 \cdot 8 \frac{E_f^{100}}{E} \quad \text{(near } E_f^{100}).$$

By taking values of θ_n and θ_{n+1} from the Brookhaven simulation, f^{100} may be plotted against E and compared with this expression. Fig. 89 shows the result of this comparison and it is clear that expression (5.67) over estimates the focusing at low energy. There are two reasons for this; first the impulse approximation over estimates ϕ because the ring atom moves out during collision. Secondly, the hard sphere approximation underestimates the angle θ_2 since A_2 moves forward early in a soft collision.

Dederichs & Leibfried (1962) have made another analysis using the Brookhaven results to check their approximations, as with the

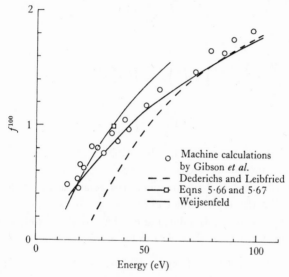

Fig. 89. The assisted focusing parameter f^{100} in Cu using the Gibson–Vineyard potential. Machine calculations by Gibson *et al.* (1960); $----$ Dederichs & Leibfried (1962); —□— equations 5.66 and 5.67; — Weijsenfeld (1965). (From Weijsenfeld, 1965.)

$\langle 110 \rangle$ sequences. Unfortunately their expressions do not lead to a simple form for f^{100} or E_f^{100} and the reader must be referred to fig. 89 and their original paper.

An alternative treatment by Weijsenfeld (1965) gives the best agreement with machine calculations (see fig. 89) and has the advantage that the expression for f^{100} is fairly easy to tabulate. The model does not assume that the ring atoms B remain fixed, but allows a free relaxation during collision. It also allows for the soft nature of the collision between A_1 and A_2.

$$f^{100} = \frac{D^{100} - 2R_{\text{eff}}}{2R_{\text{eff}}} - \frac{\frac{1}{2}D^{100} - 2R_{\text{eff}} + X}{2R_{\text{eff}}} \frac{\phi}{\theta_1} \qquad (5.68)$$

with $\qquad R_{\text{eff}} = R/(1 + 1\cdot39b/2R - 1\cdot36b^2/4R^2)$

(5.44 gives R) and $\qquad X = \tfrac{1}{2}D^{100}\theta_s - z.$

$$\theta_s = \frac{1}{2}\frac{A}{E}\frac{p}{b}K_0\left(\frac{p}{b}\right)$$

(K_0 and K_1 are modified Hankel functions)

$$z = \frac{A}{E}\rho K_1\frac{(\rho)}{b}$$

ρ being given by (4.25)

$$\frac{\phi}{\theta_1} = \frac{A}{E}\frac{2}{b}\left\{\left(\frac{\rho}{b}\right)K_1\left(\frac{\rho}{b}\right) - K_0\left(\frac{\rho}{a}\right)\right\}\frac{\tfrac{1}{2}D^{100} - X}{3\gamma - 2},$$

$$\gamma = \frac{p}{2b}\left(\frac{\rho^2}{p^2} - 1\right) + \frac{p}{\rho}.$$

Since $D^{hkl} > 4R$ when assisted focusing operates, the assisted sequences are inevitably focused replacements. Along the $\langle 100 \rangle$ lines of a f.c.c. crystal, if one neglects the relaxation of B atoms during collision the replacement energy is given by equating the kinetic energy of A_1 in G coordinates $\{\tfrac{4}{5}E\}$ to the difference in potential energy between A_1 and O in fig. 90, i.e.

$$E_r^{100} = 5A\exp(-D^{110}/b\sqrt{2}). \tag{5.69}$$

Values of this calculated with Brinkman's potential are shown in table XX. The effect of introducing relaxation during collision would presumably be to reduce E_r^{100} by a small amount.

In the f.c.c. structure it is also possible for assisted focusing to occur along $\langle 111 \rangle$ directions as shown in fig. 90. A_1 passes through two rings of three atoms before striking A_2 and these produce the focusing. The $\langle 111 \rangle$ collision sequences may be treated in the same manner as above using the impulse approximation which leads to

$$E_f^{111} = \sqrt{\left(\frac{6}{19}\right)}\frac{A(D^{110})^2}{b^2}\exp\left\{-\frac{D^{110}}{2b}\sqrt{\left(\frac{19}{12}\right)}\right\}. \tag{5.70}$$

Because the radius of the rings is only $D/\sqrt{3}$ in this case, and there are two rings, the focusing is much stronger than for $\langle 100 \rangle$ sequences and E_f^{111} tends to be large. For example, with Vineyard's Cu potential expression (5.70) gives $E_f^{111} = 490\,\text{eV}$.

Again, one has a focused replacement sequence and calculating the replacement energy by the methods above gives

$$E_r^{111} = 4A \exp(-D^{110}/b\sqrt{3}). \qquad (5.71)$$

Because the B atoms are more closely spaced, the value along $\langle 111 \rangle$ directions is greater than for $\langle 100 \rangle$, as shown in table XX.

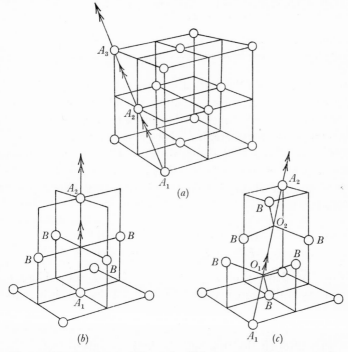

Fig. 90. The three types of focused collision sequence in f.c.c. crystals.

The replacement energies E_r^{110}, E_r^{100} and E_r^{111} must be closely related to the mean displacement energy E_d, discussed in § 5.5.1 but allowance must be made for the extent of the unstable region along the different crystal axes. From table XX it appears that $\langle 100 \rangle$ is the easiest direction for replacement and $\langle 111 \rangle$ the most difficult. The same order is followed by E_d^{hkl}, the displacement energy, although this energy is of course larger since the interstitial end vacancy must be sufficiently separated for them to be stable.

In addition to providing most of the focusing, the momentum

TABLE XX. *Calculated values of the replacement energies in f.c.c. metals using* (5.48), (5.69) *and* (5.71). *Also shown are the values of* E_d^{hkl} *found by the Brookhaven method in* Cu

Potential	E_r^{110}	E_d^{110}	E_r^{100}	E_d^{100}	E_r^{111}	E_d^{111}
Cu (Brinkman 1962)	10	—	13	—	38	—
(Gibson *et al.* 1960)	17	28	11	25	50	85
Ag (Brinkman 1962)	22	—	13	—	78	—
Au (Brinkman 1962)	155	—	63	—	350	—
(Thompson 1968)	70	—	42	—	180	—

transfer to atoms in the rings also constitutes an energy loss. For a $\langle 100 \rangle$ sequence in the f.c.c. lattice the first equation in §4.5.3 gives the momentum transfer ΔP to one of the B atoms in the ring, when the impact parameter is p. If one assumes the atom A_1 to pass through the centre of the ring, then $p = D^{110}/\sqrt{2}$ and the energy transfer per B atom may be calculated from $(\Delta P)^2/2M_2$. Then the energy loss per A_1–A_2 collision, which involves four A_1–B collisions is

$$\Delta E = 2(\Delta P)^2/M_2$$

or

$$\frac{(\Delta E)}{E_f^{100}} = \frac{(D^{110})^2 A^2}{b^2 E_f^{100} E} \exp\left\{-\frac{3D}{2b}\right\}. \qquad (5.72)$$

Because ΔP depends on $1/\sqrt{E}$ the energy loss increases like $1/E$ as the energy of the sequence attenuates. This behaviour is in marked contrast to that of mechanism 1 in simple collision sequences where ΔE is almost independent of E. Note that in simple focusing A_1 does not pass through the ring and the impulse approximation there would be inappropriate.

To obtain an order of magnitude one may take the value of expression (5.72) at E_f^{100} whence

$$\frac{\Delta E}{E_f} > 9b^2/(D^{110})^2$$

using (5.66), to express E_f^{100} in terms of the Born–Mayer constants. Now, since $b/D^{100} \sim 10^{-1}$ then $\Delta E/E_f^{100} > 10^{-1}$ and one might expect the order of 10 collisions to be the range in a static lattice. But remembering that the impulse approximation tends to over-estimate the energy loss the range is likely to be larger than this.

Because this basic attenuation process is much more severe than mechanism 1 in simple focusing, the scattering effect of lattice vibration is not likely to affect this estimate drastically.

The Brookhaven calculations provide direct confirmation of this order of magnitude but do *not* show any dependence of ΔE on $1/E$. Indeed they indicate that the energy loss is almost independent of energy. For example a typical $\langle 100 \rangle$ sequence in Au, using the potential of table X (Erginsoy & Thompson, 1964) had maximum energies on successive atoms: 150, 120, 98, 75, 52, 29, 8 eV. Thus, with the exception of the first collision, about 22 eV was lost in each collision. Evidently the outward relaxation of the rings during collision prevents the energy loss from rising at lower energies.

In the above sequence we notice a larger energy loss in the first collision. This is a characteristic of *all* focused collision sequences and arises from the potential energy stored in compressed bonds, when starting the sequence. It is given roughly in magnitude by the interstitial formation energy since a replacement sequence has a similar instantaneous configuration as the interstitial. In fact replacement sequences are sometimes referred to as *dynamic interstitials* or *dynamic crowdions*.

5.5.5. *Focusing in b.c.c. crystals.* In the b.c.c. structure, illustrated in fig. 91, the close packed direction is $\langle 111 \rangle$ and this is a direction of simple focusing. In the $\langle 100 \rangle$ although one has rings of B atoms their spacings are such that their focusing action is negligible compared with the simple focusing effect. In the $\langle 110 \rangle$ one has effective rings but it will be clear from fig. 91 (*c*) that they are unsymmetrical, being compressed along the $\langle 001 \rangle$ direction. This leads to anisotropy in the focusing and although a focusing energy can be defined for motion in the $\{1\bar{1}0\}$ plane deviations in the $\{100\}$ plane may be large. Following the methods used in § 5.5.4, Nelson (1963) derived the following expressions for focusing energies:

$$E_f^{111} = 2A \exp\{-D^{111}/2b\}, \tag{5.73}$$

$$E_f^{100} = 2A \exp\{-D^{111}/\sqrt{3}b\}, \tag{5.74}$$

$$E_f^{110} = \frac{4\sqrt{2}(D^{111})^2}{15b^2} A \exp\{-\sqrt{5}D^{111}/2/3b\} \tag{5.75}$$

(in the $\{1\bar{1}0\}$ plane only).

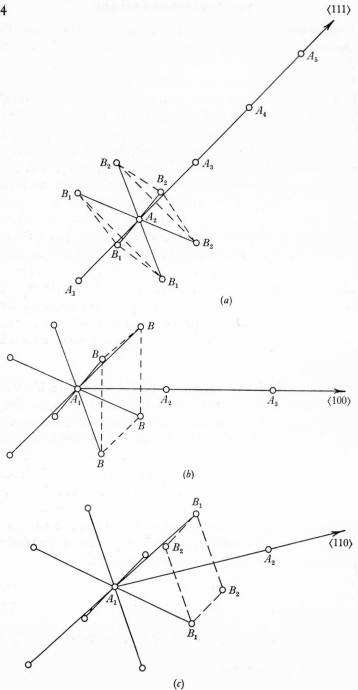

Fig. 91 (a), (b) and (c). Focusing in the b.c.c. lattice.

Erginsoy has adapted the Vineyard simulation to a b.c.c. lattice with a potential to represent Fe (Erginsoy 1964; Erginsoy, Vineyard & Englert 1964). This confirms the simple analytical model qualitatively, but as before the simulated focusing energies are about one half of those predicted by the expressions above, showing that the impulse and hard-sphere approximations predict over strong focusing.

Because simple focusing operates along $\langle 111 \rangle$ and $\langle 100 \rangle$ directions, the replacement energy is given by (5.48). Along $\langle 110 \rangle$, because assisted focusing operates, the methods of §5.5.4 give

$$E_{\mathrm{r}}^{110} = 3A \exp\{-D^{110}/2b\}. \tag{5.76}$$

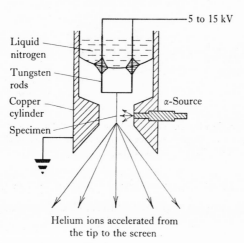

Fig. 92. Formation of a field-ion micrograph during alpha bombardment.
(From Brandon, Southon & Wald, 1962.)

5.5.6. *Experiments involving the collision cascade.*

The heavy ion penetration experiments provide a means of checking our ideas on the primary recoil track, though one must have some reservations because the heavy ion does not start out from a lattice site. The secondary recoils and all subsequent stages must be studied in other ways.

In the field ion microscope it is possible to observe atomic sites on the tip of a point metal specimen (see fig. 92). The radius of curvature of the tip is about the same as typical cascade dimensions and by bombarding such tips with energetic ions several research groups have seen surface damage which they claim to result directly from a single collision cascade as in fig. 93. Plate XIII shows a typical pair of micrographs from W bombarded with 5 MeV α-particles in which a number of atoms have simultaneously been either knocked out of the surface or up on to the surface.

In the experiments shown in figs. 92 and 93 and plate XIII the irradiation took place at a temperature too low for vacancies in W to have migrated, hence the holes on the surface must have resulted directly from the collision cascade. The same is not true in the case of the extra atoms however and these could either have been pro-

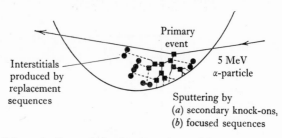

Fig. 93. Schematic diagram showing how the collision cascade produced by a 5 MeV alpha particle causes surface damage visible in the field-ion microscope. (From Brandon, Southon & Wald, 1962.)

duced by migration of interstitials from the interior, or by the arrival of replacement sequences at the surface (Brandon, Southon & Wald 1962).

Such experiments provide valuable confirmation of our ideas on the size of collision cascades.

In chapter 7 the electron microscopy of point defect clusters will be discussed in detail. At this point however we may anticipate one of the important results by looking at a micrograph in plate XIV which purports to show a group of vacancy clusters in Au apparently formed at the core of a single collision cascade (Merkle, 1966). The specimen had been irradiated with fission fragments. It is easy to see a parallel between this and Beeler's computed cascade of fig. 57.

Unfortunately all forms of microscopy rely on investigation of debris after the cascade has subsided and after some rearrangement has occurred. Sputtering experiments, in which one observes the atoms ejected from the surface by collision cascades, do not suffer

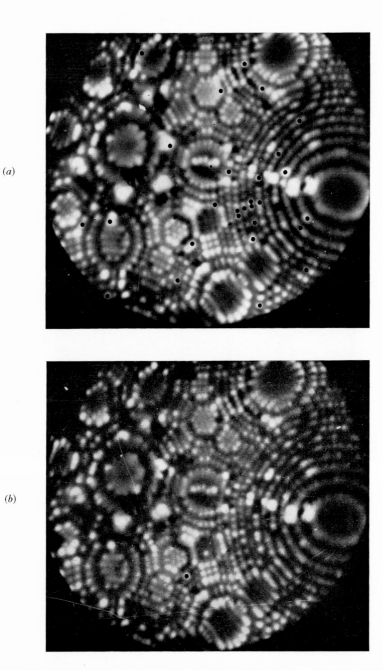

(a)

(b)

Plate XIII Field-ion micrograph of a W specimen bombarded with 5 MeV α-particles at 77 °K. (a) Before the event with atoms subsequently ejected marked by dots. (b) After the event with extra atom dotted in. (From Brandon, Southon & Wald, 1962.)

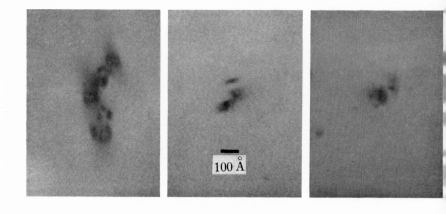

Plate XIV Electron micrographs of multiple vacancy clusters caused by individual cascades in Au under fission fragment bombardment. (From Merkle, 1966.)

from this disadvantage and have provided most of the experimental data that exists today. The remainder of this section will be devoted to such experiments.

The essential features of a sputtering experiment are a beam of energetic ions which impinge on a solid target causing ejection of atoms from the surface which are then detected. In order for the results to be easily interpreted in terms of radiation damage theory the energy of the ions should be high enough for their penetration

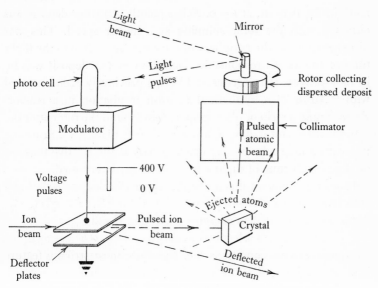

Fig. 94. The experiment to determine the energy spectrum of sputtered atoms. (From Thompson, 1963.)

depth to exceed the dimensions of the collision cascades and the maximum recoil energies should be comparable to recoils from, say, fast neutron collisions. We generally find ourselves working, therefore, with heavy ions at energies of the order of E_a (i.e. 10 to 100 keV). Such particles are easily produced by low energy accelerators.

Perhaps the most fundamental measurement one can make is of the energy spectrum of the atoms that are ejected, or sputtered. Fig. 94 shows a suitable apparatus developed for this purpose. (Thompson, 1963; Thompson, Farmery & Newson, 1968).

Atoms were sputtered from a single crystal target of either Cu or

Au by an ion beam from an accelerator and the time of flight t of the atoms between the target and a collector was measured.

After being formed into a collimated beam the ejected atoms were collected on the cylindrical surface of a rotor, magnetically suspended in vacuum and capable of spinning at speeds up to 5×10^3 rev/sec. The ion beam was pulsed in synchronism with the rotor, the pulse length being about 1 % of the period of rotation. The rotor was first exposed to the atomic beam whilst rotating very slowly so that all the atoms landed at the same point, providing a reference mark in the deposit, at $t = 0$. Subsequently a further deposit was obtained with the rotor spinning at a higher speed. This was dispersed around the periphery of the rotor, according to the flight time of the atoms responsible. Detection of the deposit was by its radioactivity, the target crystal having previously been activated with ^{198}Au or ^{64}Cu by thermal neutron irradiation in a reactor. By wrapping a strip of photographic film around the rotor after the experiment an autoradiograph was made (see plate XV). Scanning this with a microphotometer enabled $N(t)\,dt$ the relative number of atoms with time of flight t to $t + dt$ to be determined.

From this function it is a simple matter to calculate $p(t)\,dt$ and $p(E)\,dE$ the probabilities of ejection in dt at t or dE at E respectively, or the mean and median energies of ejection $\bar{E}$ and E_m.

Figure 95 shows a selection of energy spectra, $p(E)$ *versus* E, for Au under various conditions. Perhaps their most striking feature is an approximate dependence on $1/E^2$ and we shall see that this confirms an important result from the theory of collision cascades presented in § 5.1. We saw there that when we have an infinite solid in which there is a density $\dot{q}(E_2)\,dE_2$ primary recoils per unit volume per sec with energy in dE_2 at E_2, each of these produces a cascade with $\nu(E_2, E)$ atoms slowing down through the energy level E. These atoms take a time dt to slow down through the interval dE at E and

$$dt = \frac{dE}{dE/dt} = \frac{dE}{v(dE/dx)},$$

where v is their velocity dx/dt. Thus the density of atoms in dE at E due to recoils in dE_2 at E_2 is:

$$\frac{\dot{q}(E_2)\,\nu(E_2, E)\,dE}{v(dE/dx)}.$$

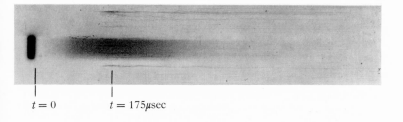

$t = 0$ $t = 175\mu\mathrm{sec}$

Plate XV An autoradiograph from the rim of a rotor exposed to a radio-active Au target whilst spinning at 1190 rev/sec.

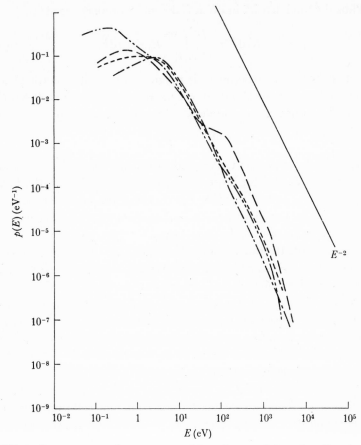

Fig. 95. Energy spectra of atoms sputtered from Au crystals in various directions.
— — — 43 keV A$^+$ on Au $\langle 100 \rangle$; —·—·— 41 keV A$^+$ on Au $\langle 110 \rangle$ (1);
— — — — 41 keV A$^+$ on Au (poly 1); —··—··— 45 keV Xe$^+$ on Au (poly 1).
(From Thompson, 1968.)

The behaviour of atoms in the cascade below $E_d\,(\gtrsim 10^3\,\text{eV})$ is such that their mean free path is roughly $D/2$ and their mean energy loss per collision is about $E/2$. Then

$$\frac{\mathrm{d}E}{\mathrm{d}x} \simeq \frac{E}{D}.$$

The function $\nu(E_2, E)$ is just the damage function of §5.1 with E replacing $2E_d$, i.e.

$$\nu(E_2, E) \simeq \frac{E_2}{E}.$$

Thus the total density in dE at E due to all primary recoils is

$$\rho(E)\,dE = \frac{D}{vE^2}\int_E^{\Lambda E_1} E_2\,\dot{q}(E_2)\,dE_2. \tag{5.77}$$

The sputtering experiment is roughly equivalent to cutting the infinite solid in half and observing the flux of atoms $\Phi(E)\,dE$ emerging. But the flux and density are simply related by

$$\Phi(E) = v\rho(E),$$

hence if we can neglect the dependence of the integral above on its lower limit E, and for most cases $\dot{q}(E_2)$ will permit this, the flux $\Phi(E)$ is proportional to $1/E^2$. The fact that the observed spectra behave in this way gives us some confidence in the theoretical model of the collision cascade developed earlier.

This simple treatment does not take account of focused collision sequences and channelling for which $\nu(E_2, E) \equiv 1$, or of the potential energy barrier which ejected atoms must surmount at the surface. The effect of introducing a binding energy E_B into the model is to cause $\Phi(E)$ to pass through a maximum near $E = E_B$ and, as E decreases to zero, to approach zero linearly (Thompson 1968). In most cases the spectra of fig. 95 have such a peak near 5 eV, which is a very reasonable value for E_B.

It is apparent from fig. 95 that there are some departures from the simple $1/E^2$ spectrum deduced from a random collision cascade model and these become even more pronounced when the experimental data are presented as time-of-flight spectra, $p(t)$ versus t.

Figure 96 shows $p(t)$ versus t for ejection from a Cu crystal in a direction well removed from focusing or channelling axes. Here one might expect the random cascade model with surface binding to fit the spectrum rather well, and indeed it does.

On the other hand if one looks at atoms ejected near simple crystallographic directions the spectra of fig. 97 shows that there are considerable differences. In the $\langle 121 \rangle$ case, where channelling might be possible, but not focusing, a peak appears at high energies which indicates a decrease in dE/dx below the value suggested in the random cascade model. Clearly this is just what would be expected if some atoms were being channelled out of the crystal with small dE/dx. Furthermore this peak increases in magnitude as the primary recoil energies in the experiment are increased, just as one would expect from the theory of channelling.

Nor can the $p(t)$ spectrum from a $\langle 110 \rangle$ direction be fitted by a random cascade theory. Figure 97 shows that an extra contribution is required in the form of a curve which rises from zero at a lower time limit t_p to a peak at about 5 t_p. One might expect such a contribution from simple focused collision sequences in Au with t_p corresponding roughly to the focusing energy.

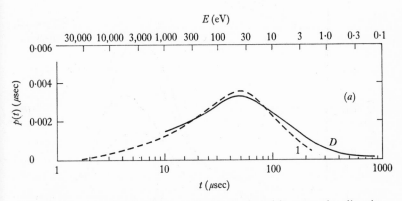

Fig. 96. The time of flight spectrum for atoms ejected in a complex direction from a Cu crystal under 47 keV A$^+$ ion bombardment (solid curve) compared with the prediction of a random cascade model.

More precisely we should make it correspond to an upper energy limit for propagation of the sequences E_p^{hkl}, which may be somewhat larger than E_f^{hkl} because defocused sequences can also make a small contribution to sputtering.

It is possible to calculate the focused collision contribution to the spectrum in the following way. (See Thompson, 1968, for a full discussion.)

The density of primary recoils per second, $\dot{q}(E_2)\,dE_2$, has been defined already and can obviously be calculated if the differential cross-section for the primary collision is known. For example,

$$\dot{q}(E_2) = n\Phi_1 \frac{d\sigma}{dE_2}. \tag{5.78}$$

Suppose that each recoil at E_2 produce $X(E_2, E'')\,dE''$ atoms with energies in dE'' at E'' and moving in the right direction to create a

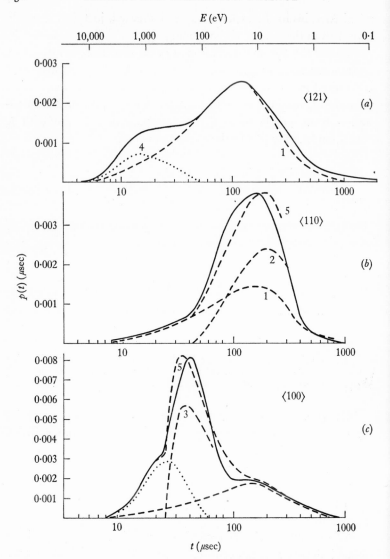

Fig. 97. Time of flight spectra from Au under bombardment with 42 keV A^+ ions. (a) Ejection in a $\langle 121 \rangle$ direction, experimental data (solid curve) compared with the spectrum predicted by a random cascade model. (b) Ejection in a $\langle 110 \rangle$ direction (solid curve) fitted with a random (1) plus a focused (2) contribution with $E_t^{100} = 170$ eV. (c) Ejection in a $\langle 100 \rangle$ direction (solid curve) fitted with a random (1) plus a focused collision (3) contribution with a maximum propagation energy of 500 eV.

focused collision sequence in a particular direction. In transit to the surface, a distance r away, this sequence will lose energy and arrive with E'. Then if ΔE is the energy lost per collision E', E'' and r are related by the equation

$$r = \int_{E'}^{E''} \frac{D}{\Delta E} dE. \tag{5.79}$$

Suppose E_p is the maximum energy with which the sequence can propagate, then the maximum distance $\hat{r}$ from which a sequence can reach the surface with E' is:

$$\hat{r} = \int_{E'}^{E_p} \frac{D}{\Delta E} dE. \tag{5.80}$$

On ejection the surface atom loses the binding energy E_B and the energy E with which it is detected is:

$$E = E' - E_B. \tag{5.81}$$

With equation (5.79) this allows us to relate E with E'' at distance r.

The contribution from a slab of thickness dx at depth x ($x = r \cos \phi$, where the focusing direction makes an angle ϕ to the surface normal) to ejection into dE at E is

$$\int_{E_2} \dot{q}(E_2) X(E_2 E'') \, dr \cos \phi \, dE_2 \, dE''.$$

Then using the relations above to put E'' in terms of E and r, the total flux crossing the surface into dE at E in the solid angle $d\Omega$ can be calculated from

$$\Phi(E) \, dE \, d\Omega = \int_{E_2} \dot{q}(E_2) \int_0^{\hat{r}} X(E_2, E'') \, dr \cos \phi \, dE_2 \, dE \, \frac{d\Omega}{\Omega_f}, \tag{5.82}$$

where Ω_f is the effective solid angle into which atoms are ejected.

Further progress demands details of the functions ΔE and $X(E_2, E'')$. The first of these has already been considered for simple and assisted focusing and we know that reasonable approximations for $E < E_f$ are:

$$\Delta E \begin{cases} = \alpha DE & \text{simple focusing}, & (5.83\,a) \\ = \beta D & \text{assisted focusing}, & (5.83\,b) \end{cases}$$

where α and β are constants (5.61).

The function $X(E_2, E'')$ can be calculated following methods developed by Leibfried (1959) and Sanders & Thompson (1967).

In section 5.1 we saw how to calculate the number $N(E_2, E)\,\mathrm{d}E$ of recoils in $\mathrm{d}E$ at E generated by each primary at E_2. In the case of an inverse-square potential between primary and secondary recoil this has the form

$$N(E_2, E)\,\mathrm{d}E = \frac{E_2}{2E_\mathrm{f}^{\frac{1}{2}}E^{\frac{3}{2}}}\,\mathrm{d}E.$$

Here, we have substituted E_f for E_d since this is more appropriate as the lower energy limit of the integrals involved.

In § 5.5.3 we calculated the probability $P_\mathrm{f}(E)$ of a randomly directed recoil starting a simple focused sequence. By multiplying these two relations together one obtains the function $X(E_2, E)\,\mathrm{d}E$, giving the number of focused sequences, started in $\mathrm{d}E$ at E, in a particular direction, by a primary of energy E_2.

$$X(E_2, E)\,\mathrm{d}E = \frac{b}{2D}\frac{E_2}{E_\mathrm{f}^{\frac{1}{2}}E^{\frac{3}{2}}}\log\frac{E_\mathrm{f}}{E} \quad \text{for} \quad E < E_\mathrm{f}. \qquad (5.84)$$

This falls to zero as E approaches E_f, as one should expect. In order to simplify the calculation of a spectrum we shall identify E_f with E_p for the present.

When one is interested in the number of focused collision sequences, irrespective of starting direction, this function is simply multiplied by the number of focusing directions leaving one lattice site (12 in the case of f.c.c. simple $\langle 110\rangle$ focusing).

Now by using assumption (5.83a) to give $\hat{r}$ with equation (5.80), the general expression (5.82) can be evaluated for simple sequences giving

$$\Phi(E)\,\mathrm{d}E = \frac{b\cos\phi}{3D\alpha E_\mathrm{f}^2\,\Omega_\mathrm{f}}\left[\frac{E_\mathrm{f}}{E+E_\mathrm{b}}\right]^{\frac{3}{2}}$$

$$\times\left\{\log\frac{E_\mathrm{f}}{E+E_\mathrm{b}}+\frac{2}{3}\left[\frac{E+E_\mathrm{b}}{E_\mathrm{f}}\right]^{\frac{3}{2}}-\frac{2}{3}\right\}\int_{E_\mathrm{f}}^{\Lambda E_1}E_2\,\dot{q}(E_2)\,\mathrm{d}E_2. \qquad (5.85)$$

This can easily be converted into a time-of-flight spectrum $\Phi(t)$, using the fact that

$$\Phi(E) = \Phi(t)\frac{\mathrm{d}t}{\mathrm{d}E},$$

which is convenient for comparison with experiment. The simple focusing $\Phi(t)$ corresponding to (5.85), rises from zero at a time t_f, equivalent to the focusing energy, and passes through a peak near $5\,t_f$. Thus, ejection by low-energy sequences is strongly favoured, as one might expect from the theory leading to (5.83a) which suggests that sequences near E_f are strongly attenuated by thermal vibration.

Figure 97(b) shows how the experimental $\Phi(t)$ curve can be fitted by adding a contribution from random cascades with surface binding to one from simple focused collision sequences.

The fit is extremely good if one takes t_f corresponding to an energy $E_f = 170\,\mathrm{eV}$.

There is some doubt as to the exact meaning of E_f in this context since one is really measuring an upper energy limit to propagation E_p and this could be rather greater than E_f due to the contribution of defocused sequences.

The $\langle 100 \rangle$ spectrum is shown in fig. 97(c). This requires a focused contribution of a different type since the peak is much sharper with fewer low energy ejections and the assumption of equation (5.83b) $\mathrm{d}E/\mathrm{d}x = \beta$ appears to be appropriate since it discriminates against low energy ejections (Thompson 1968). Such behaviour is consistent with the theoretical models of assisted focusing where the energy lost per collision is expected to be almost independent of energy. The upper energy limit for propagation of $\langle 100 \rangle$ sequences is evidently at about $500\,\mathrm{eV}$ in this case.

In the case of Cu generally similar results have been obtained with focused energy limits as shown in table XXI. A comparison of spectra at $2\frac{1}{2}°$ and $10°$ from $\langle 110 \rangle$ shows an interesting feature since the apparent propagation limits are at 57 and $90\,\mathrm{eV}$ respectively. If one looks at a large angle to the $\langle 110 \rangle$ axis a relatively larger contribution must be expected from defocused sequences travelling above E_f^{110}, hence a larger energy of propagation E_p^{110} is possible. Extrapolating the data to zero angle one sees that E_f^{110} must be about $50\,\mathrm{eV}$ in Cu, which is consistent with the Brookhaven computation in table XX, and gives us further confidence in the Brookhaven method and the potential chosen by Gibson $et\ al.$ (1960) for Cu.

In §5.5.3 we saw that in simple focusing directions the replacement energy should be about half the focusing energy. Hence E_r^{110} in Cu should be about $25\,\mathrm{eV}$ and in Au about $85\,\mathrm{eV}$. The former is

TABLE XXI. *Characteristics of focused collision sequences in* Cu *and* Au *deduced from time-of-flight spectra* (*Thompson*, 1968, *Farmery &* *Thompson*, 1968)

		Cu	Au
Maximum energy of propagation (eV) $\begin{cases} E_{\text{p}}^{110} \\ E_{\text{p}}^{100} \end{cases}$		50	170
		320	500
$\langle 110 \rangle$ attenuation constant α (Å^{-1}) $\left(\dfrac{\text{d}E}{\text{d}x} = \alpha E\right)$		—	< 0·07
$\langle 100 \rangle$ attenuation constant β (eV/Å^{-1}) $\left(\dfrac{\text{d}E}{\text{d}x} = \beta\right)$		—	5·5
Maximum range of $\langle 110 \rangle$ sequences (Å)		—	> 56
Maximum range of $\langle 100 \rangle$ sequences (Å)		—	100

clearly consistent with the computed values of 17 eV for E_{r}^{110} in table XX.

The value of E_{f}^{110} deduced for Au from these experiments may be used in conjunction with the compressibility to give an interatomic potential for Au–Au. This is the one quoted in table X (Thompson, 1968).

The relative magnitudes of the random and focused contributions to the time-of-flight spectra must depend on the range of focused collision sequences and the constants α and β in the approximations to $\text{d}E/\text{d}x$. For instance where the range of simple sequences is large and α is small the focused collision peak should be large in comparison with the random peak. A rough estimate may be made of the constants α and β for Au by comparing theoretical expressions with observed spectra and some values are shown in table XXI.

These values are entirely consistent with the theoretical estimates made earlier.

Returning to the energy spectra of fig. 95 we can see how the presence of focused collision sequences modifies the $1/E^2$ expected for a random cascade. In the energy range near 100 eV the spectrum behaves like $1/E$, and this is consistent with our model of focused collision sequences for which $\nu(E_2, E) = 1$, *since no multiplication of the number of moving atoms occurs once a focused sequence is formed.* Thus, in (5.77) $\rho(E)$ becomes proportional to $1/vE$ and $\Phi(E)$ to $1/E$.

If the angular distribution of ejected atoms is determined it is found that pronounced maxima occur near crystal directions where focused collision sequences are expected. For example in the experiment illustrated in fig. 98 ejected atoms are collected on a glass plate and form ejection patterns of the type seen in plates XVI, XVII,

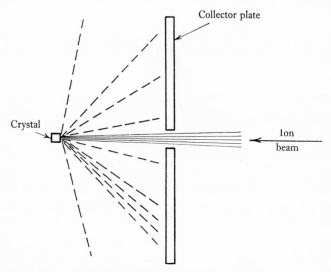

Fig. 98. The formation of an ejection pattern.

XVIII and XIX. Such effects were first reported by Wehner (1956) with ion bombardment at a few hundred eV. Since then experiments have been conducted over a wide range of energies, extending up to several MeV with qualitatively similar results (e.g. Thompson, 1959; Yurasova, Pkeshivtsev & Orfanov, 1959; Koedam 1959; Thompson & Nelson, 1961).

One might think that these experiments provided rather direct evidence of focused collision sequences, but this is not necessarily the case (Lehmann & Sigmund, 1967). Even a cascade that contains no long range focused sequences will cause an anisotropic distribution of ejected atoms because when $E < E_D$ atoms collide with near neighbours the surface atom is ejected as a result of an impulse received from a near neighbour in an adjacent layer of atoms. There will always be a tendency therefore to move off in a cone of directions around the near-neighbour axes of the crystal. However, if focused collision sequences are present one must expect a narrower range of angles, and smaller spots in the ejection pattern. This is found to be so. We have seen above that the energy spectra in high energy sputtering can only be understood completely if some ejection is due to focused collision sequences and the same is true of ejection patterns.

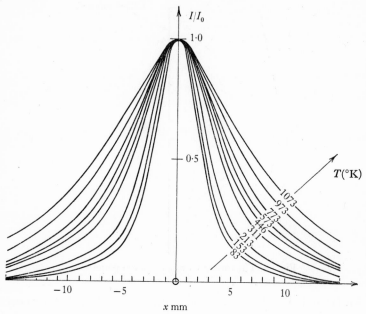

Fig. 99. The intensity profiles of ejection spots observed in the case of Au {111} bombarded with 40 keV A$^+$ ions at various temperatures. (From Nelson, Thompson & Montgomery, 1962.)

Plates XVI to XIX show that spots are observed for all the principal focusing directions considered in the previous section. The width of the ⟨110⟩ spots from Au has been measured as function of temperature and fig. 99 shows how the spots are affected by going from 83 to 1073 °K. The angular width ψ can be estimated theoretically using (5.57) and (5.58) to give $\overline{\psi}_N^2$ the mean squared angle of deviation for a sequence at energy E, and averaging this over the energy spectrum of ejected atoms.

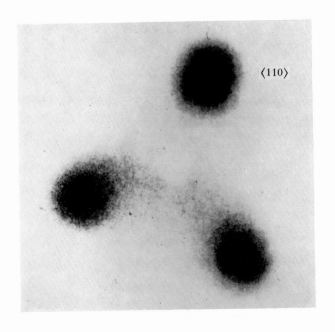

Plate XVI Ejection pattern from the {111} face of an Au crystal under 40 keV A[+] ion bombardment. (From Nelson, Thompson & Montgomery, 1962.)

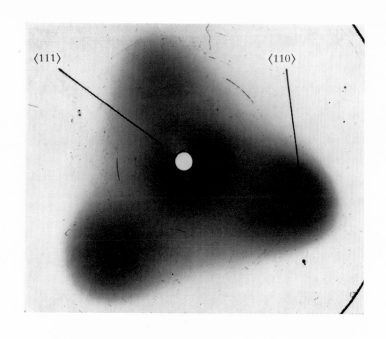

Plate XVII Ejection pattern from Al {111} under 40 keV A⁺ bombardment. (From Nelson & Thompson, 1962b.)

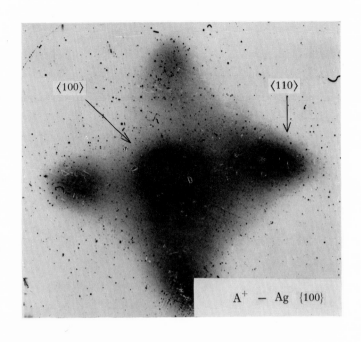

Plate XVIII Ejection pattern from Ag {100} under 10 keV A⁺ ion bombardment.
(From Thompson & Nelson, 1961.)

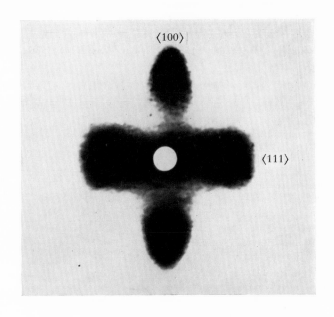

Plate XIX Ejection pattern from W {110} under 40 keV A⁺ bombardment. (From Nelson, 1963.)

Figure 100 compares the theoretical widths, as a function of temperature, with those deduced from ejection patterns by Chapman & Kelly (1967). These experimental widths agree quite well with a theory which includes focused collision sequences.

Ejection patterns made under conditions where primary recoil energies were about 50 keV show evidence of channelling in the

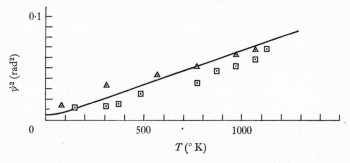

Fig. 100. Angular widths of spots, such as those in fig. 99 as a function of temperature, compared with theory assuming ejection by focused collision sequences. (From Chapman & Kelly, 1967.)

cascade. Figure 101 shows a contour map of the intensity in a spot pattern from a {110} face of a Cu crystal under A$^+$ ion bombardment at three energies. If a section is taken through the mid point of the streak joining two ⟨110⟩ spots, the intensity distribution of fig. 102 is found. The relative intensity of this streak clearly increases as the starting energy of the cascade increases. The energy spectrum for this direction of ejection, which is close to a ⟨121⟩ axis, shows an increase in the high energy ejections. This must indicate that some Cu recoils in the cascade are channelled. They are relatively few in number, however, and even in the ⟨121⟩ direction cause less than 1 % of all ejections.

240

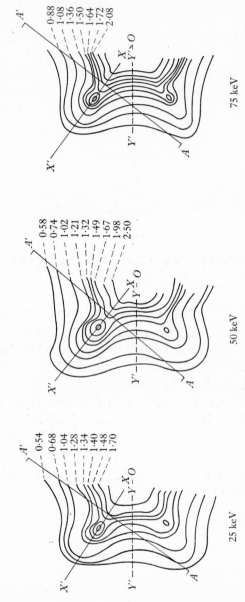

Fig. 101. Contour maps showing the relative density of Cu atoms in three deposits obtained by bombarding a {110} crystal plane with A⁺ ions of 25, 50 and 75 keV. The densities have been scaled to make the three distributions fit along the line YY′. (From Nelson & Thompson, 1962a.)

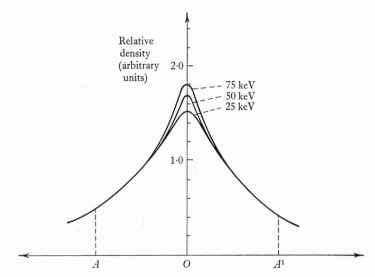

Fig. 102. Relative density, after scaling, plotted along AA' to show the enhancement of the {111} streak at higher energies of bombardment. (From Nelson & Thompson, 1962 a.)

5.6. Revision of the damage function $\nu(E_2)$ to account for all dynamic effects

The effect of focusing is to prevent further multiplication of the cascade, for once the momentum is focused, collisions are either head-on, leading at the most to replacement, or else they are glancing and not energetic enough to create displacements. Both the computer and the sputtering experiments show that below E_f^{cpd} much of the energy in a cascade is focused into the closest-packed directions (denoted by superfix cpd). In the simple cascade theory one should therefore use either $2E_d$ or E_f^{cpd} as the limit of multiplication, whichever is greater. If $E_f^{\text{cpd}} > 2E_d$ then $\nu(E_2) = E_2/E_f^{\text{cpd}}$, otherwise the original equation will hold. For the present we neglect the effect of channelling and electron excitation.

In the case of Cu or Fe we have seen that $E_f^{\text{cpd}} \simeq 2E_d$ and the numerical value of $\nu(E_2)$ is little affected. In Au however, E_d and E_r^{100} are about 35 eV, whereas $E_f^{110} = 170$ eV, and $\nu(E_2)$ must be about half the value predicted by the simple theory. To generalize, one expects little effect on $\nu(E_2)$ in the light or medium-weight elements but in heavy elements a reduction of 50 % should be typical.

The range of replacement sequences determines the separation between vacancy and interstitial. One expects a maximum of $30D^{110}$ in Au and $15D^{110}$ in Cu. Average values will obviously be much less than this, but the trend with Z_2 clearly favours wider separation in heavy elements.

Lattice vibration, and hence temperature, exerts an important influence on both $\nu(E_2)$ and i–v spacing. In the first place it reduces the effective value of E_f by reducing the focusing effect, and $\nu(E_2)$ becomes closer to $E_2/2E_d$ the higher the temperature. Secondly it reduces the range, directly reducing the i–v spacing. Also, by its defocusing effect it may permit subsidiary replacements to branch out from the track of the main collision sequence, making the damage even more concentrated and $\nu(E_2)$ closer still to $E/2E_d$.

Another effect which can be important is the interaction of focused collision sequences with defects in the crystal. This will make $\nu(E_2)$ depend on both the natural defect structure and also on the accumulated radiation damage. Perhaps its main importance lies not in the effect on $\nu(E_2)$ but on the likelihood that i–v pairs will form preferentially near other defects.

We have now reached the point where the dynamic effects, focusing, channelling and electron excitation, can be accounted for together in a generalized damage function. We take $E_2/2E_d$ as the basic expression, replacing $2E_d$ by E_f^{cpd} where appropriate to allow for focusing. Electron excitation is introduced by replacing E_2 with the collision energy E_c, defined in § 5.3. Channelling is introduced in a crude way by raising the whole expression to the power $(1-2c)$, as described in § 5.4.4. We then have:

$$\left(\frac{E_c}{E_f^{\text{cpd}}}\right)^{1-2c} < \nu(E_2) \leqslant \left(\frac{E_c}{2E_d}\right)^{1-2c} \quad \text{for} \quad E_f^{\text{cpd}} > 2E_d. \quad (5.86)$$

When $E_f^{\text{cpd}} < 2E_d$ the equality applies.

The effect of lattice disturbance, whether by vibration or defects, will shift $\nu(E_2)$ towards an upper limit of $(E_c/2E_d)$.

5.7. The end of the cascade

5.7.1. *Athermal rearrangements in a single cascade.* The processes by which atoms are displaced have been considered and we now have some idea of the number of displaced atoms $\nu(E_2)$ and the distribution of interstitials relative to vacancies. The next stage is to determine whether local rearrangements can occur as the disturbed region settles down. We shall not be concerned here with the rearrangements that occur by thermal activation at the ambient temperature of the crystal, but only with processes that occur within a short time, say 100 lattice vibration periods or 10^{-11} sec., of the primary event. The cascade stops displacing atoms at about 10^{-13} sec. this being the approximate lifetime of a long-range focused collision sequence (see § 5.5.5). For convenience the rearrangements are divided into two categories, *athermal* and *thermal*, the second of which is discussed in the next section.

An athermal rearrangement occurs whenever two defects are formed close enough together for coalescence to occur, without any thermal activation being required. If the two are interstitial and vacancy, annihilation results, otherwise a small cluster is formed. In the cascade calculation we did not count unstable i–v pairs because E_d was defined as the energy to produce a *stable* i–v configuration. But we did not introduce the possibility of different branches of the cascade overlapping and, for instance, an interstitial of one branch falling within the unstable zone of a vacancy from the other. At first sight this is most likely in the higher energy cascades and Beeler (1964b) has shown the magnitude of the effect. In a 10 keV cascade in Fe he found that almost half of the point defects were lost in athermal annihilations. This can only be taken as a rough guide, for the result depends critically on the size of the unstable zone and on the initial separation, which is underestimated in his model. There is no adequate treatment of the problem at present, either by computer or analytical methods, but a simple treatment follows which illustrates the physical principles and draws attention to some of the difficulties.

Take a single cascade in which i–v pairs are produced initially within a volume V. We suppose that the unstable zone around each vacancy has a volume v_0.

As the vacancies are being produced, consider the behaviour of the total unstable volume v when their number increases from n_r to $n_r + dn_r$ assuming a uniform distribution of vacancies. A volume $v_0 \, dn_r$ is associated with the dn_r new vacancies but a fraction v/V of this is already unstable, making the net increase of unstable volume

$$dv = \left(1 - \frac{v}{V}\right) v_0 \, dn_r$$

Integrating this from $v = 0$ to V_0 and $n_r = 0$ to ν we obtain:

$$\frac{V_0}{V} = 1 - e^{-\nu v_0/V}.$$

This is the fraction of the total volume V which is unstable. If the interstitials are now introduced randomly and we neglect the possibility that more than one is within the unstable zone of a particular vacancy, then this is also the fraction of interstitials which are annihilated. Hence the number of interstitials or vacancies which survive is

$$\nu' = \nu \, e^{-\nu v_0/V}. \tag{5.87}$$

It must be emphasized that the assumptions are not strictly justified. Because the interstitials and vacancies tend to form in separate zones (5.87) will overestimate the athermal effect. The fact that two or more interstitials may compete for a particular vacancy will also lead to an overestimate by this method. However, it will serve to give the magnitude and the rough behaviour of ν' with E_2.

In a typical 10 keV cascade in Fe we have seen that $\nu \sim 10^2$ and the cascade affects a volume containing $\sim 10^5$ atoms. In chapter 2 we saw that the unstable volume contains about 10^2 atoms, making the ratio $v_0/V \sim 10^{-3}$. Then ν' is of the order of 10 % less than ν, in agreement with Beeler's result. Next we consider the energy dependence of the exponential factor. Two extreme assumptions can be made, first that the linear extent of the cascade is proportional to energy, hence that $V \propto E_2^3$. This might be approached in lighter elements where the energetic recoil ranges are much longer than focused collision sequences and therefore determine V. Secondly, one could assume that V is independent of energy, which might be approached in the heavy elements where the size of the cascade depends on the range of focused collision sequences which are larger than energetic recoil ranges. In both cases we can take $\nu = E_2/E_f$ in the exponent. Then:

(1) $V = C_1 E_2^3$ (light elements)

$$\nu' = \nu \exp\left(-\frac{v_0}{C_1 E_f}\right) \cdot \frac{1}{E_2^2}.$$

As recoil energy increases the exponential approaches unity and the effect should eventually disappear. This is perfectly reasonable for as the cascade spreads out further the chance of branches overlapping diminishes.

(2) $V = C_2$ (heavy elements)

$$\nu' = \nu \exp\left(-\frac{v_0}{C_2 E_f}\right) E_2.$$

Here, the more defects one puts into the limited volume, the greater the fraction annihilated athermally.

5.7.2. *Thermal spikes.* In the early days of radiation damage much attention was devoted to the local heating effect which must follow as a collision cascade subsides. The question was, whether or not this could produce any damaging effects or rearrangements.

Before talking about 'heating' and 'temperature' in this situation one must first establish their applicability on such a small physical scale. If an assembly of atoms is isolated from its surroundings and the atoms are able to exchange energy with each other, a state of dynamic equilibrium will be reached in which the energy distribution is described by a Maxwell–Boltzmann function. In this function the temperature appears as a parameter which determines the mean energy by $\bar{E} = \frac{3}{2}kT$.

Suppose we draw an imaginary surface around the region to be filled by a collision cascade. In its earlier stage the energy distribution is nothing like the Maxwell–Boltzmann, as all the energy is concentrated on a few atoms. Once the energy has been spread by the focused collision sequences though, the sharing process can begin properly and, once each atom has had time to interact with its neighbours several times, the Maxwell–Boltzmann distribution will be approached.

We must investigate some orders of magnitude before proceeding further. The lifetime of a focused collision sequence will provide a magnitude for the cascade duration. This is roughly the time required for an atom of E_f to travel a distance $n(E_f) D$ and is hence given by

$$n(E_f) D \sqrt{(M_2/2E_f)}.$$

This ranges between 10^{-12} and 10^{-13} sec, the smallest values being expected in light elements. Once this violent stage is over, the atoms communicate in times of the order 10^{-13} sec; the atomic vibration period. Thus, after a time between 10^{-12} and 10^{-13} sec. the energy distribution will approximate the Maxwell–Boltzmann and it makes sense to think in terms of temperature and heating. The region is then referred to as a *thermal spike* (Seitz & Koehler, 1956).

The next question to consider is the rate at which heat leaks out of the spike into the rest of the crystal. We suppose that both electrons and ion cores receive energy in the cascade and that the ensuing disorder justifies a liquid model being used. We assume that energy is transmitted in two-body collisions of three types: ion–ion, electron–electron and ion-electron. As the first two are between particles of equal mass, the transfer is very efficient and up to 100 % can be exchanged in a single collision. The third type, however, is very inefficient, due to the widely different masses, the maximum fraction transferred being $4m_0/M_2$.

The electrons and ion cores can therefore be regarded as separate systems, loosely coupled by ion–electron collisions. The mean time between collisions for an electron with velocity v_e, in a material with interionic spacing D, will be $\sim D/v_e = 5 \times 10^{-16}$ sec for 1 eV electrons; whereas that between ion–ion collisions is

$$\sim D/v_i = 10^{-13} \text{ sec}$$

at 1 eV. The spike electrons must therefore dissipate their energy into the surrounding crystal long before ion–ion collisions take effect. Collisions between 'hot' ions and 'cool' electrons would require more than $(4m_0/M_2)^{-1}$, i.e. between 10^4 and 10^5, collisions to cool the ion system. The relaxation time for energy to pass between the two systems is therefore greater than 10^{-11} sec. This time will be smallest in light elements, largest in heavy elements, in proportion to M_2.

For the moment, we neglect the loss of ion-core energy to the electron system and consider their cooling by ion–ion collisions only. If we can show this cooling time to be less than 10^{-11} sec the electrons can be disregarded. From classical theory of heat conduction it is well known that a hot region of radius r_0 in an infinite medium cools with a characteristic relaxation time of the order τr_0^2; where $\tau = c\rho/4K$, c being the specific heat, ρ the mass density and K the thermal conductivity. Because of the local disorder the appropriate value of K will here be intermediate between that of a glassy solid ($\sim 10^{-3}$ cal.cm^{-1}sec^{-1}deg^{-1}) and that due to a crystal lattice ($\sim 10^{-1}$ cal.cm^{-1}.deg^{-1}). The use of the macroscopically deter-mined value for a metal would be quite wrong, since this is domi-nated by electronic heat transport and this we have shown not to remove heat from the ion cores in times less than 10^{-11} sec. Taking $K = 10^{-2}$ and $r_0 = 100$ Å the decay time is of the order 10^{-11} sec. Since the loss to the electron system is slower than this it seems justifiable to neglect ion-electron losses as a first approximation. It is important to notice that the spike duration depends on r_0^2 and because the thermal concept cannot be applied for times less than about 10^{-12} sec, small cascades cannot be said to generate thermal spikes and the concept is only valid either for energetic recoils, or crystals where focusing ranges are long. For example, with the numerical values used above, the duration is 10^{-12} sec when $r_0 = 30$Å and *this is about the smallest cascade for which the thermal concept is applicable.* An estimate of the temperature in a spike T_1 is made by assuming the recoil energy E_2 to be shared amongst atoms in the volume $\frac{4}{3}\pi r_0^3$. Then if n is the atomic density the mean energy increase of spike atoms is

$$E_0 = E_2/\tfrac{4}{3}\pi r_0^3 n$$

and, if the ambient temperature of the crystal is T_0, the spike temperature is

$$T_1 = T_0 + E_0/\tfrac{3}{2}k. \tag{5.88}$$

With $r_0 \simeq 100$ Å and $E_2 = 40$ keV, E_0 is about 0.17 eV and the temperature rise about 1300 °C. This is a fairly extreme example, for only in a heavy element with a short recoil range could the energy of a 40 keV recoil be contained in such a small region (see table XVI).

In general the high spike temperature always requires this com-bination of a small recoil range and a large recoil energy. The behaviour of recoil range shown by equation (5.18) is clearly to reduce r_0 as Z increases, then since E_0 depends on r_0^{-3} the spike temperature must depend on a high inverse power of Z. On the other hand, the duration of the spike depends on r_0^2 and we can expect large cool spikes in light elements that persist for times of

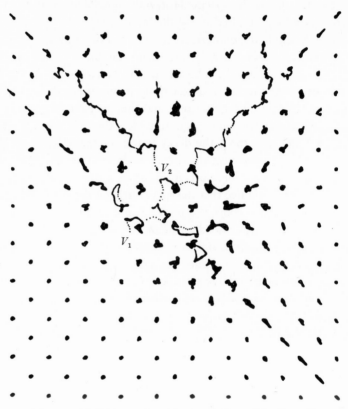

Fig. 103. (a) Orbits produced by a 100 eV recoil in Cu at V_1. Time runs from zero to $6 \cdot 5 \times 10^{-13}$ sec.

10^{-11} sec or more. Conversely heavy elements will have small, short-lived, hot spikes.

A more sophisticated approach to this problem has been made by the Brookhaven computer method (Vineyard, 1961). A recoil was started with 100 eV in a model representing Cu and the events followed after the cascade was over. The effective temperature was found for each atom by dividing its mean kinetic energy by $\frac{3}{2}k$ and

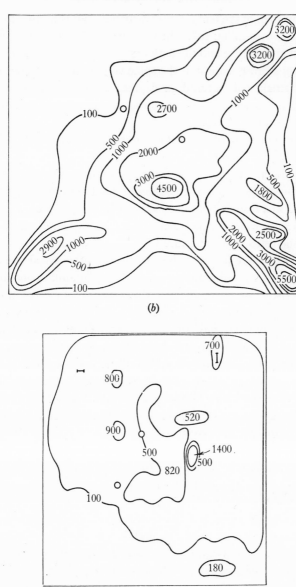

Fig. 103. (b) Isothermals at $t = 3\cdot27 \times 10^{-13}$ sec following the cascade shown in (a). (c) Isothermals at $t = 9\cdot9 \times 10^{-13}$ sec following the cascade shown in (a). ((a), (b) and (c) from Vineyard, 1961.)

a contour map of the temperature distribution was plotted for three times, as shown in fig. 103. The structure of the cascade is retained by the thermal spike and the paths of $\langle 110 \rangle$ collision sequences can be clearly seen. For most of the spike volume the temperature rise is quite small, only a hundred degrees or so, and there is good quantitative agreement with the spike picture presented in order of magnitude terms above.

Experimental evidence for thermal spikes is hard to obtain. Many attempts have been made to observe physical changes due to the heating effect, such as phase transformation. But in these experiments there is generally the possibility that the changes could have occurred as a result of defect migration and aggregation.

Because sputtering experiments deal with dynamic effects rather than a modified solid structure they are better suited for the task. If a thermal spike forms at the free surface of a crystal one might expect some evaporation to occur which should contribute a characteristic low-energy peak to the spectrum of ejected atoms.

Using the techniques described in § 5.5.5 the spectra shown in fig. 95 were obtained from Au. These compare the effect of 40 keV A^+ and 40 keV Xe^+ ion bombardments, which give respectively a 20 keV and 40 keV maximum recoil energy. If ejection were due solely to random cascades, focused collision sequences or channelling, one would expect the low recoil energy to give the most intense spectrum at low energy. In fact the reverse is observed and the peak moves to lower energy, which is consistent with the high-energy recoil producing the hottest spike and hence the most evaporation. Perhaps it should be pointed out that the spectrum below 1 eV accounts for less than 10 % of the total at room temperature and the conclusions of § 5.5 about focused collision sequences are not invalidated.

A theoretical model has been developed to predict the shape of the thermal spectrum from Au and its magnitude (Thompson & Nelson, 1962). The assumption is made that r_0 is independent of E_2, which is only for heavy elements where one expects the recoil ranges to be smaller than focused collision ranges which must therefore play the greatest part in determining r_0. Because of the strong exponential dependence of evaporation on temperature, only the very hottest spikes caused by recoils with energy approaching $\hat{E}_2$ contribute significantly. The position of the peak, which corresponds to temperature, is then determined by $\hat{E}_2$ and r_0, and the area under it depends on E_B, the binding energy.

Using r_0 and E_B as fitting parameters, the predicted spectrum matches the observed when $r_0 = 110$ Å and $E_B = 4 \cdot 2$ eV, when Au is bombarded with 43 keV Xe^+ ions. The observed peak corresponded to a temperature rise of 1250 °C. With 43 keV A^+ ions the maximum recoil energy was reduced from 42 to 24 keV. The same model then

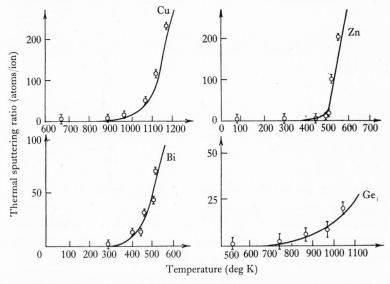

Fig. 104. Theoretical curves of sputtering ratio fitted to experimental data for 45 keV Xe^+ ion bombardment of Cu, Zn, Bi, Ge. (From Nelson, 1965.)

gave $r_0 = 92$ Å, justifying the assumption of a constant radius spike for Au.

An independent confirmation of this interpretation came from an experiment in which sputtering rate was observed as a function of temperature (Nelson, 1965). Near room temperature we have seen that focused collision sequences and random cascades are the dominant mechanisms of sputtering. Although the focusing range is slightly affected by temperature above Θ_D one should not expect any dramatic effect, and none is found. But as T_0 rises, there should come a point where T_1 is great enough for evaporation to overtake collisions. The sputtering rate should then rise rapidly with temperature. Figure 104 shows curves for a series of metals which all exhibit this behaviour. By curve fitting, the temperature rise $(T_1 - T_0)$, radius and duration can be deduced with the results shown in the table XXII. The values of E_B are taken as sublimation energies.

TABLE XXII. *Thermal spike parameters deduced from sputtering experiments with* 45 keV Xe$^+$ *ions (after Nelson, 1965)*

Metal	E_b (eV/atom)	$\hat{E}_2$ (keV)	Temperature rise (°K)	Radius (Å)	Duration (10^{-12} sec)
Au	3·7	43	910	110	3
Ag	3·0	45	530	134	6
Cu	3·5	38	49	130	5
Zn	1·4	39	150	250	1
Bi	1·9	42	600	128	9
Ge	3·9	41	1060	95	3

It will be seen that the values deduced for Au are very similar to those in the spectrum experiment which gives confidence in the interpretation.

We now go on to examine the significance of thermal spikes in radiation damage, and particularly whether they can cause migration of the defects formed in the collision cascade. Consider a defect with an activation energy for movement of U_m. The number of jumps it could make in a spike lifetime of τr_0^2 would be

$$\nu \tau r_0^2 \exp\left(-U_m/kT_1\right)$$

with $\nu = k\Theta_D/h$. Using values appropriate to 40 keV recoils in Au, which will be a fairly extreme case, the number of jumps is greater than one only if $U_m < 0.6$ eV. The quenching experiments described in chapter 2 suggest that $U_m^v > 0.6$ eV and hence that no vacancy migration could occur during the spike, unless it was the collapse of a closely spaced cluster. In the next chapter, however, we shall see that electron-irradiation experiments suggest that the interstitial can move with less than 0·1 eV. We should therefore expect some i–v recombination, and possibly some interstitial clustering, to occur in the spike lifetime.

So far as the damage function is concerned, we do not know enough about spike parameters and migration energies to predict the amount of recombination from first principles. All we can do is to note that the number of surviving interstitials and vacancies will be less than $\nu(E_2)$, and some clustering may occur due to spike heating in the 10^{-11} sec that follow the primary event.

5.8. Saturation at high doses

5.8.1. *Irradiations producing isolated i–v pairs.* The simplest forms of radiation damage occur when the primary recoil energy is close to the threshold and i–v pairs are formed singly or in small groups dispersed through the crystal. A model in which newly formed defects can be annihilated in the unstable zone of existing defects seems to describe the situation, and should be a good approximation for electrons or gamma rays in the MeV range or slow neutrons inducing an (n, γ) reaction, and a fair approximation for light charged particles. Of course we assume that the irradiation proceeds at a low enough temperature to prevent any thermally activated migration of defects.

The rate of formation of i–v pairs is given by (5.1) as

$$\dot{C}_d = \sigma_d \Phi$$

with σ_d the effective cross-section for one displacement. Since interstitials and vacancies are formed in equal numbers and the only process by which they can disappear at low irradiation temperatures is mutual annihilation, their concentrations must at all times be equal. For example, $$C_{iv} = C_i = C_v.$$

Let the unstable zone for either defect contain m lattice sites, with $m = v_0/\Omega$ in the notation of § 5.7.1. Then as C_d increases by dC_d in a crystal of N sites, the $N\,dC_d$ new interstitials have the choice of NmC_v unstable sites or $N(1 - mC_v)$ stable sites. By simple proportion it follows that the fraction formed on stable sites is $(1 - mC_v)$, and the increase in interstitial concentration is:

$$dC_i = (1 - mC_v)\,dC_d.$$

Putting $C_i = C_v$ and integrating leads directly to

$$C_i = \frac{1}{m}(1 - e^{-mC_d}). \tag{5.89}$$

An exactly similar expression holds for C_v or C_{iv}. As one might expect intuitively, the saturation concentration is $1/m$. In the next chapter we shall see how this behaviour has been observed in cases of electron or light charged particle bombardment, and that the saturation concentration is of the order 10^{-3}, implying that $m \sim 10^3$.

Lück & Sizmann ($1964a, b$) have modified this simple model to allow for the fact that two or more similar defects may lie close enough to share an unstable zone, which then contains *less than m* unstable sites per defect. This overlapping is more likely to happen at high concentrations, hence the effective value of m is a function

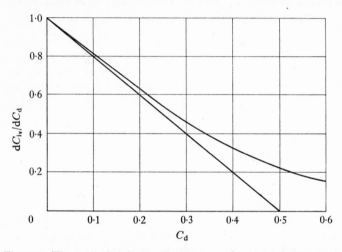

Fig. 105. The rate of defect production as a function of the concentration of displaced atoms C_d, from the calculations of Lück & Sizmann (1964). The straight line shows the behaviour of a simple saturating exponential like (5.89). The curve takes account of the possibility that several defects share an unstable zone. (From Lück & Sizmann, 1964.)

of C_d. They show that (5.89) is replaced by a function that saturates less rapidly, as one would expect. This is shown in fig. 105. We shall see in § 6.3.1 that such a curve fits the observations for a noble metal better than the simple exponential.

Sizmann also shows that the unstable zone leads to athermal clustering, even though i–v pairs are randomly introduced. This is because a defect in a cluster has a reduced effective value of m. Since m is a measure of the probability of annihilation, defects in clusters are less likely to be annihilated than those that are remote from other defects. Thus when a cluster forms due to statistical fluctuation there will be a tendency for it to persist and grow.

This model is probably realistic for gamma and electron irradiation where one might have thought that no initial clustering would occur. In the case of neutrons and charged particles we have seen in §5.6 how the structure of the cascade favours athermal formation of clusters, with the possible assistance of the thermal spike. Thus in *all* cases of irradiation there is a tendency for clusters to form in the earliest stages.

5.8.2. *Irradiations producing extended cascades.* We now consider saturation due to irradiation with particles such as fast neutrons, which form extended collision cascades in which some annealing occurs, both athermally and in the thermal spike. Saturation will be due to overlap of cascades and rigorous treatment would require consideration of further athermal annealing and the effect of repeated applications of a thermal spike to a region. This would clearly be very difficult but with the help of a simplified model some insight can be gained.

Suppose that each cascade completely saturates a volume V with defects but leaves the rest of the crystal unaffected. At a given stage of irradiation let there be N of these damaged volumes per unit volume, then

$$N = n\sigma\Phi t \tag{5.90}$$

with σ the cross-section for the primary event.

Let F be the fraction of the crystal volume affected. Increasing N by dN affects a volume $V\,dN$, of this only a fraction $(1-F)$ is **not** already saturated. The net increase in F is therefore

$$dF = (1-F)\,V\,dN,$$

hence

$$F = 1 - e^{-VN}. \tag{5.91}$$

If the assumptions are justified then one could deduce V from the shape of saturating damage *versus* dose curves. We shall see some attempts in this direction in §§6.3 and 6.4.

It is interesting to relate (5.91) to the corresponding expression (5.89) in the last section. If the saturation concentration is $\hat{C}_{1v}$ then

$$F = C_{1v}/\hat{C}_{1v}. \tag{5.92}$$

Furthermore the number of defects in V is

$$v' = nV\hat{C}_{1v}.$$

Putting this with (5.90) we find

$$VN = v'\sigma\Phi t/\hat{C}_{1v}.$$ (5.93)

Then remembering that $\hat{C}_{1v} = 1/m$ and putting $C'_d = v'\sigma\Phi t$, substitution of (5.93) into (5.91) gives:

$$C_{1v} = \frac{1}{m}(1 - e^{-mC'_d}).$$ (5.94)

This is just the same as (5.89) except for C'_d, which is simply the net concentration of displaced atoms after allowing for annealing processes within a single cascade.

So far as defect concentrations are concerned, these simple models predict no difference in the approach to saturation, once the difference between C_d and C'_d has been recognized. Any big difference between the two classes of irradiation is likely to show in the degree of clustering. Here, by concentrating defects into small volumes the probability of clustering is greatly increased and one should expect many more clusters in an irradiation with fast neutrons, for instance, than with electrons; even though C_d was the same in each case.

CHAPTER 6

POINT DEFECTS
IN IRRADIATED METALS

6.1. Introduction

The next problem is to follow the events occurring after the collision cascade has subsided, leaving its debris of lattice defects. Chapters 4 and 5 showed that these defects are often interstitials and vacancies and in chapter 2 they were shown to be capable of migration under thermal activation. To proceed with a study of damage from this stage, it is essential at first to prevent migration of defects during the irradiation, thus preserving the initial configuration for annealing experiments. This may be done by holding the specimen at a sufficiently low temperature during irradiation. At the end of this chapter some effects of migration during irradiation will be discussed.

As a criterion to decide the irradiation temperature one requires that no jumps be made by any defect in the duration of the experiment, t. Then, since the rate of jumping is approximately

$$\nu \exp\left(-U_{\mathrm{m}}^{\mathrm{r}}/kT\right)$$

by (2.9), the average time per jump is the inverse of this, which must greatly exceed the time t. For example,

$$t \ll 1/\nu \exp\left(-U_{\mathrm{m}}^{\mathrm{r}}/kT\right)$$

or
$$T \ll \frac{U_{\mathrm{m}}^{\mathrm{r}}}{k \log \nu t}. \tag{6.1}$$

With $U_{\mathrm{m}}^{\mathrm{i}} \sim 0.1\,\mathrm{eV}$, which appears to be the lowest activation energy expected in Cu, $t = 10^{6}\,\mathrm{sec}$ and $\nu = 10^{13}\,\mathrm{sec^{-1}}$; one has $T \ll 20\,°\mathrm{K}$.

One is therefore faced with a severe technical problem in making irradiations at the temperature of liquid He in order to preserve the defect configuration that exists immediately after the damage event. Historically, the earliest experiments were performed with deuterons or neutrons, but because electrons and gamma rays should produce isolated interstitial vacancy pairs, the simplest form of damage, it is advantageous to consider these first.

6.2. Electron and gamma irradiation

6.2.1. *Property changes.* Gamma rays are included here with electrons since they produce damage by forming fast electrons. However, the majority of experiments are concerned with electrons, and relatively little space is devoted to γ-irradiation.

A good irradiation experiment with any charged particles must satisfy the following requirements:

(1) The specimen must be maintained close to the temperature of liquid He (4·2 °K) during irradiation.

(2) There must be a means of raising the temperature after irradiation, for annealing at an accurately controlled temperature, after which the property change can be measured again at 4·2 °K.

(3) In order to have a well defined energy in the sample, the particles should have a range much greater than the sample thickness.

(4) The flux of particles Φ must be accurately known and be uniform over the specimen. It is generally most convenient to measure the current passing through a hole of known area and to integrate this electronically to give the dose Φt. (1 μA $= 6 \times 10^{12}$ singly charged particles per second.)

The method used by Sosin & Neely (1961) is rather typical and although it specifically refers to an electrical resistance experiment, the apparatus is easily modified for other measurements. In fig. 106 a schematic diagram shows the working principles. The specimen was in good thermal contact with a copper cooling block on which was wound an electrical heater of manganin wire. This block plugged the end of a thin-walled stainless steel tube that connected to a reservoir of liquid He, and when the tube was full of liquid the block was at 4·2 °K. In order to raise the block's temperature for an annealing experiment, a valve was closed in the reservoir to shut off the supply of liquid and current passed through the heater. When the liquid in the tube had evaporated through the vent pipe, the block was in poor thermal contact with the reservoir on account of the high thermal resistance of the thin-walled tube. The heater could then raise the temperature to any value up to 350 °K in a time less than 20 sec. On switching off the heater and opening the valve the temperature quickly returned to 4·2 °K.

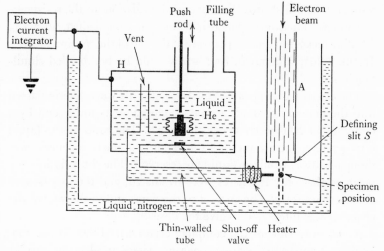

Fig. 106. Schematic view of Sosin's cryostat.

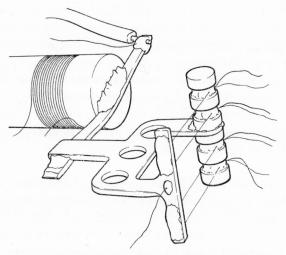

Fig. 107. Detail of the holder for resistance specimens.
(From Sosin & Neely, 1961.)

For an electrical resistance measurement thin wires are an advantage and this also satisfies requirement (2) above. Sosin stretched his wires between a sapphire post and a Cu support, both in good thermal contact with the cooling block as shown in fig. 107. Sapphire was chosen as it has a good thermal conductivity at low temperature whilst providing electrical insulation for the resistance specimens. Thermocouples using pure Cu wire against Au 3·2 % Co alloy measured the temperature at the end of the specimen wires. In the holder illustrated, four wires could be bombarded simultaneously.

The electron beam from a Van de Graaf 2 MeV accelerator passed into the cryostat through a pipe A, closed at its lower end by a defining slit S, directly above the target. The tube was at liquid nitrogen temperature and thus the majority of the beam energy was dissipated away from the liquid He reservoir. The slit S limited the beam so that only the wires were irradiated, and the flux Φ over these was uniform. The current to the wires and to the wall of the chamber was integrated to give the dose Φt.

A cut-away view of the complete cryostat is shown in fig. 108. With this apparatus, using 0·0053 cm wires, an electron current density of 5 μA cm^{-2} produced a temperature rise in the wire of less than 2 °K. The energy lost by 1 MeV electrons in 0·005 cm of Cu is about 0·1 MeV, so that condition (3) was fairly well satisfied. (Range-energy curves for various particles will be found in USAEC Nuclear Data Tables, part 3, or Bethe & Ashkin, 1953.) Some compromise was necessary here, because cooling of the wire increases with its cross-sectional area, hence its thickness.

Many experiments have been carried out with Cu, the majority using electrical resistance at 4 °K as an index of damage (see footnote* for references).

The first stage in most experiments has been to follow the property change as irradiation proceeds at a constant electron energy E_1. In all cases the change is effectively linear with dose up to $\sim 10^{18}$ electron cm^{-2} which corresponds to a concentration of displaced atoms C_d of the order of 10^{-5}. The measurement is repeated for several energies and the rate of property change plotted as a function of E_1. Figure 109 shows the results of Corbett $et\ al.$ for Cu.

In principle, an extrapolation to zero rate of increase should give

* Corbett $et\ al.$ (1957); Corbett, Smith & Walker (1959a,b); Corbett & Walker (1958, 1959); Walker (1962 (a review)); Meechan & Sosin (1959a); Luccasson & Walker (1962); Sosin (1962a,b); Sosin & Neely (1962); Sosin & Rachal (1963); Lomer & Niblett (1962); Lomer (1963).

Electron
beam

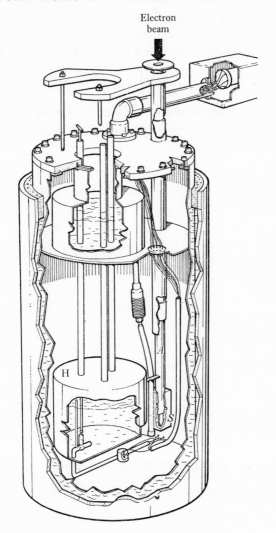

Fig. 108. Cutaway view of target box. (From Sosin & Neely, 1961.)

$\breve{E}_1$, the electron energy just sufficient to transfer the threshold displacement energy E_t in a head-on collision and thus displace an atom. Then by (4.56)

$$E_t = 2\breve{E}_1(\breve{E}_1 + 2m_0 c^2)/M_2 c^2$$

and the threshold displacement energy E_t could be determined. But because the graph is not linear the extrapolation is not straight-forward and a better procedure is to fit a theoretical curve, proportional to the concentration of displaced atoms, given by (4.58) and (5.1) as

$$C_d = t\Phi \int_{E_t}^{\hat{E}_2} \nu(E_2) \frac{d\sigma}{dE_2}\, dE_2$$

using E_t as fitting parameter. Figure 109 shows how the best fit to Corbett & Walker's results is obtained with a mean displacement energy $E_d = 22\,\mathrm{eV}$.

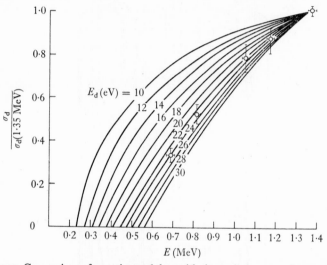

Fig. 109. Comparison of experimental data with theoretical curves of normalized displacement cross-section versus bombarding electron energy. All the curves were calculated under the assumption that the displacement probability is a step-function and the numbers refer to the assumed value of the threshold energy. Error limits include the uncertainty in normalization. (From Walker, 1962.)

The simple theory assumes a sharp threshold with $\nu(E_2) = 1$ for $2E_d > E_2 > E_d$ and $\nu(E_2) = 0$ if $E_2 < E_d$. In fact $\nu(E_2)$ must increase from 0 to 1 over a range of E_2 starting at E_t. Various forms of the function have been assumed in order to improve the fit but their choice has been somewhat arbitrary and it is not possible to find the correct function from this experiment. However, it does appear that the best fit is obtained by taking $22\,\mathrm{eV}$ as the energy at which $\nu(E_2) = \frac{1}{2}$ with $\nu(E_2) = 0$ at about $19\,\mathrm{eV}$. It seems that the mean displacement energy E_d of § 5.1 is best identified with $22\,\mathrm{eV}$ in Cu.

Then using the results of §§ 4.4 and 5.1 with this measured value of E_d one can calculate C_d, which should then be equal to the concentration C_{iv} of interstitial-vacancy pairs (hereafter referred to as i–v pairs). Knowing the resistivity change found in the experiment, one can then estimate the resistivity of i–v pairs. For Cu assuming a sharp threshold to calculate C_{iv} one obtains $1\cdot3\,\mu\Omega$ cm per $1\,\%$ i–v.

But if a gradually increasing $\nu(E_2)$ is used, the calculated C_{iv} is a little smaller, giving resistivity between $2\cdot2$ and $2\cdot5\,\mu\Omega$ cm per $1\,\%$ i–v. This seems in good agreement with an estimate made by adding the vacancy and interstitial resistivities from § 2.3, remembering that the resistivity of a closely spaced i–v pair may not be exactly the sum of the individual resistivities. Other metals have been studied in the same way and a summary of the results is contained in table XXIII. The case of Au illustrates the danger of using a step function $\nu(E_2)$ as this procedure gives a value of $0\cdot89\,\mu\Omega$ cm per $1\,\%$ i–v, whereas we know from § 2.3.10 that the vacancy alone has a resistivity of $1\cdot54\,\mu\Omega$ cm per $1\,\%$.

TABLE XXIII. *The mean displacement energy E_d and the resistivity of i–v pairs in various metals, determined in electron irradiation experiments*

Metals	E_d (eV)	$\Delta\rho_0$ ($\mu\Omega$ cm per $1\,\%$ i–v)
Al	32 (1)	3·4 (1)
Cu	22 (1)	1·3 (1)
	23 (4)	2·2 (2)
		2·5 (3)
Ag	28 (1)	1·4 (1)
Au	> 40 (1)	—
	35 (5)	0·89 (1)
Fe	24 (1)	12·5 (1)
		20 (2)
Ni	24 (1)	3·2 (1)
Mo	37 (1)	4·5 (1)
	51 (6)	—
W	> 35 (1)	—
Ti	29 (1)	42 (1)

(1) Lucasson & Walker (1962), electrical resistance, step function $\nu(E_2)$
(2) Lucasson, Lucasson & Walker (1962) electrical resistance, gradual $\nu(E_2)$
(3) Sosin (1962a)
(4) Lomer & Niblett (1962), internal friction, step function $\nu(E_2)$
(5) Bauer & Sosin (1964), electrical resistance, step function $\nu(E_2)$
(6) Lomer & Taylor (1967), internal friction, step function $\nu(E_2)$

If one proceeds with irradiations up to doses of the order of 10^{20} electron cm^{-2} the tendency to saturate, predicted in § 5.8.1, can be observed (Dworschak, Schuster, Wurm & Wollenburger, 1967). Accepting the interpretation in terms of an unstable zone for annihilation surrounding a point defect, the (5.89) may be used to fit the experimental results. Hence the value of m, the number of lattice sites in the unstable zone as defined in § 5.8.1, has been deduced for Al, Cu and Au to be 360, 620 and 270 respectively.

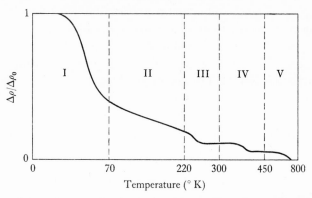

Fig. 110. Schematic curve for the isochronal annealing of resistivity showing Cu annealing stages. (From Holmes, 1964.)

6.2.2. *Recovery.* Annealing of electron irradiation damage in Cu occurs over a very wide range, from 14 to about 600 °K. Recovery can be divided into a number of stages, labelled I to IV, whose approximate temperatures are shown in the schematic recovery curve of fig. 110. In electron irradiation I and III are the most important, accounting for roughly 90 and 10 % of the recovery in resistivity, respectively. In metals other than Cu similar stages can be identified by their reaction kinetics and their temperature relative to the melting point, but we shall return to this point later, in § 6.4.2.

In the case of electron irradiated Cu, many annealing experiments similar to those described in § 2.3.8 have been carried out. A typical isochronal recovery curve shows in fig. 111 that Stage I contains five substages, Ia, Ib, Ic and Id and Ie. Their properties and behaviour under various conditions have been exhaustively studied in several laboratories and some of the results are summarized in table XXIV.

We already know from equilibrium and self-diffusion experiments (§ 2.3.10) that U_m^v must be near 0·9 eV in Cu. Furthermore, in § 4.4 we found that electrons around 1 MeV can transfer only enough energy to produce isolated interstitial-vacancy pairs. Vacancy migration cannot possibly be responsible for Stage I, which strongly suggests interstitial migration. The observed activation energies near 0·1 eV are certainly in accord with the predicted values of U_m^i from chapter 2.

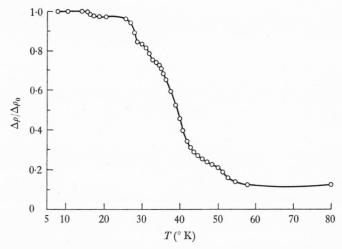

Fig. 111. Isochronal annealing. The experimental points were obtained by pulse-heating the specimen and holding it for ten-minute periods at successively higher temperatures—each ten-minute annealing being followed by a resistivity measurement at 4·2 °K. (From Walker, 1962.)

Stages I a, I b and I c behave similarly and since $j \sim 1$ it is attractive to interpret them as recombination of close i–v pairs, with slightly different spacings. This is consistent with their first order reaction kinetics, since an isolated close pair can be regarded as a single defect which, when activated, disappears without involving any other defect in crystal. Hence the rate of disappearance depends on the concentration to the first power, as shown in § 2.2.2. If this interpretation is correct the activation energies in I a, I b and I c must be rather less than U_m^i, due to the attractive strain field between interstitial and vacancy.

TABLE XXIV. *Recovery stages in electron irradiated* Cu

	Ia	Ib	Ic	Id	Ie	II	III	IV	V
Approximate temperature (°K)	16 (1)	28 (1)	32 (1)	39 (1)	53 (1)	not in pure Cu (1)	300 (7, 8) / 235 (4)	350–600(7)	—
Activation energy (eV)	0·050 (1)	0·085 (1)	0·095 (1)	0·12 (1)	0·12 (1)	—	0·60 (4, 7)	—	—
Order of reaction (x)	1 (1)	1 (1)	1 (1)	1 (1)	2–3 (1, 2)	—	2 (7)	—	—
Number of jumps (j)	1 (1)	1 (1)	1 (1)	10 (1)	10^4 (1)	—	10^6 (7)	—	—
Effect of increased dose		none (1)			reduces temp. (1)		—		
Effect of increased impurity	increase (1, 3)	small reduction (1, 5)	reduction (1, 3)	large reduction (1, 5)		introduces peaks (8)	affects magnitude (8)		
Effect of increased electron energy		reduction (1, 3)		increase (1, 3)					
Change in dislocation pinning due to stage		none (4, 6)			increase (4, 6)		decrease (9)		
Percentage recovery of resistivity	90 % (1, 2)					9 % (1, 2)	1 % (1, 2)		

(1) Corbett *et al.* see review by Walker (1962)
(2) Meechan, Sosin & Brinkman (1960)
(3) Sosin (1962a)
(4) Sosin (1962b); Keefer & Sosin (1964)
(5) Sosin & Neely (1962)
(6) Lomer (1963)
(7) Meechan & Brinkman (1956)
(8) Martin (1961)
(9) Keefer & Sosin (1964)

The influence of impurities and dose ($\equiv C_{iv}$) on the first three substages is small, which supports this picture. Ib and Ic are reduced in relative importance when higher bombarding energies are used, presumably due to the interstitial being thrown further from its vacancy, resulting in fewer close pairs. Ia, however, increases slightly at higher energies, here the interpretation is less straightforward and will be dealt with later (Sosin, 1962a).

Stages Id and Ie are different from the earlier stages and although they share an activation energy of 0·12 eV each has characteristic properties. The fact that they occur at different temperatures with the same activation energy shows that their reaction kinetics must differ. Closer investigation shows Id to require of the order of 10^2 jumps and to be roughly first order, whereas Id needs $\sim 10^6$ jumps and an order of reaction of 2 or 3, depending on the method of analysis. Both substages are sensitive to impurities and Id, having a high order of reaction, is sensitive to damage concentration and hence dose.

It has been suggested that they are due to free migration of interstitials through the lattice with $U_m^i = 0.12$ eV; these being the interstitials that were thrown well clear from the vacancy, perhaps by a focused replacement sequence. Because of its small number of jumps and first order kinetics, Id could be due to the return of the interstitial to its own vacancy: *correlated recombination*. The properties of Ie are consistent with a long range migration, probably leading to annihilation of a vacancy other than its own. Since $C_1 \propto C_v$ at all times in this simple model, and the reaction rate depends on $C_1 C_v \propto C_1^2$, one expects second order kinetics.

Confirmation of this model comes from measurements of elastic modulus and internal friction during the annealing through Stage I. The modulus measurements were made by Sosin & Neely (1962), and by Keefer & Sosin (1964). The decrement due to internal friction was measured by Lomer (1963). Both experiments used frequencies in the kilocycle region and amplitude dependence of the decrement showed that breakaway was the operative loss mechanism (see §§ 3.1.6 and 3.1.7). No change in decrement or modulus occurred until Stage Ie was reached when the measurements indicated an increase in pinning points, presumably due to the arrival of point defects. This is clearly consistent with long range migration of interstitials in Stage Ie and correlated recombination in earlier substages.

Stage II is very weak in pure Cu and only appears when foreign atoms are introduced. Martin (1961) has shown that the recovery spectrum in Stage II is a series of peaks (see fig. 112). Their number,

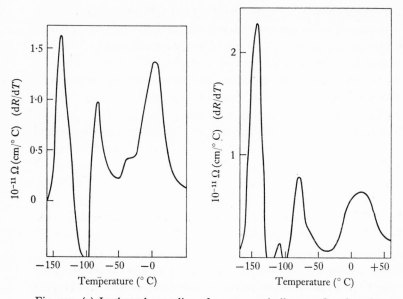

Fig. 112. (*a*) Isochronal annealing of spectroscopically pure Cu after electron bombardment at -196 °C. Dose $= 3 \times 10^{17}$ electrons/cm². (*b*) Isochronal annealing of 0·05 atomic % Ag–Cu after electron bombardment at -196 °C. Dose $= 3 \times 10^{17}$ electrons/cm². (From Martin, 1961.)

positions and magnitudes depend on the nature and concentration of solute atoms. This can be explained in terms of interstitials from Ie becoming trapped on solute atoms, by the mechanisms described in §2.6, to be released in Stage II when sufficient activation energy is available. This model is supported by the fact that Stages Id and Ie are reduced in dilute alloys, presumably because trapping prevents annihilation of vacancies.

Even in pure Cu, interstitials will disappear from the crystal without annihilating vacancies. Either they could escape to the boundaries, become absorbed into dislocation lines or perhaps form clusters. An indirect result of this will be to leave an excess of vacancies, providing an obvious mechanism for Stage III; the migration of vacancies to boundaries, network dislocations, trapped interstitials, or to form clusters. There are two objections to this, first the activation energy of 0·6 eV is rather low, since $U_m^v = 0·94$ eV from § 2.3.10. Secondly, any migration to a large fixed concentration of sinks, such as boundaries or dislocations, would proceed by a first order reaction whereas the observed order is 2. The kinetics of clustering would certainly be more complex than first order, but if such clusters are formed after electron bombardment they have so far eluded detection in the electron microscope, setting their maximum size below 20 Å. Vacancy migration to trapped interstitials should show second order kinetics, but there remains the first objection about activation energy.

An alternative scheme, originally proposed by Seeger (1958) and developed amongst others, by Meechan, Sosin & Brinkman (1960), and Bauer, Seeger & Sosin (1967) apparently resolves this difficulty, but only at the expense of considerable complication. Two stable forms are postulated for the interstitial with greatly differing migration energies. One is supposed to migrate freely in Stage Ie with $U_m^i = 0·12$ eV and the other in Stage III with $U_m^i = 0·6$ eV. These are tentatively identified in the earlier models with the $\langle 110 \rangle$ crowdion and the $\langle 100 \rangle$ dumb-bell respectively, mainly because no Stage I recovery is observed after cold-work at 4·2 °K, and it is difficult to see how such a treatment could produce crowdions, though dumb-bells are a possibility. Then, Stage III is explained as dumb-bell migration to vacancies, with second order kinetics. If the majority of vacancies are annihilated in this process, very few are left for Stage IV, where they migrate with $U_m^v = 0·94$ eV.

Stages Ib and Ic are attributed to close dumb-bell-vacancy pairs and Ia to a close crowdion-vacancy pair. The first two types of pair are supposed to result from $\langle 100 \rangle$ focused replacement sequences depositing their interstitial at two different distances from the vacancy, whilst Ia is due to a $\langle 110 \rangle$ focused replacement. Then if one makes the plausible assumption that $E_r^{110} > E_r^{100}$, the fact that Ia increases with electron energy whilst the other two decrease can be understood. Stages Id and Ie are explained as mixtures of other close pair configurations and long range crowdion migration.

One objection to this scheme is the high migration energy

required for the second form of interstitial, but for this there is some theoretical justification (Bauer, Seeger & Sosin, 1967) and it may be substantiated by other experiments, to be described here and in the next chapter. Following from this, it is difficult to understand why there should be such a large gap between the migration energies associated with the close dumb-bell-vacancy pairs and the free dumb-bell. Another difficulty is that for recovery to be 99 % completed after Stage III, nearly all the vacancies must be annihilated by

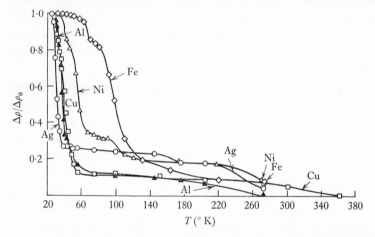

Fig. 113. Recovery of several metals bombarded at 20·4 °K with 1·5 MeV electrons. The recovery curve is of the isochronal type obtained by pulsing the samples to the specified temperatures and holding then there for 10 min. The samples are then quenched to 20·4 °K for measurement. (From Walker, 1962.)

dumb-bells. This permits only a surprisingly small fraction of interstitials to escape into other types of sink.

In general these higher temperature stages are very inadequately understood and much more work is required.

Several other metals besides Cu have been annealed after electron irradiation and resistivity recovery curves are shown for some of these in fig. 113. In all cases the majority of the resistivity recovers below 100 °K, and in this low temperature region substages are generally present. Au and Al have been more thoroughly studied than the rest, and their recovery is summarized in table XXV, in comparison with Cu.

In Al only two substages at 18 °K and 35 °K are resolved. Figure

TABLE XXV. *A comparison of resistivity recovery stages in Al, Cu and Au after electron bombardment*

	Stage I			Stage II			Stage III		
	Al	Cu	Au	Al	Cu	Au	Al	Cu	Au
Approx. temperature (°K)	14–50	16–55	<45	140	Absent	45–260	140–270	300	—
Percentage recovery	85	90	28	5	0	38	10	9	—
Activation energy (eV)	—	0·05–12	0·02–0·15	0·22	—	—	0·45	0·6	—
Order of reaction	—	1–3	1?	1	—	—	2	2	—
Number of jumps	—	1–10⁶	—	—	—	—	10⁶	10⁶	—
	(1, 11)	(1, 2)	(12, 13)	(11)	(1)	—	(11)	(7)	—

	Al	Cu	Au
$U_{\mathrm{m}}^{\mathrm{v}}$ from Table II (eV)	0·7	0·94	0·83

Additional references to those in table XXIV.

(11) Sosin & Rachal (1963)
(12) Ward & Kauffman (1961)
(13) Bauer *et al.* (1962); Bauer & Sosin (1964*a*, *b*, *c*)

114 shows the complete recovery curves for pure Al and several dilute alloys. It is clear that the behaviour of Al is rather similar to Cu since impurities evidently transfer recovery from Stage I to Stage II, without affecting the early part of Stage I.

In Au Stage I is very complicated and at least ten substages exist. Bauer and colleagues analysed their set of isotherms to produce the *activation energy spectrum* shown in fig. 115 plotting $d\Delta\rho_0/dU_m$ versus U_m, where $d\Delta\rho_0$ is the resistivity recovered in the interval

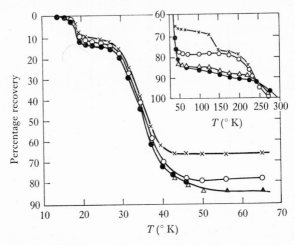

Fig. 114. Isochronal recovery of the electrical resistivity of Al following irradiation near 4·2 °K, the time at each temperature was 5 min. Dose $= 4.05 \times 10^{17}$ electrons/cm²; ● pure Al; × Al+0·1 % Zn; ○ Al+0·1 % Zn; Δ pure Al. (From Sosin & Rachal, 1963.)

dU_m at U_m. This appears to be a more valuable way resolving substages than the normal resistivity recovery spectrum, which resolves only five substages in Au over the same range.

In contrast to Cu, both Al and Au show very marked recovery in Stage II even when their purity is apparently just as high. As before, the effect of adding foreign atoms in solution is to enhance recovery in this stage.

Stage III is identified by its second order kinetics in Al and in both Al and Au it is clear that the activation energy does not correspond to U_m^v, found in quenching or from equilibrium and self-diffusion experiments. There is therefore the same problem of interpretation as was found in Cu, and one is again lead to postulate two forms of the interstitial.

We turn now to an entirely different type of recovery experiment, from which the formation energy of interstitials in Cu has been deduced. A crystal that contains defects has a higher internal energy than one that is well-annealed. During the process of annealing each defect releases its formation energy U_f, which generally appears as heat in the sample. A valuable class of experiment in radiation damage measures this *stored energy release* as annealing proceeds.

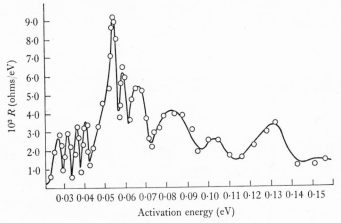

Fig. 115. Activation energy spectrum for gold.
(From Bauer *et al.* 1962.)

It is generally a very small quantity, as a simple calculation will show. Suppose a crystal contains a concentration C_{iv} of i–v pairs. We know that $U_f^{iv} = U_f^i + U_f^v$, $\simeq 5\,\mathrm{eV}$ from calculations, and for a concentration $C_{iv} = 10^{-5}$ the total stored energy should be $\sim 5 \times 10^{-5}\,\mathrm{eV}$ per atom, corresponding to only 0·03 cal. gm^{-1}.

A typical experiment to measure the stored energy release is shown schematically in fig. 116(a). Two specimens, 1 and 2 of identical dimensions, are suspended in symmetrical positions within an enclosure with which they have poor thermal contact. Specimen 1 has been irradiated and contains stored energy. The temperature of the enclosure is raised steadily while thermocouples measure the temperature T of 1 and the temperature difference ΔT between 1 and 2. The stored energy release becomes apparent in ΔT, since the temperature of 1 will rise slightly more than 2 if they are identical in all other respects. In practice there are always small differences in the heat transfer to each specimen and in their masses. To correct for this a second warm-up is carried out immediately afterwards, without disturbing the specimens in any way, obtaining a second curve of ΔT versus T, without stored energy. Subtracting the two curves give θ, the temperature difference that is due to stored energy

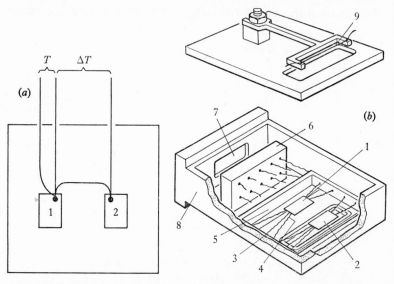

Fig. 116. (*a*) Schematic diagram of a stored energy experiment. (*b*) Target chamber with stored energy specimens in position: 1, dummy specimen; 2, irradiated specimen; 3, heater filament; 4, thermocouple wires; 5, Lavite frame; 6, terminal block; 7, helium reservoir access hole; 8, copper chamber; 9, resistivity wire. (From Meechan & Sosin, 1959*b*.)

alone. In the example given above, if the specific heat is 0·03 cal . gm^{-1}. degC^{-1}, the difference will be $\sim 1\ ^{\circ}$C.

If θ increases by dθ in the interval dT at T the stored energy release dU is given by

$$\frac{\mathrm{d}U}{\mathrm{d}T} = mc(T)\frac{\mathrm{d}\theta}{\mathrm{d}T} \qquad (6.2)$$

where m is the mass of 1 and $c(T)$ the specific heat. Then a graph of dθ/dT will be proportional to dU/dT and should show recovery stages as peaks, in the same way as an isochronal recovery spectrum.

The total energy release from temperature T_0 to T_1 is

$$U(T_0, T_1) = m \int_{T_0}^{T_1} c(T) \frac{d\theta}{dT} dT. \tag{6.3}$$

Using tabulated data for $c(T)$ and finding $d\theta/dT$ from the experiment one then obtains $U(T_0, T_1)$.

Meechan & Sosin (1959*b*) performed such an experiment for electron bombarded Cu, and their apparatus is shown in fig. 116(*b*). The two specimens were thin foils mounted in a copper box with thin windows top and bottom through which the beam entered and left. A resistivity specimen in the form of a thin Cu wire was placed above the target foil in order to relate the stored energy release to the resistivity increment. During irradiation the box was filled with liquid He, which kept the target below 20 °K. This was pumped away before starting the first warm-up. After annealing to 56 °K the box was cooled down again and the second warm-up carried out. The two curves are shown in fig. 117, whilst fig. 118 shows $U(20, T_1)$.

Some evidence of the substages can be seen, though the accuracy of this difficult experiment is not so good as the resistivity recovery. The irradiation temperature was probably too high for I*a* to be retained.

Expressing the total energy release in Stage I, $U(20, 56)$, as a ratio to resistivity recovered one obtains $5 \cdot 4 \pm 0 \cdot 8$ cal.gm^{-1} for 1 $\mu\Omega$ cm. If one accepts that recovery in this temperature range is due to i–v recombination with an energy release ($U_f^i + U_f^v$) for each pair, and knowing that the resistivity of 1 % i–v pairs is about $2 \cdot 5$ $\mu\Omega$ cm from the experiments in § 6.2.1, it follows that a concentration of 1 % i–v pairs is associated with $13 \cdot 5$ cal. gm^{-1}. Converting this to atomic units gives

$$U_f^i + U_f^v = 4 \text{ eV}$$

but from the equilibrium experiments described in § 2.3.5

$$U_f^v = 1 \cdot 2 \text{ eV},$$

hence
$$U_f^i = 2 \cdot 8 \text{ eV}$$

which bears out the theoretical predictions for Cu rather well.

Thus, the electron irradiation experiments have added three

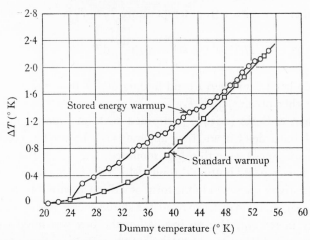

Fig. 117. Curves showing temperature difference between specimen and dummy as a function of absolute temperature for two runs. (From Meechan & Sosin, 1959*b*.)

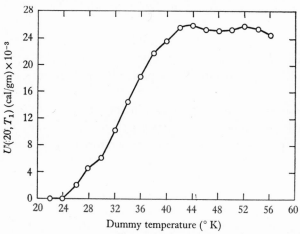

Fig. 118. Curve showing total energy release after correction for heat losses. (From Meechan & Sosin, 1959*b*.)

basic defect properties in Cu to the list in table II, §2.3.10; $U_{\mathrm{m}}^{\mathrm{i}}$, $U_{\mathrm{f}}^{\mathrm{i}}$ and $\Delta\rho_{0}$ for 1 % i–v.

In b.c.c. metals relatively few electron irradiations have been carried out, but the limited data available bears a qualitative similarity to the noble metals. For instance, in Mo Stage I occurs in three main substages at 15, 30 and 42 °K. (Lucasson & Lucasson, 1963). Below 40 °K annealing causes little change in the internal friction, but a large increase in dislocation pinning occurs between 40 and 70 °K. (Lomer & Taylor, 1967). One could interpret these observations as due to close pair recombination in I_A and I_B with long range interstitial migration beginning at about 40 °K.

6.3. Light charged particle irradiation

6.3.1. *Property changes.* Many irradiations with this class of particles, which include protons, deuterons and alphas, have been performed by the Illinois group using 12 MeV deuterons from a cyclotron. The expected form of the damage was discussed in chapters 4 and 5, where it was shown that the primary recoils are distributed according to dE_2/E_2^2 from E_d up to the maximum ΛE_1. The mean recoil energy given by (4.40a) is $E_d \log(\Lambda E_1/E_d)$. For 10 MeV deuterons on Cu the mean primary recoil energy is then 220 eV, though the maximum is 1·25 MeV. Thus, although some very energetic recoils are present, the majority produce relatively small cascades, and we may expect the damage to retain some of the characteristics found with electron irradiation.

12 MeV deuterons have a range of 20×10^{-3} cm in Cu (see Bethe & Ashkin, 1953, or USAEC Nuclear Data tables, part 3), thus to study damage that is characteristic of this energy the specimen thickness should be of the order of 10^{-3} cm. This is fairly similar to the requirements for electron irradiation and it is not surprising that the cryostats used for deuterons resemble the one described in §6.2.1. A typical experiment is described by Magnuson, Palmer & Koehler (1958). Specimens were thin foils, clamped by their ends to a liquid He cooled copper block, the deuteron beam passing through the specimen, then out of the cooled region. The copper block was connected to the helium reservoir by a helium filled chamber. For annealing the samples, this gas was pumped out in order to isolate the copper block thermally. For 9 MeV deuterons, a mean value taking account of degradation in the target, the cross-section for displacing Cu atoms is about 4×10^4 barn, about 1000 times more than for 1 MeV electrons. It is therefore possible to accumulate rather high defect concentrations $\sim 10^{-3}$ and to study

saturation effects. The changes are sufficiently large for lattice parameter and length change experiments.

The resistivity increase of Cu, Ag and Au was one of the first results to be published by the Illinois group and fig. 119 reproduces their graphs of $\Delta\rho$ versus Φt, and this illustrates the departure from linearity. This could have three possible origins; (a) a tendency of

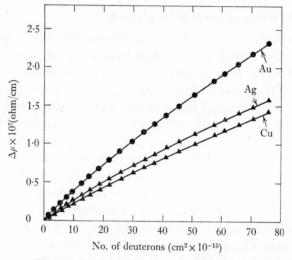

Fig. 119. Resistivity increase as a function of integrated deuteron flux. (From Cooper, Koehler & Marx, 1955.)

C_{1v} to saturate, (b) a breakdown of the proportionality between C_{1v} and $\Delta\rho$, or (c) some thermally activated annealing processes taking place at $10\,^{\circ}\text{K}$. Alternative (b) seems unlikely in these metals, for it is known that the resistivity of dilute alloys increases linearly with solute concentration up to far higher levels than C_{1v} reaches here. Any annealing that occurs during irradiation is almost certainly due to correlated close pair recombination, from the results of the previous section, and it is difficult to see how this could cause the observed curvature. The most probable explanation is that the defect concentration is tending towards a maximum $\hat{C}_{1v}$, above which any new defects are instantaneously annihilated. In §5.8.1 it was shown that this model leads to an exponential saturation, and on this basis the curves have been fitted with (5.89) allowing m, the number of lattice sites in an unstable zone, to be deduced. Values of *m* are shown in table XXVI which agree exceedingly well with those deduced from the similar experiments with electron irradiation described in §6.2.1.

TABLE XXVI. *Deuteron bombardment of* Cu, Ag *and* Au, *after Cooper, Koehler & Marx* (1955)

	(From table XXIII) $\Delta\rho_0\,\mu\Omega$ cm for 1 % iv	$\hat{C}_{iv}$	m	$\left(\dfrac{dC_{iv}}{d\Phi t}\right)_{t=0}$ (deut.$^{-1}$ cm^2)	calculated rate / observed rate
Cu	2·5	2×10^{-3}	500	$0·9 \times 10^{-20}$	4·5
Ag	2·6*	$1·5 \times 10^{-3}$	670	$1·0 \times 10^{-20}$	4·7
Au	2·5**	$2·8 \times 10^{-3}$	350	$1·6 \times 10^{-20}$	3·7

* Double the value of Lucasson & Walker (1962), to allow for their use of a step function $\nu(E_2)$.
** Assumed.

The initial slope and saturation values, with the values of $\Delta\rho$ for 1 % i–v from table XXIII, give the saturation concentration and initial rate of increase $dC_{iv}/d\Phi t$ shown in table XXVI. It is interesting to compare the latter figure with that calculated from the simple cascade theory of (5.5). It will be seen that the observed number of defects is about a quarter of that calculated, as anticipated in chapter 5. We shall see later that as the primary recoil energy becomes larger this discrepancy is magnified, and only in the case of recoils near to the threshold are predictions at all accurate.

In §2.3.5 it was shown that when a vacancy is formed by taking an atom from the interior and replacing it on the surface, the overall volume of the crystal increases, because the number of sites has increased, but this is slightly offset by the inward dilatation around the vacancy which decreases the average lattice parameter. Finally we arrived at the relation (2.37):

$$\left(\frac{\Delta l}{l}\right)_v - \left(\frac{\Delta a}{a}\right)_v = \tfrac{1}{3}C_v.$$

A similar calculation can be made for interstitials assuming them to be made by taking an atom from the surface, thus reducing the number of sites and the macroscopic volume, then inserting it into an interstice of the lattice, when the outward dilation will increase the volume per site and hence the mean lattice parameter. Finally one obtains an equation that corresponds to (2.37):

$$\left(\frac{\Delta l}{l}\right)_i - \left(\frac{\Delta a}{a}\right)_i = -\tfrac{1}{3}C_1. \tag{6.4}$$

For an i–v pair we can suppose the formation to proceed in two stages, using the atom replaced on the surface in the first stage as the atom to be inserted in the second stage. Then one adds the two equations above to obtain:

$$\left(\frac{\Delta l}{l}\right)_i + \left(\frac{\Delta l}{l}\right)_v - \left(\frac{\Delta a}{a}\right)_i - \left(\frac{\Delta a}{a}\right)_v = (C_1 - C_v)/3 \tag{6.5}$$

but since $C_1 = C_v = C_{1v}$

$$\left(\frac{\Delta l}{l}\right)_{1v} = \left(\frac{\Delta a}{a}\right)_{1v} \tag{6.6}$$

with

$$\left(\frac{\Delta l}{l}\right)_{1v} = \left(\frac{\Delta l}{l}\right)_i + \left(\frac{\Delta l}{l}\right)_v \tag{6.7}$$

and

$$\left(\frac{\Delta a}{a}\right)_{1v} = \left(\frac{\Delta a}{a}\right)_i + \left(\frac{\Delta a}{a}\right)_v. \tag{6.8}$$

When we come to consider recovery these relations show that for any recovery stage where $\Delta l/l$ and $\Delta a/a$ change by equal amounts, C_1 and C_v must have also decreased equally. On the other hand if the defects have disappeared at different rates this will be shown by a lack of correspondence between $\Delta a/a$ and $\Delta l/l$.

The length change of a Cu strip under deuteron bombardment at 17 °K has been measured by Vook & Wert (1958), whilst the corresponding lattice parameter experiment has been performed by Simmons & Balluffi (1958). The two sets of data are plotted together in fig. 120. Because their samples differed in thickness the effective bombardment energies were 8·5 and 7·5 MeV respectively. A small correction factor has been included to make both sets refer to the same energy.

It is clear that expression (6.6) is obeyed to within the experimental error ($\sim$ 10 %) which confirms that we are dealing with interstitials and vacancies in equal numbers. The precision is not sufficient to show any tendency to saturation but taking the mean slope with the data of table XXVI ($\Delta\rho = 2\cdot5\,\mu\Omega$ cm per 1 % i–v) one finds

$$\left.\begin{aligned}\frac{\Delta a}{a} &= 0\cdot34C_{\mathrm{iv}},\\[4pt]\frac{\Delta l}{l} &= 0\cdot38C_{\mathrm{iv}},\end{aligned}\right\} \quad \text{for Cu.}$$

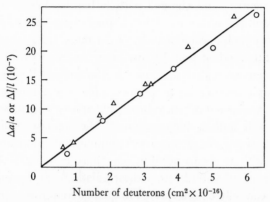

Fig. 120. The relative changes in length and lattice parameter in deuteron irradiated Cu at low temperature are equal within experimental error. ○, lattice parameter (Simons & Balluffi, 1958); △, length (Vook & Wert, 1958). (From Simmons, 1962.)

In §§2.3 and 2.4 we showed how calculations led to

$$\left(\frac{\Delta a}{a}\right)_{\mathrm{i}} = (0\cdot5\pm0\cdot2)\,C_{\mathrm{i}}$$

and

$$\left(\frac{\Delta a}{a}\right)_{\mathrm{v}} = (0\cdot11\pm0\cdot05)\,C_{\mathrm{v}};$$

which would predict

$$\left(\frac{\Delta a}{a}\right)_{\mathrm{iv}} = (0\cdot4\pm0\cdot2)\,C_{\mathrm{iv}}$$

in excellent agreement with the above results.

A very simple method of irradiating with α-particles is to store an α-emitting substance in liquid He. Such *self-irradiation* does not produce the same result as a beam of particles since, (*a*) every emitting nucleus recoils with ΛE_1, (*b*) most of the α's slow down to rest in the sample, making it impossible to define a bombarding energy, (c) both He and the daughter nuclides are introduced as impurity damage.

In the case of ^{240}Pu, which has been extensively studied by King *et al.* (1965), Wigley (1965), also Olsen, Elliot & Sandenaw (1963), the alpha energy is 5 MeV and the recoil energy 250 keV. The recoil is somewhat similar to one that might occur in high energy neutron irradiation, and for this reason α-self-irradiation is intermediate between neutron and charged particle bombardment. The main advantage of the technique is its simplicity and because ^{240}Pu has such a very high specific activity larger doses have been given at 4 °K by this method than by any other. In a period of 10^4 h for instance, one calculates $C_{1v} \sim 10^{-2}$. Unfortunately Pu is one of the most complex metals in the periodic table, having an anomalous resistivity versus temperature curve, and at least five phase changes between 0 °K and its melting point. The α-phase, which exists between 0 °K and 385 °K has a very complicated monoclinic crystal structure with 16 atoms per unit cell. Since none of these characteristics are properly understood the irradiation experiments, though valuable, can only be interpreted in qualitative terms.

The 4 °K resistivity as a function of storage time is shown in fig. 121. This follows a $(1 - \exp(-t/T))$ relationship within 2 % with $T = 1560$ h. Such a time corresponds to a calculated C_{1v} of $\sim 10^{-3}$ but we cannot here be certain of the conclusion, reached above in the context of noble metals, that saturation is only in the defect concentration.

Another ingenious technique for light charged particle bombardment was devised by Blewitt (1962) using the ^{10}B (n, α)^{7}Li reaction with thermal neutrons. He irradiated samples of Al and Cu doped with 0·1 % B^{10}, at 4 °K in a reactor. The cryostat will be described in the next section. The reaction energy of 2·7 MeV is shared by the α and the ^{7}Li in the inverse ratio of their masses. To allow for the effects of the B impurity, samples with 0·1 % ^{11}B were used as controls. Figure 122 shows how resistivity and length are effected in Cu 0·1 % ^{10}B. Only a slight tendency to saturation is observed, but the defect concentrations are rather small, $< 10^{-4}$. The ratio of length change to resistivity is $1·25 \times 10^3$ (ohm cm)$^{-1}$ which is similar to the value found in the case of deuteron bombardment by Simmons & Balluffi.

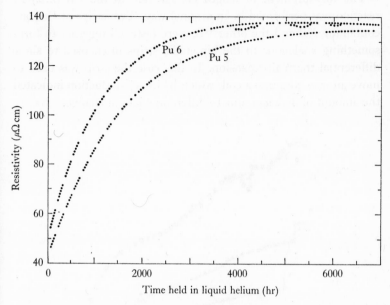

Fig. 121. Resistivities of α Pu 5 and α Pu 6 as a function of time held in liquid helium. (From King *et al.* 1965.)

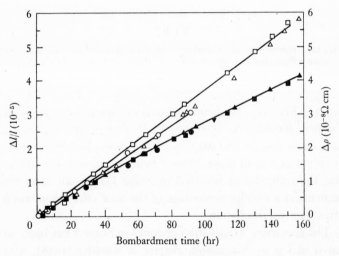

Fig. 122. Change in length as a function of dose in 0·1 % ^{10}B-doped copper. Run 1: △ $\Delta L/L$, △ $\Delta\rho$; Run 2: □ $\Delta L/L$, ■ $\Delta\rho$; Run 3: ○ $\Delta L/L$, ● $\Delta\rho$. (From Blewitt, 1962.)

The measurement of length change due to the ^{10}B (n, α) ^{7}Li reaction was performed in an original way. Two strips of Cu, one containing ^{10}B and the other ^{11}B, were fastened together to form something analogous to the bimetallic strips often used to show differential thermal expansion. In this case the strip was used to move an iron core into a coil, which by its self-induction indicated the amount of deflexion due to differential length change.

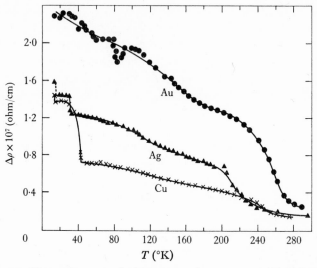

Fig. 123. Thermal recovery during warm up from 10 to 300 °K. The resistivity increase remaining at temperature T is plotted *versus* T. (From Cooper *et al.* 1955.)

6.3.2. *Recovery.* Isochronal recovery of resistivity was studied by Cooper, Koehler & Marx (1955) in their deuteron bombarded samples of Cu, Ag and Au. Figure 123 shows how Stages I, II, III are identified in all these, Stage I being greatest in Cu and least in Au, the order being reversed in Stage III. In all cases Stage I recovery is a smaller percentage of the total effect than was found after electron irradiation.

The spectrum of activation energies in Stage I has been investigated in Cu by Magnuson, Palmer & Koehler (1958), who find essentially the same peaks as in electron damage, though their relative magnitudes are different. The results seem to indicate that fewer close pairs are present after deuteron bombardment, in agreement with our picture of charged particle damage.

A most interesting property of the defects that migrate in Stage I of Al has been established by Riggauer et al. (1967) who measured internal friction as a function of temperature—following α-particle bombardment at 20 °K. In the range from 20 to 80 °K five peaks are found in the internal friction which are of the *relaxation* type discussed in chapter 3. This implies the presence of five types of defects with a symmetry different to that of the f.c.c. lattice, so that when an alternating stress is applied they try to alter their configuration in a reversible way. Knowing the frequency of oscillation and the temperature of the peak the activation energy required to change configuration can be deduced. In this case energies between 0·05 and 0·07 eV were found.

In all cases, annealing of the peaks occurred at a temperature just above the temperature at which they appeared in the internal friction *versus* temperature curve. Thus the defects evidently became sufficiently mobile to migrate, shortly after they had become sufficiently activated for the change in configuration to occur. Presumably the basic defect responsible is the interstitial in the i–v pair, and perhaps the individual peaks are due to interstitials at different distances from their vacancy.

Stored energy release in Stage I (Granato & Nilan, 1961) shows the same substages and fig. 124 shows clearly how I_d is dependent on dose, confirming its assignment to long range defect migration. The ratio of energy release to resistivity recovered in Stage I is 7·1 cal.gm^{-1}.$\mu\Omega^{-1}$.cm^{-1}. Taking $\Delta\rho_0$ for 1 % i–v as 2·5 $\mu\Omega$ cm gives 5·2 eV per i–v pair.

The recovery of $\Delta l/l$ and $\Delta a/a$ have both been followed over Stage I and fig. 125 shows that they correspond closely throughout the stage. Interstitials and vacancies must therefore be disappearing at the same rate, presumably by annihilation.

Stages II and III in deuteron bombarded Cu have also been investigated (Overhauser 1955; Dworschak, Herschbach & Koehler, 1964). An activation energy is found near 100 °K of 0·2 eV, rising to 0·7 eV with near second order kinetics in Stage III at 240 °K and this stage appears to have two substages. In both Stages II and III the stored energy release bears the same ratio to resistivity of 1·7 cal.

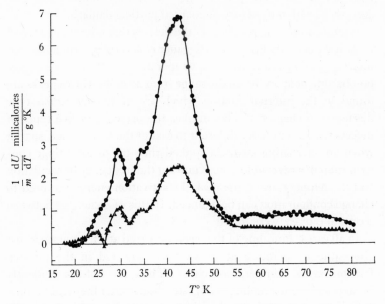

Fig. 124. Stored energy release in deuteron-irradiated Cu. ● $\Phi t = 8\cdot25 \times 10^{15}$ deuterons/cm²; $U(25, 55)/m = 92\cdot8$ mcal./gm. ▲ $\Phi t = 2\cdot89 \times 10^{15}$ deuterons/ cm²; $U(25, 55)/m = 35\cdot3$ mcal/gm. (From Granato & Nilan, 1961.)

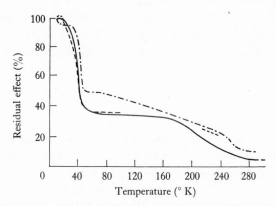

Fig. 125. Comparison of low-temperature recovery effects of resistivity —·—·— (Cooper, Koehler & Marx, 1955), lattice parameter ——— (Simmons & Balluffi, 1958), and macroscopic length (Vook). (From Vook & Wert, 1958.)

gm^{-1} per $\mu\Omega$ cm. This is so much lower than in Stage I that one suspects a quite different mechanism to be operating.

The recovery of Blewitt's Cu o·1 % ^{10}B samples showed the substages both in resistivity and stored energy. There is a ratio of 4·0 cal.gm^{-1} $\mu\Omega^{-1}$ cm^{-1} in Stage I which implies $U_f^{iv} = 3$ eV. This again seems rather low, and we shall return to this point later.

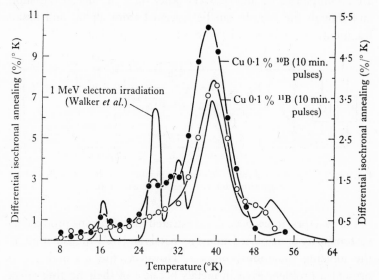

Fig. 126. The differential isochronal annealing curves of reactor-irradiated ^{10}B and ^{11}B-doped Cu and of electron-irradiated Cu in the temperature region from 6 to 65 °K. (From Blewitt, 1962.)

In fig. 126 the recovery spectra of Cu o·1 % ^{10}B and Cu o·1 % ^{11}B are compared with Walker's spectrum for electron irradiated Cu. The α-bombardment evidently produces weak substages Ia, Ib, Ic and Ie with a relatively strong Id. Substages are absent in the non-fissile, ^{11}B-doped Cu where no α's are produced. We shall see that this is characteristic of fast neutron bombardment.

6.4. Neutron irradiation

6.4.1. *Property changes.* Nuclear reactors are the most frequently used source of neutrons for irradiation experiments. A less popular alternative are the neutron generators that convert accelerated beams of deuterons into neutrons by reactions such as ^{9}Be (d, n) ^{10}B. The latter method is limited to fluxes below about 10^{11} neutron. cm^{-2}.sec^{-1} but has the advantage that the neutrons are confined to a small range of energies. The modern reactor can provide fluxes up to 10^{15} neutron.cm^{-2}.sec^{-1} but these are distributed over a wide spectrum from zero to $\sim$ 10 MeV, as described in §4.5. Also they are accompanied by almost every other form of nuclear radiation, and though the sample can be screened from alphas and fission fragments, some beta and γ-rays are always present.

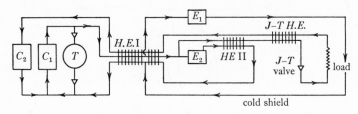

Fig. 127. Idealized refrigerator cycle for temperatures in the 2 to 4 °K temperature interval. (From Blewitt, 1962.)

For reactors, it has been demonstrated that neutrons produce the vast majority of displaced atoms in metals and only in insulators can the ionizing radiations compete. However, the β's and γ's do present a problem even in metals because of their heating effect. The so-called *gamma heating* amounts to about 15 mW.gm^{-1} in a typical graphite reactor with 10^{12} neutron.cm^{-2}.sec^{-1}, and some watts.gm^{-1} in the core of a high-flux D_2O moderated reactor with 10^{14} neutron.cm^{-2}.sec^{-1}. One must reckon with γ-heating, not only in the sample, but also in the chamber that surrounds it and the pipes that connect to the exterior. In a 10^{12} neutron.cm^{-2}.sec^{-1} flux a litre of liquid He would last less than 30 min, and one requires an irradiation lasting several days for most experiments. It is therefore necessary to build a continuously operating refrigerator to cool the sample, keeping the greater part of the circuit outside the reactor core to minimize γ-heating. This approach was pioneered at Oak Ridge by Blewitt & Coltman and a full account is contained in the book *Experimental Cryophysics* (eds.: Hoare, Jackson & Kurti (Butterworths, 1961)).

Figure 127 shows a typical flow diagram for such a device and

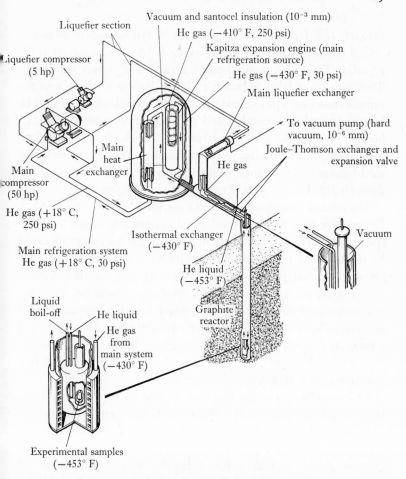

Fig. 128. Pictorial of the Oak Ridge Hole 12 facility.
(From Blewitt, 1962.)

fig. 128 illustrates the layout of an early cryostat at Oak Ridge. Helium gas at room temperature is raised to 250 p.s.i. by compressor C_1 and passes through a heat exchanger $HE.\,I$, where it is cooled to 28 °K by exhaust gas returning to the compressor. After $HE.\,I$ the gas stream divides in two, the majority passing into an expansion engine E_2, where it is cooled by the external work it performs, emerging at 30 p.s.i. and about 17 °K. It then flows into a heat exchanger $HE.\,II$, where it cools the secondary gas stream, still at

19   TDA

high pressure, to 17 °K. The primary gas stream then returns through $HE.I$ to the compressor whilst the secondary stream is cooled further in $J\text{-}T.HE$ to about 10 °K by exhaust gas from the cryostat. Then it expands through a valve $J\text{-}T$, where the Joule–Thomson effect takes it below 4 °K, and finally it passes into the sample chamber. The exhaust gas from the sample chamber may be below atmospheric pressure and requires pre-compression by C_2 before rejoining the main gas stream in C_1. The heat leakage into the cooled part of the circuit is minimized by providing a radiation shield around all pipework, cooled by gas that is diverted part-way through $HE.I$ at 70 °K, and expanded in an engine E_1.

The neutron energy spectrum in a reactor was considered in §4.5. In the core of a graphite moderated reactor one might expect a spectrum similar to that shown in fig. 49a. Such a spectrum could be modified by placing a hollow cylinder of uranium around the cryostat, obtaining a spectrum like that in fig. 49b, greatly enhanced in fast neutrons. To remove the *thermal* neutrons, below 1 eV, the uranium may be lined with B or Cd which have large capture cross-sections at low energy and act as an effective filter. Alternatively if a thermal flux is required, either fuel may be removed from the cryostat's vicinity, or the cryostat may be moved into the fuel-free moderator, or *reflector*, around the core.

The contribution of the fast and thermal parts of the spectrum to the resistivity increase of various metals has been determined by Coltman, Klabunde, McDonald & Redman (1962) in an ingenious set of experiments. Their cryostat was in the core of the ORGR graphite reactor at Oak Ridge where the thermal flux is 6×10^{11} neutron.$cm^{-2}.sec^{-1}$. They first measured the rate of resistivity increase with all the fuel rods in position, giving thermal plus fast; then with the nearby fuel rods removed and the power adjusted to maintain the same thermal flux at the cryostat, giving thermal plus residual fast; finally in the same condition but with a Cd filter, giving only the residual fast. Making suitable subtractions enabled the damage rates due to thermal and fast in the normal spectrum to be isolated. Table XXVII shows the results for a series of metals. In the case of Al the contribution of fast flux vastly outweighs that of the thermal. In Cd the position is reversed, whilst in Cu, Ag, Au and Pt the two components produce effects of comparable magnitude.

TABLE XXVII. *Rate of resistivity increase at* 4·5 °K *in ORGR with* $\Phi_{\text{thermal}} = 6 \times 10^{11}\ cm^{-2}.\ sec^{-1}$. *Units are* $10^{-11}\ ohm\ cm.\ h^{-1}$ *and values show the contribution from fast and thermal neutrons respectively*

	Fast	Thermal	Total
Al	8·70	0·001	8·70
Cu	2·37	0·567	2·96
Ag	1·92	2·958	4·88
Au	1·69	2·71	4·40
Pt	5·52	3·31	8·83
Cd	15·54	126·0	141·5

(After Coltman *et al.* 1962.)

Since thermal neutrons have insufficient energy to displace atoms in direct collision their contribution must be through nuclear reactions. An investigation of the possibilities shows that (n, γ) reactions predominate in many cases. In §4.6 we considered these reactions in some detail and table XIII*b* contains a set of thermal neutron (n, γ) cross-sections, and mean recoil energies for various metals. Bearing in mind that the cross-section for scattering of fast neutrons is about 5 barns for all nuclides, comparison of tables XIII*b* and XXVII shows how the relative magnitude of the two contributions can be related to the (n, γ) cross-section.

Our initial picture of irradiation damage in a reactor is therefore far from simple. The (n, γ) reactions, because of their small recoil energy, will produce small groups of i–v pairs. The direct collision of fast neutrons will give a wide energy spectrum of primary recoils extending up to 10^5 eV, or more. As described in chapter 5 the high energy cascade will leave an inner core of vacancies surrounded by a cloud of interstitials, and there may be some clustering due to athermal annealing or the thermal spike.

In their early experiments the Oak Ridge group measured the resistivity increase of a wide variety of metals during irradiation at low temperature. Mostly, these were carried out in a normal reactor spectrum. In the range of doses available ($< 4 \times 10^{17}$ neutron $cm^{-2}.sec^{-1}$) the resistivity increased linearly with dose and the last column of table XXVII gives a selection of the rates. The maximum dose corresponds roughly to a calculated $C_{1v} \sim 10^{-5}$. More recently, using the high flux reactor at Munchen, Berger, Meissner & Schilling (1964) have accumulated much higher doses in Al, Cu, Ag and Au which clearly show a tendency to saturation with $C_{1v} \sim 10^{-3}$.

Saturation follows an approximate exponential curve, as predicted in §§ 5.8, and analysis showed the typical cascade volume in Cu to be of the order of 10^5 atomic volumes. The size of the unstable zone around a vacancy was also deduced and the numbers of sites m are similar to those in the cases of electron and deuteron irradiation.

A calculation of the expected concentration of displaced atoms involves a detailed knowledge of the reactor spectrum and a fair amount of computation, along the lines indicated in § 4.5. Using simple cascade theory and taking account of electron excitation, D. K. Holmes (1964) has carried out such a calculation for both the thermal flux and the case of the fission flux inside a hollow uranium cylinder. In the first case the predicted value is too high by a factor 2, in the second this increases to a factor of 4·3. We note a continuing trend for the discrepancy between theory and experiment to increase with primary recoil energy.

Changes in modulus and decrement (internal friction) have also been studied in Cu by the Oak Ridge Group (see Holmes, 1962, and Thompson & Paré, 1964, for reviews). The general behaviour is shown in fig. 129 as a function of irradiation time at 25 °C. At lower temperatures the shape of the curves is similar but the effect is smaller, though even at 20 °K a small effect is found. Calculation shows its magnitude in this case to be consistent with the small amount of pinning expected for producing defects in the immediate vicinity of the dislocations. A striking fact is that these effects are observable at doses of 10^{12} neutron.cm^{-2}, whereas the change in resistivity is hardly noticeable until doses four orders of magnitude greater are reached. This emphasizes the sensitivity of dislocation pinning effects to small changes in segment length, predicted in §§ 3.1.6 and 3.1.7. We deduced there how the bowing string model predicts a dependence of modulus and decrement on segment length l_0. An excellent test of that model is provided by comparing the shape of the curves in fig. 129 with a theoretical prediction based on (3.15) and (3.21).

The segment length l_0 is inversely proportional to the number of pinning points n, and n should be a linear function of irradiation dose, Φt. For example, $\quad n = n_0(1 + \gamma \Phi t)$,

where n_0 is the number of pinning points before irradiation and γ is a constant measuring the efficiency of transmitting point defects to

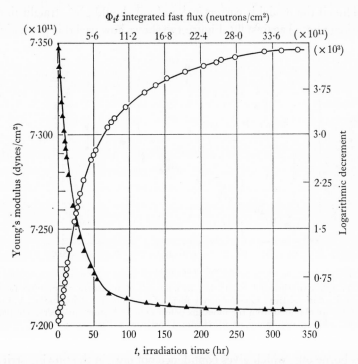

Fig. 129. Variation of decrement and modulus of a Cu crystal during irradiation. Measuring frequency about 12kc. The crystal was well annealed prior to irradiation. ○ Young's modulus, △ Logarithmic decrement. (From Thompson & Paré, 1964.)

the dislocations. Hence γ is smaller at low temperature. Using this with $l_0 \propto n^{-1}$ gives, from (3.15) and (3.21)

$$\frac{\mu_{\text{eff}} - \mu}{\mu} = \frac{k\rho d}{2n_0^2} \frac{1}{(1 + \gamma \Phi t)^2},$$

$$\Delta = \frac{\pi B \rho_{\text{d}} \omega}{\mu b^2 n_0^4} \frac{1}{(1 + \gamma \Phi t)^4}.$$

The experimental quantities are:

$$y = \frac{\mu_{\text{eff}}(\Phi t) - \mu}{\mu_{\text{eff}}(0) - \mu} = \frac{1}{(1 + \gamma \Phi t)^2}$$

and

$$z = \frac{\Delta(\Phi t)}{\Delta(0)} = \frac{1}{(1 + \gamma \Phi t)^4}.$$

Thus if the model is correct $y^{-\frac{1}{2}}$ and $z^{-\frac{1}{4}}$ should give straight lines when plotted against Φt. Figure 130 shows how well this prediction

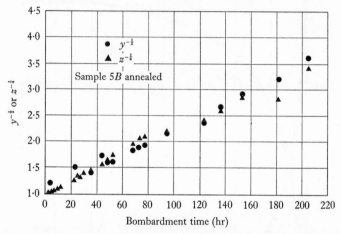

Fig. 130 Experimental irradiation dependence of measures of pinning point density derived from low-frequency form of the vibrating string model. Values of $y^{-\frac{1}{2}}$ are obtained from the modulus defect data; $z^{-\frac{1}{4}}$, from the decrement data shown in fig. 129. ● $y^{-\frac{1}{2}}$, ▲ $z^{\frac{1}{4}}$. (From Thompson & Paré, 1964.)

is borne out, which gives confidence not only in the bowing string model but also gives respectability to experiments that use it to deduce point defect behaviour. Some experiments in this category were described in § 6.2.2 and more will appear in the next section.

6.4.2. *Recovery after neutron irradiation.* We shall deal first with the f.c.c. metals. The Stage I recovery of Cu after fast neutron irradiation is well illustrated in fig. 126, which shows that the first three substages are merged into a continuum stretching down to $7\,^{\circ}\text{K}$. Recovery rises to a maximum in the vicinity of Stage Id and there is some evidence of a separate Stage Ie. Figure 131 shows the recovery curves and spectra of Cu irradiated in different types of neutron spectrum. As one progressively reduces the fast neutron content, the amount of recovery in Stage I increases and substages emerge from the continuum. Amongst other f.c.c. metals, Ag and Pt behave rather similarly, with substages appearing with thermal

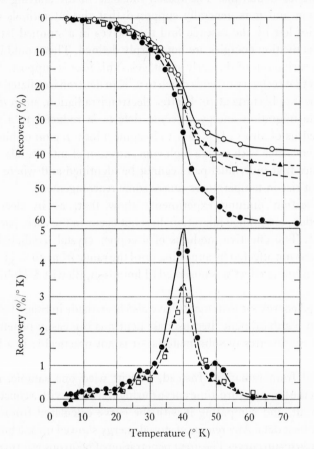

Fig. 131. Isochronal annealing of irradiated Cu. Flux and fission $\Delta\rho$ (10^{-9} Ω cm), ○ Fission (87 %) 1·01, Δ_3 — slug fast 82 % 2·36; □ normal 2·68; ● thermal (67 %) 1·27–3 min. impulses. (From Coltman et al. 1962.)

neutrons, but in Al there is no detectable change with alterations in the neutron spectrum. The behaviour of Cu, Ag and Pt seems qualitatively in accord with the ideas presented above since the thermal neutron induced (n, γ) damage should resemble electron damage. In Al, however, the (n, γ) reaction makes an insignificant contribution and damage is mainly due to fast neutrons.

There is an impressive difference between the smooth Stage I after neutron irradiation and the resolved substages found with electrons or deuterons. The reason must lie in the starting configuration and it is generally supposed that the interstitials and vacancies left by the cascade find themselves in a strained lattice where activation energies are not sharply defined. This would lead to a smearing out of the early substages. This idea is supported by the fact that recovery extends down to 7 °K in neutron irradiated Cu, whereas it only starts at 16 °K after electron irradiation, suggesting that the migration energy of some defects is reduced to a few hundredths of an eV. It is easy to imagine a local region of highly defective lattice with such a high concentration of interstitials and vacancies that individual pairs cannot be identified and where the slightest rise in temperatures causes some rearrangement.

Dislocation pinning experiments show that, as in electron irradiation, long range migration does not occur in the early part of Stage I. The effective modulus of a copper crystal irradiated at 20 °K was not affected by annealing until the range from 40 to 55 °K, there a pinning effect was observed (Thompson, Blewitt & Holmes, 1957).

Stored energy measurements have also been made in Stage I after neutron bombardment. Because fast neutrons can easily penetrate several centimetres of solid material, it is not practicable in a low temperature reactor irradiation to use two samples one of which is screened from radiation. Instead, Blewitt used one sample, observing the temperature-time curve during warm-up with a constant heat input, and comparing it with the curve calculated from the specific heat data. Any release of stored energy showed up as a bump on the warm-up curve. The great penetration of neutrons was turned to advantage because a massive specimen could be used, which minimized the effect of heat leaks relative to energy release. With the reactor running at low power the gamma heating provided a convenient and easily stabilized source of heat input for the warm-up.

A typical apparatus used by Lucas & Blewitt (1967) is shown in fig. 132. The specimen was hollow and formed the bulb of the gas thermometer used to measure specimen temperature. On the outside of the specimen a resistance sample, in the form of a thin wire, was wound. This enabled the ratio of stored energy to resistivity recovery to be measured accurately. In this state the sample was irradiated for several weeks. With the reactor off, the exchange gas

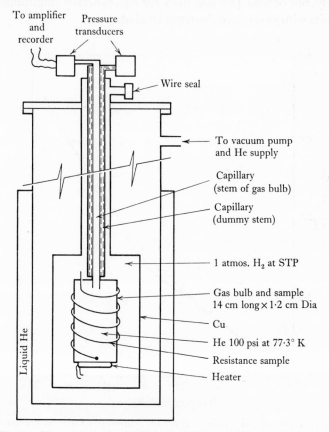

Fig. 132. The stored energy experiment of Lucas & Blewitt (1967). (From Blewitt, 1962.)

was pumped away, leaving the sample thermally isolated. The reactor power was then increased to a sufficient level to supply enough γ-heating to the sample. The temperature was followed as a function of time and a graph of dt/dT plotted against temperature. The result for Al is shown in fig. 133 and the release in Stage I is clearly evident.

Table XXVIII summarizes the experimental results in various laboratories. In Cu and Al the ratios of stored energy to resistivity recovered are about 5 cal. gm^{-1} $\mu\Omega^{-1}$ cm^{-1} and 6 cal. gm^{-1} $\mu\Omega^{-1}$ cm^{-1}. Taking 2·5 $\mu\Omega$ cm and 3·4 $\mu\Omega$ cm for 1 % i–v in Cu and Al respectively, one obtains 4 eV and 10 eV for U_i^{iv}, generally confirming the results with electron or deuteron irradiation.

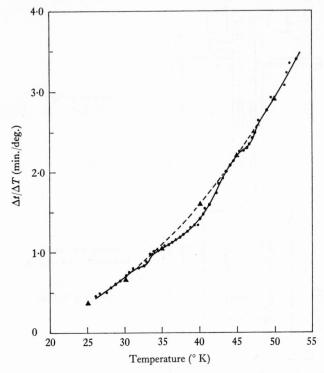

Fig. 133. Differential heating curves of neutron-irradiated aluminium normalized to the specific heat of aluminium. The bump in the curve at the highest temperature is an instrumental error. Integrated flux $\sim 4 \times 10^{17}$ nvt; time ~ 150 hr; exchange gas ~ 25 μm; $\triangle \gamma = 0.315$. (From Blewitt, 1962.)

However, the results obtained by Blewitt *et al.* give rather smaller values and this has led to suggestions that either some other process besides i–v recombination is occurring in Stage I, or that the local strains in the cascade region reduce the energy release per i–v pair. One could then attribute the disagreement between various labora-

Table XXVIII. *Properties of i–v pairs. The type of irradiation experiment is denoted by e^-, d, (n, α) and n. References are given in brackets and are collected at the foot of the table. Values of $(\Delta\rho_0)_{iv}$ are from table XIII; in the case of Al an allowance has been made for the use of a step function $\nu(E_2)$*

	Al	Cu
Stored energy/Resistivity for stage I (cal. gm^{-1}. $\mu\Omega^{-1}$.cm^{-1})		e^- 5·4 (4) d 6·4 (5) (n, α) 4·0 (6)
	n 4·4 to 9·1 (9, 11)	n 3·8 to 6·7 (6, 9, 10)
U_i^{iv} (eV)		e^- 4·0 (4, 7) d 5·2 (5, 7) (n, α) 3·0 (6, 7)
	n 4·5 to 15 (6, 8, 11)	n 2·8 to 5·0 (6, 7, 9, 10)
$\left(\dfrac{\Delta a}{a}\right)_{iv}$ for 1 %		d 0·34 × 10^{-2} (1, 2)
$\left(\dfrac{\Delta l}{l}\right)_{iv}$ for 1 %	n 0·4 × 10^{-2} (11)	d 0·38 × 10^{-2} (1, 3)
$(\Delta\rho_0)_{iv}$ for 1 %, $\mu\Omega$ cm	e^- 3·4 (8)	e^- 2·5 (7)

(1) Cooper, Koehler & Marx (1955)
(2) Simmons & Balluffi (1958)
(3) Vook & Wert (1958)
(4) Meechan & Sosin (1959a, b)
(5) Granato & Nilan (1961)
(6) Blewitt (1962)
(7) Sosin (1962a)
(8) Lucasson & Walker (1962)
(9) Lucas & Blewitt (1967), and private communication
(10) Heeger et al. (1966)
(11) Isebeck et al. (1966a, b)

tories to the different neutron spectra causing different sizes of cascade.

Table XXVIII contains a summary of the stored energy experiments for various types of experiment and also collects together the properties of i–v pairs that have been deduced in the course of this chapter. Unfortunately the critical experiment here, comparison of Δl and Δa, has not yet been carried out for Stage I due to the extreme practical difficulty. In fact it is generally much more difficult to perform measurements in a reactor than at the end of an accelerator's beam tube. Some experimenters have removed samples from the reactor to the laboratory, keeping them at low temperature. This is relatively easy for liquid N_2 but not liquid He.

The later stages in Cu have been studied in this way by Martin

(1962, 1963) who irradiated in liquid $N_2(77\,°\text{K})$ and made the assumption that his sample started in the same state as though it had been irradiated at $4\,°\text{K}$ and warmed to $77\,°\text{K}$. We shall see in the next chapter that this cannot be strictly justified on theoretical grounds but the observed similarity between Martin's recovery curve and Blewitt's data between 77 and $300\,°\text{K}$ shows it to be a good approximation in this case.

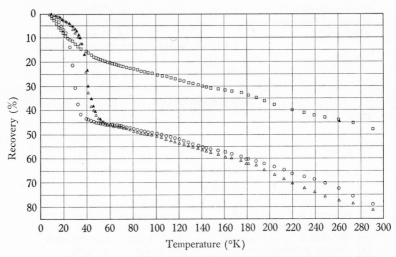

Fig. 134. Isochronal annealing curves of reactor-irradiated Cu, Ag and Au. Dose 1×10^{18} n/cm². Bombardment temperature $4 \cdot 2\,°\text{K}$. (From Blewitt, 1962.)

Figure 134 shows that a prominent Stage II is present in Cu, Ag and Au. Closer investigation of Cu gives the recovery spectra in fig. 135. Stage II is seen to consist of a series of substages superimposed on a plateau ending with Stage III. Stage IV is not at all well defined. Evidently Stage III is complex beginning with first order kinetics and ending with second (Isebeck *et al.* 1966 *a*, *b*).

The effect of impurities in Cu is most interesting and has been studied by both Martin and Blewitt. Some of their results are shown in fig. 135, 136 and 137. In general the impurities transfer recovery from Stage I to higher temperature. Even when spectroscopically pure, fig. 135 shows that Stage II is different in material from different suppliers. With a controlled amount of Ag impurity Martin has investigated the relationship between the -150 and $-80\,°\text{C}$

peaks and found that as the Ag concentration increases, the low temperature substage diminishes whilst the $-80\,°C$ one increases, as the square of the concentration.

A favoured interpretation of Stage II is the release of interstitials trapped on impurity atoms, each substage peak corresponding to a particular type of trap. This should give first order reaction kinetics, and this is indeed found (Schilling, 1967). In the Ag case the $-150\,°C$

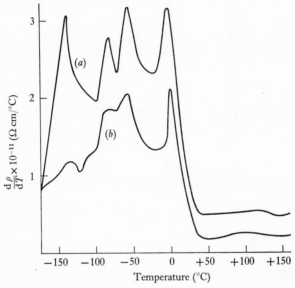

Fig. 135. Isochronal annealing of spectroscopically pure Cu after neutron bombardment at $-196\,°C$. (a) Johnson Matthey, $\rho_0 = 2 \times 10^{-8}\ \Omega$ cm. (b) American Smelting and Refining Co., $\rho_0 = 2 \times 10^{-9}\ \Omega$ cm. (From Martin, 1962 a.)

peak is supposed to be due to trapping at single Ag atoms whilst the $-80\,°C$ peak is attributed to trapping by neighbouring pairs of Ag atoms. Since the concentration of Ag pairs should increase with the square of the total Ag concentration one can then understand the observed behaviour of these two peaks.

Of the b.c.c. metals Mo and W have received the most attention. The majority of irradiations have been at $77\,°K$ and recovery data is therefore subject to the same reservations that were expressed above in the context of Martin's experiment. However, some irradiations of W have been performed in liquid helium. The result of annealing a W sample irradiated at 4 °K is shown in fig. 138. Three main stages can be identified, the first below 100 °K the second between 100 and 620 °K, and the third between 620 and 720 °K. Stage I in W has not been investigated for substructure but Stage II exhibits a series of

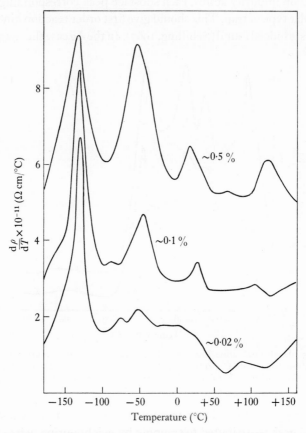

Fig. 136. Isochronal annealing of three dilute Cd–Cu alloys after neutron bombardment at −196 °C. Atomic concentrations of cadmium are 0·02, ∼ 0·1 and ∼ 0·5 % respectively. Ordinates of the upper two curves have been displaced by 2 and 4 × 10⁻¹¹ Ω cm/°C respectively for the sake of clarity. (From Martin, 1962a.)

small peaks, affected by purity in much the same way as Cu. The activation energy rises from 0·25 eV at 100 °K to 1·7 eV at 670 °K. In Mo the recovery spectrum is similar to that of W, but with an overall shift to lower temperature. The activation energy rises from

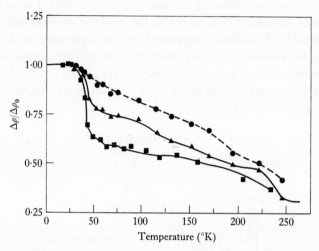

Fig. 137. The isochronal annealing curve of Au-doped Cu following a reactor irradiation at 14·5 °K; ●, 1 % Au, $\rho_0 = 5\cdot882 \times 10^{-7}$ Ω cm; ▲, 0·1 % Au, $\rho_0 = 6\cdot354$ Ω cm; ■, 99·999 + % Cu, $\rho_0 = 7\cdot5 \times 10^{-10}$ Ω cm. (From Blewitt, 1962.)

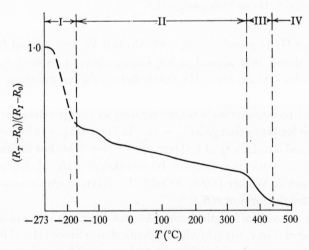

Fig. 138. A recovery curve for W irradiated for 140 hr at 10^{12} n.cm.$^{-2}$sec^{-1}. (From Thompson, 1962.)

0·25 eV at 100 °K. Figure 139 shows that Stage II contains sub-stages, these are different in materials from different suppliers, strongly suggesting that, as in Cu, the trapping of point defects is responsible. This idea is supported by the observation in W that Stage II is absent after very short irradiations (Thompson, 1960, 1959) where it can be supposed that the traps greatly outnumber the point defects and enables changing of traps to occur without significant resistivity change.

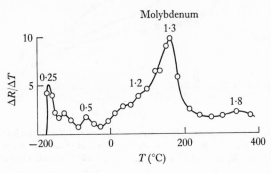

Fig. 139. Differential recovery curve for electrical resistance of Mo irradiated with 2×10^{18} n.cm.$^{-2}$ at -196 °C; figures above the curve show activation energies in eV. (From Thompson, 1962.)

Stage III is prominent in both Mo and W, and Nihoul (1962, 1964) shows that second order kinetics prevail. Stored energy release from Mo in Stage III closely follows the resistivity recovery (Kinchin & Thompson, 1958). The assumption that interstitial-vacancy recombination is occurring with an energy release of 5 eV per pair leads to a change $\Delta C_{1v} \sim 10^{-5}$ in Stage III per 10^{18} neutron. cm^{-2}, and a resistivity of $9 \mu\Omega$ cm per 1 % i–v. This last figure is in agreement with the electron bombardment data of Lucasson, Lucasson & Walker (1962) in table XXXIII, when allowance is made for the form of $\nu(E_2)$.

The recovery of length change (Adam & Martin, 1958) and lattice parameter (Gray, 1959) in Mo are identical over Stage III and follow the same pattern as resistivity or stored energy. This lends weight to the idea that i–v recombination is occurring here. If we assume that 1 % i–v pairs produce a 1 % lattice expansion then we again deduce $C_{1v} \sim 10^{-5}$ in stage III for 10^{18} n.cm^{-2}.

Amongst the h.c.p. metals only Cd has been studied after neutron irradiation at low temperature. Its behaviour, shown in fig. 140, is quite unlike anything described so far. In the first place a large stage is found at 7 °K. Secondly, the behaviour of this with dose is quite

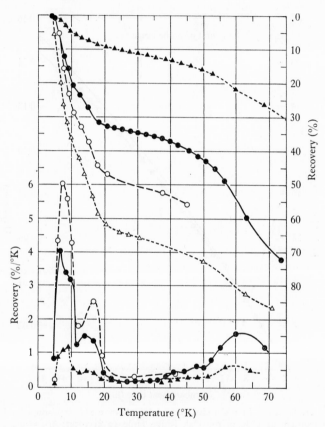

Fig. 140. Isochronal annealing of irradiated Cd. Flux and $\Delta\rho$ (10^{-9} Ω cm); ▲, 3 slug fast (92 %) 14·0; ●, 'thermal' (99·4 %) 191·7; ○, 'thermal' (99·4 %) 32·3; △, 'thermal' (99·4%) 5·9. (From Coltman *et al.* 1962.)

extraordinary as it decreases with increasing dose. It cannot therefore be first order recombination of i–v pairs; one suspects that in Cd Stage I starts below 4 °K and the observed recovery at 7 °K is due to the equivalent of a Stage Ie in Cu.

6.5. Irradiation at low temperature with fission fragments

Blewitt (1962) carried out one of the few low temperature irradiations with fission fragments using Al and Au doped with 0·1 % ^{235}U. When specimens of these materials were irradiated in a reactor the

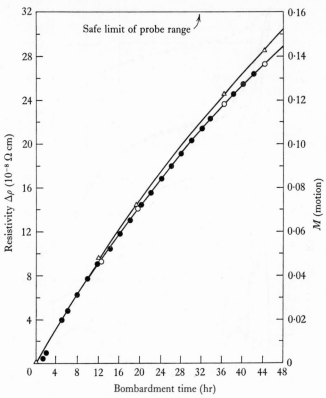

Fig. 141. Changes in resistivity and of length of 0·1 % ^{235}U-doped Al with bombardment at 4 °K in the Oak Ridge Hole 12 cryostat: $\Delta l/l = 2\cdot4 \times 10^{-3}$. △, resistivity readings; ●, position read from chart; ○, position read directly on 6-dial. (From Blewitt, 1962.)

rate of resistivity increase was two orders of magnitude higher than in the pure metal. The difference can only lie in thermal neutron induced fission of the ^{235}U and the damage is therefore due to fission fragments.

The resistivity and length change increments of Al are shown in fig. 141. Both exhibit curvature indicating a tendency of C_{1v} to saturate. Recovery of Al after fission fragment bombardment is remarkably similar to that following fast neutron irradiation. This is illustrated in fig. 142 which shows also that $\Delta\rho$ and Δl change together, suggesting that C_i and C_v maintain parity throughout. In Au the effect of ^{235}U doping on recovery is quite marked. Stage I is almost entirely absent and Stage III is apparently moved to lower temperature, about 230 °K.

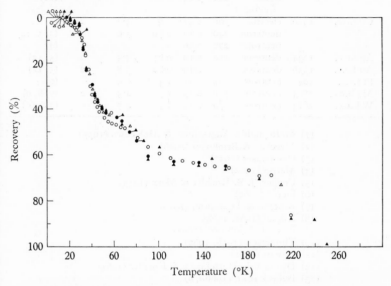

Fig. 142. Isochronal annealing of pure Al, 0·1 % ^{235}U-doped aluminium and 0·1 % ^{11}B-doped Al, ● Δl, Al −0·1 % ^{235}U; ▲, ΔR, Al −0·1 % ^{23}U; ○, ΔR, Al; △ ΔR, Al −0·1 % ^{11}B. (From Blewitt, 1962.)

In Chapter 4 we saw that fission fragments are expected to produce a very local concentration of displaced atoms. Perhaps one can understand the general similarity of neutron and fission fragment damage because both start out with a high local C_{iv}. However, in its starting condition, the Al −0·1 % ^{235}U contains a much higher average C_{iv} than does the fast neutron irradiated specimen with which it is compared in fig. 141.

Such investigations with this type of radiation are still in a very early stage and much more work will be necessary before the low temperature recovery can be understood. It is possible to raise

TABLE XXIX. *Comparison of Stage III recovery in various f.c.c. and b.c.c. metals*

Metal	T_m (°K)	Radiation	T (°K)	$\dfrac{T}{T_m}$	U_m (eV)	$\dfrac{U_m}{T_m} \times 10^4$	x	Reference
Al f.c.c.	933	neutron	190	0·20	0·55	5·9	2	(1, 13)
		deuteron	230	0·24	0·58	6·2	1–2	(12)
		fission fragment	220	0·24	—	—	—	(11)
Cu f.c.c.	1356	electron	300	0·22	0·6	4·4	2	(2)
		deuteron	240	0·18	0·70	5·0	2	(3, 5, 12)
		neutron	270	0·20	—	—	—	(4)
Ag f.c.c.	1234	deuteron	230	0·19	0·67	5·5	2	(5, 12)
Au f.c.c.	1336	deuteron	260	0·19	0·80	5·8	2	(5, 12)
Pt f.c.c.	2047	neutron	500	0·25	1·4	6·9	—	(6)
Mo b.c.c.	2893	neutron	420	0·15	1·3	4·5	2	(7, 8, 9)
W b.c.c.	3643	neutron	670	0·18	1·7	4·7	—	(7 & 10)

(1) McReynolds, Augustiniak & McKeown (1955)
(2) Meechan & Brinkman (1956)
(3) Overhauser (1953)
(4) Martin (1962)
(5) Cooper, J. S. Koehler & Marx (1955)
(6) Piercy (1960)
(7) Kinchin & Thompson (1958)
(8) Nihoul (1962, 1964)
(9) Peacock & Johnson (1963)
(10) Thompson (1958, 1960)
(11) Blewitt (1962)
(12) Dworschak, Herschbach & Koehler (1964)
(13) Isebeck *et al.* (1966*a, b*)

objections to the doping technique on the grounds of sample purity and it is to be hoped that future work will use accelerated beams of 100 MeV heavy ions.

6.6. Summary of recovery behaviour in irradiated metals

In making a comparison between recovery in b.c.c. and f.c.c. metals one is more aware of general similarities than of differences. This impression is reinforced when the absolute melting temperature T_m is introduced to provide a scale of reference. One then finds that Stage III, identified initially by near-second order kinetics, occurs near $0·2T_m$. In table XXIX the properties of Stage III in various metals are compared. A remarkable correlation exists between the ratios U_m/T_m which are close to 5×10^{-4} eV . deg^{-1} in all cases. In Chapter 2 we saw how a similar correlation existed between self-diffusion activation energies, which was taken to imply a common mechanism. The similarity here could lead one to the same conclusion. From the data presented so far it will be clear that Stage I behaviour cannot be so classified with melting temperature.

To summarize the recovery models for irradiated cubic metals, Stage I is fairly well understood, the early part being due to close pair recombination and the later parts to long range migration of interstitials. Stage II probably contains a contribution from trapped interstitials. Stages III and IV are not at all well understood but at the time of writing models based on two types of interstitial and the vacancy appear to have some attractive features. It is depressing to find how little progress has been made here despite the intense activity in this field over a period of 15 years. Further attempts to clarify the situation will be deferred to the next chapter, where clustering is considered.

6.7. Irradiation at temperatures where migration occurs

6.7.1. *The accumulation of point defects.* Now consider the case where irradiation is at a high enough temperature for one defect, say the interstitial to be mobile. The treatment is based on work by the author (1960) and R. M. Walker (1962). At first the crystal will be assumed infinite, with an initial concentration C_{st} of *saturable* trapping sites, which once occupied cannot accommodate any new interstitial. The dose will be taken small enough for $C_d \ll 1/m$ and the saturation due to unstable zones to be neglected. Clustering will also be neglected. As irradiation proceeds the concentration of trapped interstitials C_{sti} will increase, and the number of free trapping sites will decrease to $(C_{st} - C_{sti})$. In this model, either the interstitial is trapped, or it suffers annihilation with a vacancy. It follows that for each trapped interstitial there is one vacancy left. For example,

$$C_v = C_{sti}. \tag{6.9}$$

As C_d increases by dC_d the interstitials have the choice of C_v vacancies or $(C_{st} - C_{sti})$ unfilled traps. Making the assumption that the trap and the vacancy are surrounded by equally attractive strain fields, the proportion going to traps is $(C_{st} - C_{sti})/(C_{st} - C_{sti} + C_v)$. Using (6.9) this becomes $(1 - C_{sti}/C_{st})$ and

$$dC_{sti} = (1 - C_{sti}/C_{st})\, dC_d$$

and $$dC_v = (1 - C_v/C_{st})\, dC_d.$$

On integration these become

$$\left.\begin{array}{c} C_{sti} \\ C_v \end{array}\right\} = C_{st}(1 - \exp(-C_d/C_{st})). \qquad (6.10)$$

The saturation is therefore governed by trap concentration. If the saturable traps are associated with impurities then $C_{st} \sim 10^{-5}$ in a reasonably pure sample, and saturation should occur at much lower dose than in the low temperature case, where the unstable zone produces saturation for $C_d \sim 10^{-3}$. Walker (1962) has found in an electron irradiation at 80 °K, that a graph of $\Delta\rho$ versus Φt, is strongly curved at doses where the 4 °K irradiation shows no curvature. The magnitude of $\Delta\rho$ is dependent on purity, which once more confirms the impurity trapping model.

A more realistic model must take account of other sinks for interstitials and we shall now consider some of the more obvious possibilities. First, if interstitials can condense on the network dislocations, causing them to climb, to a first approximation the number of such trapping sites remains constant. Thus we can regard them as *unsaturable traps*, with concentration C_{ut} unaffected by the concentration C_{uti} of interstitials that have passed to them.

Secondly, the escape of interstitials to crystal boundaries can be represented by a constant concentration of sinks, which are again unsaturable, and which we shall lump together with C_{ut} in this model. As C_d increases by dC_d in a crystal containing only unsaturable traps the concentration of vacancies is always equal to the concentration of interstitials that have passed to the traps. For example, $$C_v = C_{uti}.$$

Using similar arguments as those above,

$$dC_{uti} = \frac{C_{ut}\, dC_d}{(C_{uti} + C_{ut})}$$

which on integration gives

$$C_{uti}^2/2C_{ut} + C_{uti} = C_d,$$

solving this quadratic in C_{uti} gives

$$\left.\begin{array}{c} C_{uti} \\ C_v \end{array}\right\} = [\sqrt{(1 + 2C_d/C_{ut})} - 1]\,C_{ut}. \tag{6.11}$$

Thus the presence of unsaturable traps will not produce saturation in the defect concentration, but simply a negative curvature of the damage *versus* dose curve. This is intuitively reasonable, for as C_v increases there is an increasing tendency for interstitials to annihilate vacancies rather than to become trapped. When C_d is very small one expects this effect to be small and $C_d \simeq C_{uti}$ and therefore $C_v \simeq C_d$, in accord with the small-C_d behaviour of expression (6.11).

In a crystal containing both types of trap the defect concentrations can be shown to increase according to

$$C_{uti} = [\sqrt{(1 + 2C_{ut}C_d/(C_{ut} + C_{st})^2)} - 1]\,(C_{ut} + C_{st}), \tag{6.12}$$

$$C_{sti} = C_{st}[1 - \exp(-C_{uti}/C_{ut})] \tag{6.13}$$

and

$$C_v = C_{uti} + C_{sti}. \tag{6.14}$$

The interstitials on the unsaturable traps can probably be assumed not to contribute to electrical resistivity. If we take a_{sti} as the resistivity increase due to unit concentrations of interstitials on saturable traps, and a_v as the corresponding resistivity for vacancies then the total resistivity will be

$$\Delta\rho = (a_{sti} + a_v)\,C_{sti} + a_v\,C_{uti}. \tag{6.15}$$

Thus the increase can be analysed into two terms, one of which saturates as the saturable traps become filled, the other increasing steadily with a similar square root dependence to that seen in (6.12). At very large doses there will, of course, be an ultimate saturation due to the unstable zone effect. Figure 143 shows how this sort of behaviour is observed in Mo and W irradiated with neutrons, at 300 °K.

Before pursuing the study of clustering in the next chapter, one must consider the various ways in which clusters might be nucleated.

First, there is the possibility that nuclei are present in the unirradiated crystal and that as point defects become available, some will gather together around these pre-existing imperfections. Alternatively, the nuclei might be formed in the spike region so that their density would increase in proportion to Φt as irradiation proceeds. Or there may be small clusters formed by the statistical means considered in § 5.8.1. Any of these mechanisms would be examples of *heterogeneous nucleation*.

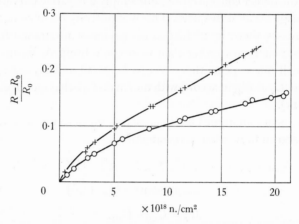

Fig. 143. The increase in resistance with neutron dose for molybdenum ○ and tungsten +, irradiated at 300 °K, measured at 273 °K. (From Kinchin & Thompson, 1958.)

A third possibility is *homogeneous nucleation*, in which clusters form as a result of two or more point defects meeting during their migration and forming a nucleus for further condensation. Such a process would clearly depend on the ambient concentration of mobile defects and, in the case where two are involved, the formation of nuclei would be a second order reaction. Here we have an example of an accumulation process which depends therefore, on Φ and T. But this feature is not confined only to homogeneously nucleated clusters, for it is possible that the growth of any cluster will be the outcome of a competition between growth and dissociation under thermal activation. Here again growth will depend on Φ and T. A further source of complication is the surrounding strain field which will clearly change as the cluster grows. It is quite possible that once a certain concentration of clusters is reached it will be impossible for new ones to nucleate due to the long range attractive force from their larger neighbours. The simple model above is inadequate to cope with these effects, but in a crude way the clusters can be regarded as traps for interstitials whose effective concentration C_{ct} depends on the density of clusters and their size distribution. C_{ct} will thus be a complicated function of Φ, t and T. Further development of the theoretical model must await the account of observed cluster behaviour in Chapter 8.

6.7.2. *The influence of irradiation temperature on the form of radiation damage.* Two general results of great importance can be deduced from the ideas presented above. The first concerns the validity of recovery experiments performed after irradiation at elevated temperature. The question to be answered is whether an irradiation at low temperature followed by an anneal to temperature T produces the same result as an irradiation at temperature T. Clearly any process that is favoured by a high ambient concentration of defects, such as clustering, operates more efficiently in the sample irradiated at the low temperature and then annealed at T when a large concentration is released suddenly. It is not so obvious that the trapping process is also different in the two cases and this will now be demonstrated.

Take the first model considered above of an infinite crystal with saturable traps. Irradiation at a temperature where only interstitials are mobile leads to the situation described by (6.10). If the irradiation had been at low temperature followed by annealing at higher temperature, the starting condition would be a concentration C_d of interstitials confronted with the choice between C_d vacancies and C_{st} saturable traps. When C_i falls by dC_i the choice is between $(C_{st} - C_{sti})$ unfilled traps and C_v vacancies, therefore the concentration going to traps is

$$dC_{sti} = -\frac{(C_{st} - C_{sti}) \, dC_i}{C_{st} - C_{sti} + C_v}$$

but the number of vacancies remaining must equal the total number of interstitials either trapped or remaining free. For example,

$$C_v = C_i + C_{sti}$$

then it follows that

$$\int_0^{C_{sti}} \frac{dC_{sti}}{C_{st}-C_{sti}} = \int_0^C \frac{dC_i}{C_{si}+C_i}$$

which gives

$$\left.\begin{array}{c} C_{sti} \\ C_v \end{array}\right\} = \frac{C_d \, C_{st}}{C_{st}+C_d}. \tag{6.16}$$

This is clearly different to the previous result of (6.10) and the ratio of either concentration in (6.10) to the corresponding one here is

$$\left(1 + \frac{C_{st}}{C_d}\right)[1 - \exp(-C_d/C_{st})]. \tag{6.17}$$

This ratio tends to unity for either small C_d or large C_d, but for intermediate values is greater than one. Thus the concentrations C_{sti} and C_v are larger in the case of low temperature irradiation followed by annealing to T, than in the case of irradiation at T. Although an irradiation at high temperature can give qualitative information about subsequent recovery stages, the magnitude of these stages will not be the same as in the low temperature irradiation experiment. Walker's work (1962) compares the result of irradiating a Cu sample at $4\,^\circ$K followed by an anneal to $80\,^\circ$K, with an irradiation at $80\,^\circ$K. The former case gives the largest defect concentration in accordance with (6.18).

6.7.3. *A relation between flux and irradiation temperature.* A second general result concerns the relation between the irradiating flux Φ and the irradiation temperature T. We have already seen that the accumulation process may depend on the ambient concentration of point defects and this will be determined by a balance between the rates of their production and removal. The first rate is proportional to Φ and the second to $\exp(-U_m/kT)$, where U_m is the migration energy of the defect concerned. Damage will accumulate in the same form in different irradiations if the ratio between these two rates is the same. Thus for irradiations at T_1 and T_2, up to the same dose $\Phi_1 t_1 = \Phi_2 t_2$, to produce the same effect we require

$$\Phi_1 \exp(U_m/kT_1) = \Phi_2 \exp(U_m/kT_2)$$

or

$$T_1 = \left[\frac{1}{T_2} - \frac{k}{U_m}\log\frac{\Phi_1}{\Phi_2}\right]^{-1}. \tag{6.18}$$

If T_1 and T_2 satisfy this relation the irradiations should be equivalent. In practical problems this is often a convenient way to relate the irradiation effect observed in a materials testing reactor, to the expected effect in a reactor under design. Note, however, that this condition can only be applied where one defect with a single value of U_m is migrating during irradiation.

CHAPTER 7

POINT DEFECT CLUSTERS IN IRRADIATED METALS

The previous chapter was devoted mainly to the properties and migration of point defects produced by irradiation. In this chapter the emphasis is on the formation and properties of point defect clusters. First we draw attention to the conclusion of chapter 5 that there is a natural tendency for point defects to be formed in clusters, even before any migration occurs. Here it will be shown how migration leads to a consolidation of existing clusters and, possibly, formation of some extra ones. The major part of the chapter is devoted to experiments to study clusters, which fall into two classes, direct and indirect. The first rely mainly on electron microscopy or field ion microscopy, and the second on either the interaction of clusters with dislocation lines or on the macroscopic distortion that they cause in some crystals.

7.1. The direct observation of point defect clusters

7.1.1. *Introduction.* Following the successful observation of vacancy clusters in quenched material by means of the electron microscope, it was natural to examine irradiated metals in the same way. Amongst the first in the field were Hirsch and Silcox at Cambridge, and Barnes, Makin, Mazey, Smallman and Whapham at Harwell. In all cases thin foils ($< 10^3$ Å) were prepared by electro-polishing thicker foils of irradiated material. This is preferable to irradiating thin foils, firstly because the atoms in such a foil are never more than a few hundred Å from a free surface, which could influence the clustering process, and secondly because severe oxidation of the foil may occur during the irradiation.

A typical micrograph of neutron irradiated Cu is shown in plate XX*a* in which one can see a number of dark spots, some of which can be resolved as small rings a few hundred Å in diameter. The number of defects per unit area varied in proportion to the thickness of the foil, which suggested defects distributed throughout the volume rather than a surface effect. The fact that the contrast altered as the foil was tilted showed that the defects were visible because of an electron diffraction effect associated with lattice rotation in a strain field (see § 3.2.3). On tilting, the rings changed their aspect, becoming more or less elliptical, suggesting that they were dislocation loops.

We have already seen one of the smaller dark spots in plate XIV, where it appeared to indicate the pattern of debris from a single cascade.

The presence of the defects was found to depend critically on the ambient concentration of point defects. This is well illustrated by the case of neutron irradiated Al. If the irradiation is at room temperature no defects are observed up to doses of the order 10^{20} n.cm^{-2}. But if the irradiation is carried out at 77 °K followed by a warm-up to room temperature, defects are clearly visible (Smallman & Westmacott, 1959).

Alternatively, by bombarding Al with protons and He^{+} at room temperature Beevers & Nelson (1961) showed a strong dependence of defect density on the rate of displacement $\dot{C}_d$, and hence on the ambient point defect concentration. These experiments strongly suggest that the defects observed are clusters of point defects. However, it is evident that $\dot{C}_d$ is not the only factor controlling their formation since an electron irradiation of Cu which gave $\dot{C}_d$ an order of magnitude greater than the neutron case, with comparable C_d, did not produce any visible effect. (Makin, Whapham & Minter, 1962). It seems likely that nucleation is the missing factor, and we shall return to this point later.

For the study of very small defect clusters the field-ion microscope is in principle a valuable complement to the electron microscope, able to resolve individual lattice sites on the surface. However, it suffers from two disadvantages, apart from its limitation to surfaces. First, the specimen is under extreme mechanical stress, due to the electric field applied, during observation, and this may affect the configuration of defects. Second, its application has so far been limited to refractory metals, most of which are b.c.c., and nothing has been learnt of the metals Al, Cu, Ag and Au by this means. Nevertheless valuable general information has been gained in the cases of Ir (f.c.c.) and W (b.c.c.) and this will be presented in §§ 7.1.2

and 7.1.5 below. Plate XX*b* shows a vacancy cluster observed in the surface of irradiated W.

It would be pointless to attempt a complete review of cluster phenomena in various metals because the field is expanding too fast for the catalogue to survive publication. However, it is possible to illustrate some of the important aspects of cluster behaviour and to emphasize some of the differences between metals. The proceedings of a Symposium on Point Defect Clusters, edited by Makin (1966), contains a representative set of papers.

7.1.2. *Clusters in Cu.* A detailed analysis of the size distribution of visible defects in irradiated Cu was carried out by Makin and his colleagues. In order to express the results in numbers per unit volume, they took great care to establish the thickness of the area of foil under investigation. This was done by leaving the foil in the microscope after the defects had been photographed and watching for movements of glissile dislocation segments that extended between the top and bottom surfaces. These leave a transient shadow, or *slip trace*, behind them which makes the slip plane clearly visible. The projected width of the slip plane on the plane perpendicular to the direction of viewing can then be measured. The crystal orientation is determined from the electron diffraction and, knowing that slip planes in Cu are generally {111}, the angle between the slip plane and the direction of viewing can be established. From this angle and the projected width, the foil thickness is calculated.

The results of this investigation are shown in fig. 144 and 145. The curves in fig. 145 suggest that the density of the larger defects (> 50 Å) tends to saturate, but the density of smaller defects (< 25 Å) does not. Even more striking is the difference in behaviour between small and large defects near to a grain boundary. It is generally observed, in both charged particle and neutron irradiation, that the grain boundaries are surrounded by a denuded zone for large clusters, whose density decreases markedly as the boundary is approached. If the small defects are affected at all it is a very small effect. The denuded zone is clearly visible in plate XX*a*.

This separation into two classes becomes even more evident after a short anneal. Plate XXI shows a micrograph of Cu irradiated with

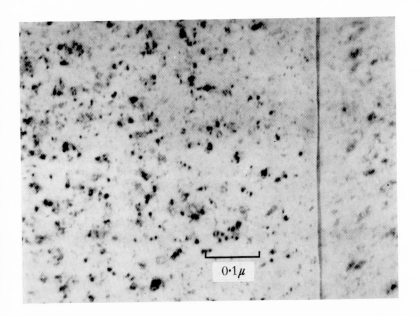

Plate XX (*a*) Neutron damage in Cu after 3.79×10^{18} fast n.cm^{-2}. A denuded zone can be seen near to the grain boundary on the right. $\times 150{,}000$ (from Makin, Whapham & Minter, 1962). (*b*) Vacancy cluster on the surface of a W specimen in a field-ion microscope (by courtesy of B. Ralph).

(*Facing p.* 318)

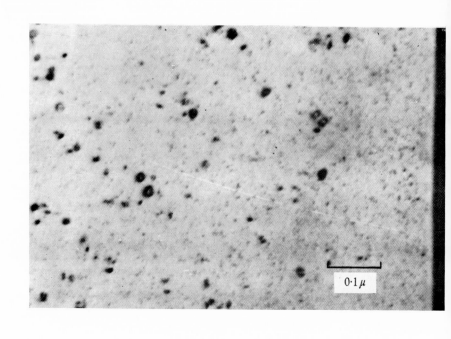

Plate XXI Copper irradiated with 2×10^{18} fast n.cm^{-2} after annealing at 306 °C for 63 min. showing small defects (< 50 Å diameter) and large loops. The difference in nature of the two types of defect is clearly shown by their different response to the grain boundary which runs along the right-hand side of the photograph. $\times 150{,}000$. (From Makin, Whapham & Minter, 1962.)

2×10^{18} n.cm^{-2} at 30 °C and annealed for 63 min at 306 °C. After annealing, the electrolytic thinning process gives better quality specimens in which greater resolution is possible. The larger defects are clearly resolved as loops and the denuded zone is even wider than before. The smaller defects can sometimes be resolved into

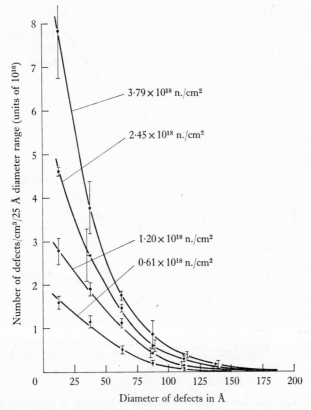

Fig. 144. The density of defects as a function of diameter after various neutron doses. (From Makin, Whapham & Minter, 1962.)

tetrahedral forms, which appear triangular when viewed along a $\langle 111 \rangle$ axis and square along a $\langle 100 \rangle$.

It is tempting to postulate that the small and large types of defect are clusters of vacancies and interstitials respectively. The rate of formation of the small clusters is about a tenth the rate of primary recoil initiation and their uniform density suggests that they may have formed during the more energetic thermal spikes. If this is so it is more reasonable to attribute them to vacancies than to interstitials since the centre of the spike contains a core of vacancies. The fact that they sometimes take up the tetrahedral form, associated with vacancy clusters in quenched Au, lends weight to this idea.

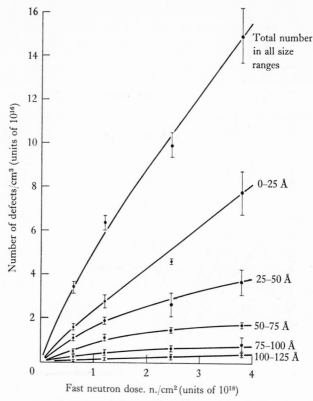

Fig. 145. The density of defects of various diameters as a function of the neutron dose. (From Makin, Whapham & Minter, 1962.)

The denuded zone of the larger clusters, and their saturating density, suggests formation by gradual aggregation of a mobile defect migrating over considerable distances at the irradiation temperature and for which the grain boundary represents a competing sink. It has been suggested that this defect is the interstitial. We shall see below that further experiments confirm this idea.

A calculation of the number of point defects in a given cluster depends critically on its form. If one assumes that the smaller defects are vacancy clusters in the form of small voids and the loops are interstitial clusters in the form of flat discs one atom thick, then there are apparently ten times more clustered vacancies than interstitials. But, if the small defects are either loops or tetrahedra of stacking fault, the numbers of clustered interstitials and vacancies are roughly equal. This second case leads to the reasonable figure of 50 clustered vacancies or interstitials per 2 MeV neutron hit. The assumption of voids would give an implausibly high number of clustered vacancies.

It is interesting to consider the possible means of nucleation of the clusters at this stage. For the small ones, heterogeneous nucleation on pre-existing defects, such as impurities, is impossible since one would then expect their numbers to saturate. Nor can it be homogeneous nucleation, for a calculation by Varley (1961) shows that their numbers should saturate after a few minutes in a flux of $10^{12}\,\mathrm{n.cm^{-2}.sec^{-1}}$. One must therefore agree with the inference above that the nuclei are formed in the thermal spike and that only one spike in ten is successful in nucleating a visible cluster. The slight decrease in rate of formation of the small clusters seen in fig. 145 may result from an increasing competition between neighbouring nuclei as their density increases.

The behaviour of the defects on annealing, emphasizes the difference between the small defects and the loops. In Cu, the temperature range from 250 to 350 °C produces a gradual shrinkage of both types and they eventually disappear.

The loops go more rapidly than the small defects and as they do so, their denuded zone at grain boundaries increases in extent. The activation energy associated with the loop shrinkage is between 1 and 1·5 eV, that for the small defects is around 2 eV. Makin suggests that the low energy process is dissociation of submicroscopic (< 10 Å) vacancy clusters into mobile vacancies which then attack the interstitial loops reducing their size. The second process is attributed to dissociation of the larger vacancy clusters, which presumably have a higher binding energy for vacancies as discussed in §3.2. The dispersal of these vacancies contributes further to the annihilation of the interstitial clusters. At the highest temperatures one is entering the range of normal self-diffusion when vacancies enter the crystal from sources such as edge dislocations and grain boundaries; these will clean up the remaining interstitial clusters and any deeply trapped interstitials that are still present.

An experiment with alpha-irradiated Cu provides convincing evidence that the loops are interstitial clusters and the small defects vacancy clusters (Barnes & Mazey, 1961). We shall see in chapter 8 that when alpha-particles come to rest in a metal, becoming He impurity atoms, they combine with vacancies and cluster together to form small gas-filled bubbles. Because He has a high positive energy of solution (see Chapter 2) these impurity atoms are extremely efficient traps for vacancies, since the vacancies enable them to come out of solution and reduce the energy of the crystal. They will therefore compete strongly with other vacancy sinks, such as interstitial clusters, in any process that involves migration of vacancies through the crystal.

When such He doped Cu is examined after irradiation both loops and small defects are observed, just as in pure Cu. But after annealing their behaviour is quite different, and although the small defects shrink as before, the loops grow, eventually joining together to form a dislocation network. If the loops are interstitial clusters and the small defects are vacancy clusters, this behaviour can be readily explained. The small defects will dissociate as before, but the vacancies released will become trapped by He to form bubbles rather than annihilate interstitial clusters. At the annealing temperatures involved ($\sim 350\,°C$) there will be a tendency for vacancies to be evolved from any dislocation loop with strong edge characteristics, such as one expects the interstitial clusters to be. In pure material the loss of vacancies would be exceeded by arrival of vacancies from dissociating vacancy clusters and other sources, and the loops shrink. In the He doped Cu however, the balance is reversed by strong vacancy trapping and the loops should therefore grow by a net loss of vacancies.

So far the nature of the loops and small defects has been inferred from circumstantial evidence, and this lacks the weight that a direct determination of Burgers vector would carry. Unfortunately, experiments of the type described in § 3.2.3 are very difficult in Cu because the defects are rather small, and none has so far been reported, but using the field-ion microscope, Ralph, Fortes &

Bowkett (1966) have made an interesting study of Ir after irradiation by fission fragments, from which some general inferences relevant to other f.c.c. metals may perhaps be drawn. They increased the electric field at the tip of the specimen until the Ir atoms were themselves being removed by field emission at a sufficiently small rate for each vacant site appearing on the surface to be recorded. After peeling off hundreds of atomic layers in this way a three-dimensional picture of the vacancy distribution could be determined. In several cases clusters were found that could only have resulted from a collision cascade.

From Chapter 6 we know that these vacancies cannot have migrated far, since the specimen was irradiated and stored near room temperature, thus the pattern of vacancies should show the skeletal structure of a cascade and also the early stages of a vacancy cluster.

The shape of clusters containing about 200 vacancies or less was approximately spherical and the region could be likened to a sponge with about 30 to 40 % of the lattice sites being vacant.

Larger clusters than this had always collapsed to form a dislocation loop of the type described in § 3.2.1, confirming that this is indeed a lower energy configuration than either the sponge or the spherical void, when the cluster is large enough.

Thus it appears almost certain that in Cu the small defects seen in the electron microscope are vacancy clusters formed during the collision cascade—these probably start out as spongy regions of the lattice but, as annealing proceeds, they collapse into dislocation loops or tetrahedra.

We are now in a position to consider the cluster formation in relation to the recovery schemes discussed in Chapter 6, from which we attempted to infer the migration energies of interstitial and vacancy defects. The experiments that link these topics together are those in which a sample is bombarded and examined at low temperature. Makin, Whapham & Minter (1961) irradiated Cu at $77\,°\text{K}$ in the reactor, prepared thin foils at $200\,°\text{K}$ and transferred these to the electron microscope without warming up. The quality of the pictures was not good, owing to contaminating films that condensed on the sample, but it was clear that both the loops and the small defects were present at $200\,°\text{K}$. No obvious changes occurred during a warm up to 300 °K, but the experimental difficulties prevented a really careful study of the defect size distributions.

Howe, Gilbert & Piercy (1963) went further, and cooled a thin Cu sample to below 30 °K on a specially constructed cold stage in the electron microscope. The electron gun of the microscope was then used as an ion accelerator to bombard the sample with 100 keV O⁻ ions following a technique suggested by Menter, Pashley & Presland (1962). From Chapters 4 and 5 it will be clear that these ions should produce collision cascades rather similar to those generated in a neutron irradiation. The doses in their experiments were calculated to be equivalent to about 10^{19} n . cm⁻², and from this point of view should be comparable to Makin's experiments. However, there is the important difference that these specimens were already thinned at the time of bombardment and this may influence the behaviour of migrating defects. Immediately after bombardment they found loops and small defects similar to those after neutron bombardment. On warming to 273 °K the small defects certainly remained and were apparently unaffected by the warmup. This would clearly be consistent with the view that they are vacancy clusters formed during the displacement spike. The behaviour of the loops is less well established, they certainly survive the warm up, but whether they increase in size or in numbers is not certain. If they are interstitial clusters, then either they have formed in the spike, or the interstitial must be able to migrate at temperatures below 30 °K. The latter alternative seems in conflict with the assignment of close pair recombination to the recovery below 30 °K, but Nelson (1964) suggests that there may be an alternative explanation. He has estimated the probability of an interstitial receiving the migration energy as a result of collisions between the 100 keV electrons in the microscope during normal observation, and neighbouring Cu atoms. With apparently reasonable assumptions he deduces that the interstitial would receive energy in excess of 0·1 eV at least 10 times per second. This may explain how the interstitial clusters can form by *stimulated diffusion*, even though thermally activated diffusion is unlikely. Because of its higher activation energy stimulated diffusion of vacancies should not be so important.

Although difficult to test in the electron microscope, electrical resistivity measurements on electron bombarded Au have shown that such sub-threshold collisions *can* induce migration of inter-stitials. Dworschak *et al.* (1967) bombarded a Au specimen at 9 °K first with electrons at 3 MeV then at 1·4 MeV and measured the rate of resistivity. The second bombardment was at too low an energy to produce Frenkel pairs, but it apparently produced a decrease in the number created by the previous bombardment. This is clear evidence of stimulated diffusion in the case of close i–v pairs. Whether it occurs in the electron microscope remains to be proved.

Finally, an experiment by Bowden & Brandon (1963) is worthy of attention. They also bombarded already thinned Cu foils, but with 100 eV A^+ ions. These are not thought to be capable of penetrating more than one or two atomic layers beneath the surface, but should be capable of generating focused replacement sequences that deposit an interstitial $\sim 10^2$ Å below the surface. They bombarded at -30 °C and 20 °C, transferring the sample to the electron micro-scope subsequently. In the -30 °C case they observed many defects close to the bombarded surface which, from the diffraction contrast, were deduced to be regions of compressive strain. This would clearly be consistent with interstitial clusters, but might possibly be due to inert gas atoms in association with vacancy clusters. After the 20 °C bombardment the effect was very much less. Their interpretation suggested that the interstitial defect from which the clusters are formed is less mobile at -30 °C, allowing a higher concentration to build up and giving a higher probability of cluster formation. In the two-interstitial recovery scheme (see § 6.2.2) this defect would be the dumb-bell interstitial. It is also possible that small interstitial clusters, too small to be resolved, were formed during the bombard-ment and that these migrated, possibly by gliding, to form the visible clusters.

7.1.3. *Clusters in Au.* Au is well suited for the study of clustering since so much is known of vacancy clusters in quenched material. Cotterill & Jones (1962) took quenched gold containing tetrahedra of stacking fault, as described in § 3.2.1, and irradiated with α-particles below 50 °C. After a dose of about $10^{13} \alpha$ cm^{-2} they found that the tetrahedra had completely vanished and been replaced by small defects similar to those found in neutron-irradiated Cu. (See plate XIV for a good example of these small clusters (Merkle, 1966)).

The small defects increased in number in direct proportion to the dose as the tetrahedra disappeared. They continued to appear when all tetrahedra had gone. This appears to show that mobile interstitials are produced in Au at 50 °C and that they annihilate clustered vacancies in the form of tetrahedra. The irradiation-produced vacancies are probably contained in the small defects, these being clusters nucleated in the displacement spike. The experiment was conducted in a < 1000 Å foil and one would expect that the surfaces provided a competing sink for interstitials. In a large crystal one might expect annihilation to be even more efficient, perhaps leading to fewer small defects. It appears to be true in general that where one bombards a thin foil the small vacancy clusters are easier to produce, presumably because interstitials can escape in this way (Hesketh & Rickards, 1966).

An obvious next step was to irradiate at low temperature and Howe & Gunn (1964) repeated their Cu experiment, described above, with a previously quenched Au foil containing tetrahedra. As bombardment proceeded at 30 °K little effect was seen until the density of small clusters reached 10^{10} cm^{-3}, when they outnumbered the tetrahedra in the ratio 25:1. Then the tetrahedra began to disappear, presumably as interstitials were produced close to them. If the irradiation was terminated before the tetrahedra had gone and the foil warmed up, it was found that nothing happened until about 0 °C. It would appear that in Au the migration of interstitials over long distances is not possible at low temperature.

Venables & Balluffi (1965 a, b) performed an elegant experiment, taking a quenched Au foil containing tetrahedra, and bombarding with 200 eV A$^+$ ions at 140 °K. These ions should displace atoms only within $\sim 10^2$ Å of the bombarded surface. The effect was to produce loops near this surface, but the majority of the tetrahedra were not affected. On warming up through 0 °C the tetrahedra disappeared, suggesting that interstitials are unable to migrate over distances $> 10^2$ Å below this temperature. If this is the case the measurements suggest that $U_m^i = 0.75 \pm 0.06$ eV. An obvious difficulty of this interpretation is the formation of clusters at the irradiation temperature. Venables & Balluffi suggest that there is a high enough concentration of interstitials produced for them to cluster spontaneously due to their close proximity. Annihilation is not thought to be an important competitor because interstitials will mainly be produced. There could also be a stimulated migration in the electron beam of the microscope. Annealing of the specimens between 250 and 300 °C produced a re-arrangement of the dislocation loops and lines. By 350 °C they had mostly disappeared.

Bowden & Brandon (1963) have also carried out their experiments with Au using 100 eV A^+ ions. The results are essentially similar and show the loops to be interstitial in character.

All these experiments can be explained with a model where $U_m^i = 0.75$ eV in Au. But apart from the means of loop formation at the low temperatures, there are other doubts about this interpretation. For instance the interstitial defect that migrates to the tetrahedra could possibly be a small cluster of interstitials. Alternatively, small clusters could be dissociating to single interstitials at this temperature. Or single vacancies could be migrating to partially annihilate interstitial clusters, partially annihilating them until they are small enough to become mobile.

Low-energy ion bombardment in combination with electron microscopy has also been used to attempt a measurement of the displacement energy. Ogilvie, Sanders & Thomson (1963) bombarded polycrystalline Au foils with monoenergetic Ne^+, A^+ and Xe^+. Observable damage was produced when the recoil energy exceeded 22, 25 and 20 eV respectively, which, though a little low in comparison with other techniques, seems a reasonable set of values.

Using oriented single crystals of Au, Ag and Cu Ogilvie (1965) extended these measurements to look for an orientation dependence. Table XXX summarizes the results.

TABLE XXX

Metal	Au			Ag			Cu
Direction of incidence	$\langle 110 \rangle$	$\langle 100 \rangle$	$\langle 111 \rangle$	$\langle 110 \rangle$	$\langle 100 \rangle$	$\langle 111 \rangle$	$\langle 110 \rangle$
Xe^+ ion energy for damage to appear (eV)	16	23	20	15	22	19	16

The thresholds are always lowest when bombardment occurs along $\langle 110 \rangle$, and if the other values are multiplied by the (cosine)^2 of the angle to the nearest $\langle 110 \rangle$ direction they all become approximately equal. These facts strongly suggest that $\langle 110 \rangle$ is the direction in Au, Ag and Cu along which interstitials are most easily introduced, in qualitative agreement with the ideas developed in chapter 5.

7.1.4. *Clusters in Al.* In aluminium a wide range of observations have been carried out in both irradiated and quenched samples, and this has made it possible to investigate the behaviour of clusters in some detail.

We have already seen that Beevers & Nelson (1961), using thin foil samples, found a strong dependence of clustering on the damage rate. They also showed that as irradiation proceeds up to a level where each atom has been displaced once ($C_d \sim 1$) the loops join up to form a tangled network of dislocations, as shown in plate XXII (Beevers & Nelson, 1963). It may be that this is the essential feature of a metal that is saturated with damage at high temperatures.

In foils prepared from Al that had been bombarded with 38 MeV alphas at ~ 250 °C, and which therefore contained He atoms as an impurity, Mazey, Barnes & Howie (1963) have found some very clearly resolved loops that are suitable for Burger's vector determination. Using a diffraction contrast analysis similar to that described in § 3.2.3, they found that the larger loops (> 125 Å) lie in planes close to the $\{110\}$ and have $\mathbf{b} = +\frac{1}{2}a \langle 110 \rangle$, close to the loop normal. The sign and magnitude of $\mathbf{b}$ show clearly that these loops are equivalent to the insertion of one extra layer of atoms. These must therefore be interstitial clusters. As we saw in § 3.2.1 the stable position for such loops should be on the $\{120\}$ plane, which is rather close to $\{110\}$. Makin (1964) suggests that the loops observed by Mazey *et al.* may have been on these planes. The smallest loops (< 125 Å) apparently lie on $\{111\}$ planes. It is presumed that they also have $\mathbf{b} = \frac{1}{2}a \langle 110 \rangle$ and that as they grow the loop plane rotates to a more stable position. Occasionally the small loops are seen with stacking fault enclosed, but this frequently disappears during observation. It is supposed that the faulted loop on a $\{111\}$ plane is the first stage of clustering, the next stage is the dissociation and sweeping across of two $\frac{1}{6}a \langle 121 \rangle$ type partials, and finally the loop rotates onto a plane close to $\{120\}$, as described in § 3.2.

In these experiments with α- irradiated Al there are apparently no

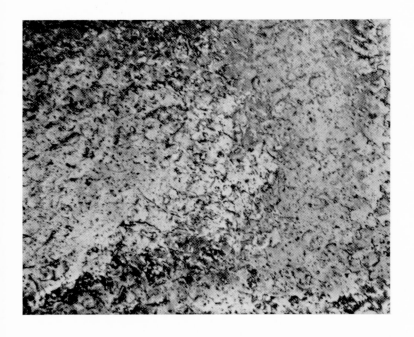

Plate XXII Dislocation networks which were formed in an Al foil after bombardment by 1.5×10^{15} Xe ions $cm^{-2} \times 36,800$. (From Beevers & Nelson, 1963.)

(*Facing p.* 328)

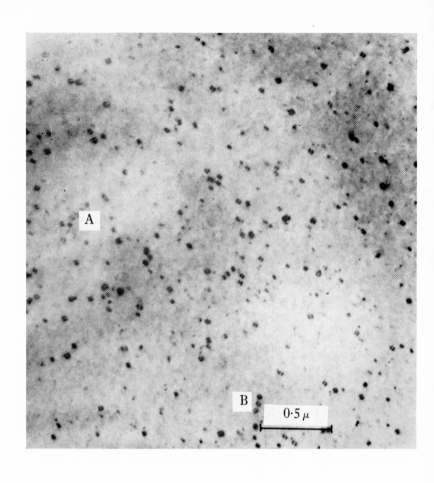

Plate XXIII Dislocation loops formed by fission fragments in Al. Rows of loops may be seen at *A* and *B*. (From Westmacott, Roberts & Barnes, 1962.)

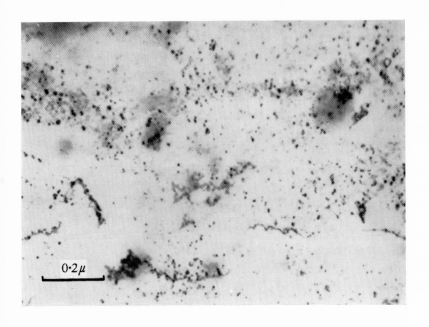

Plate XXIV Dislocation loops and point defect clusters in a thin foil of molybdenum bombarded with $\sim 1\cdot5 \times 10^{17}\,He^{2+}$ ions . cm^{-2} at 38 MeV. $\times$ 80,000. (From Beevers & Mazey, 1962.)

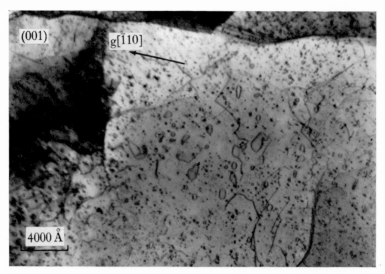

Plate XXV Initial stages of rapid loop growth in neutron irradiated Mo after annealing for one hour at 800 °C. (From Downey & Eyre, 1965.)

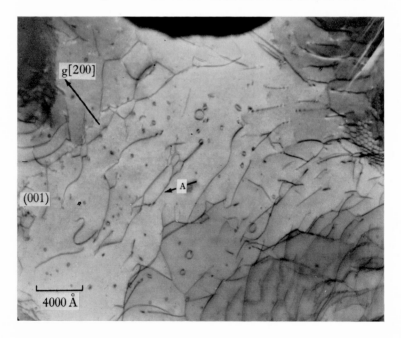

Plate XXVI Initial stages of dislocation network formation in neutron irradiated Mo after annealing for 1 h at 1935 °C. (From Downey & Eyre, 1965.)

loops that could have formed by vacancies clustering, nor are there any of the small defects found in neutron-irradiated Cu. It is likely that this is a special case in which the He impurity captures vacancies and inhibits clustering except in the form of gas-filled bubbles.

Using fission fragments to bombard thin Al foils, Westmacott, Roberts & Barnes (1962) found both types of loop. Here the ratio of displaced atoms to gas atoms is three orders of magnitude greater, and the effect of impurity is correspondingly less. In some cases one loop was found within a larger one. Occasionally these touched and the two segments of line were annihilated, leaving a horseshoe-shaped loop. Evidently the inner and outer loop had Burger's vectors of equal magnitude but opposite sign, suggesting that one was the vacancy cluster and the other the interstitial. This demonstrates that vacancy clusters can also be formed by irradiation; it could suggest that the nucleation and growth of one type of cluster is possible in the vicinity of a loop of the opposite kind. Alternatively the loops might have been moved together after formation.

We saw in §3.2 that loops can move bodily, either by slip of glissile segments or by *self-climb*, that is, pipe diffusion of atoms or vacancies from one side to the other. For movement to occur, there must also be a driving force, and we have seen that the elastic inter-action energy between a pair of loops can provide this. Depending on their relative positions, loops may then line up in rows or join together to form irregular-shaped loops that may eventually link up, forming a dislocation network. Evidence of such movements is abundant in the fission-fragment irradiated Al, and plate XXIII shows some rows of loops. In general, the larger the loop the more likely it is to move by slip rather than diffusion processes. The effect of increasing temperature will be to enhance diffusion processes. Bowden & Brandon (1963) confirm this general conclusion in their experiments with low energy ions.

7.1.5. *Clusters in the b.c.c. transition metals.* Of these metals, Mo has received the most attention in the electron microscope because it is relatively easy to prepare thin foils by electro-polishing, and the observations appear to be reproducible. An experiment by Beevers & Mazey (1962) showed the existence of small loops in Mo that had been bombarded, as a thin foil, with 85 keV H^+ and He^{++}. They also found similar effects in foils prepared by thinning bulk Mo that had been irradiated with 38 MeV α's (see plate XXIV).

Though roughly the same number of atoms were displaced in each of the lower energy bombardments, the He^{++} case produced very much larger loops. Many of the He^{++} ions must have come to rest in the foil, whereas the H^{+} ions, having a longer range, should have mostly passed through. It seems very likely that, as in the Barnes & Mazey experiment described in §7.1.2, the He combines with vacancies, thus allowing a larger fraction of interstitials to escape annihilation and form clusters.

An interesting feature of the 38 MeV α-irradiation is shown in plate XXIV where several helical dislocation lines are seen. These have presumably formed from originally straight screw dislocations by point defect condensation, as described in §3.2.1. This lends weight to our earlier supposition that dislocations act as sinks for point defects.

Neutron-irradiated Mo has been studied in great detail by Downey & Eyre (1965). They irradiated up to doses of 2×10^{20} n.cm^{-2} at 60 °C and found that many small loops were present in thin foils prepared from large irradiated samples. From loop counts, about 1 % of the calculated point defect concentration was condensed into observable clusters. Annealing in stages up to 800 °C produced progressively fewer, but larger, loops, as shown in plate XXV. Throughout this range the number of point defects in observable clusters remained roughly constant, even in the region around 170 °C, where resistivity shows a prominent recovery peak. Annealing to temperatures above 800 °C apparently causes the loops to join up to form a network, an example of which is shown in plate XXVI at A.

After annealing had made the loops large enough, detailed analysis was possible. One set lay on {111} planes, were interstitial in character, did not enclose a stacking fault and had $\mathbf{b} = \frac{1}{2}a\langle 111\rangle$. The lines in the network had either this vector or $a\langle 100\rangle$. When two $\frac{1}{2}a\langle 111\rangle$ loops with $\mathbf{b}$ in different $\langle 111\rangle$-type directions intersect to form a segment of line in a network, this will have $b = a\langle 100\rangle$ according to:

$$\tfrac{1}{2}a(111) + \tfrac{1}{2}a(1\overline{1}\overline{1}) \to a(100).$$

Thus, the observed vectors in the network could have arisen in this manner.

Eyre & Bullough (1965) have considered how such loops might be nucleated. It is thought from the theoretical work of Erginsoy, Vineyard & Englert (1964) and Johnson (1964) that the single interstitial in b.c.c. adopts a split configuration along the $\langle 110 \rangle$ direction. If a number of these cluster together with their axes parallel, forming a platelet on the $\{110\}$ plane, a loop results which has $\mathbf{b} = \frac{1}{2}a\langle 110 \rangle$ and contains a sticking fault. Consider the reaction in which a partial with $\mathbf{b} = \frac{1}{2}a\langle 100 \rangle$ dissociates from one side of the loop and sweeps across it, removing the stacking fault.

$$\tfrac{1}{2}a[110] + \tfrac{1}{2}a[00\bar{1}] \rightarrow \tfrac{1}{2}a[111].$$

The energy associated with the dislocation lines may be estimated by taking the squared magnitude of the Burger's vector involved, since $U_f^d \simeq \mu b^2 l$ from (3.2). For example,

$$\tfrac{1}{2} + \tfrac{1}{4} \rightarrow \tfrac{3}{4}.$$

These energies balance, so that by including the stacking fault energy on the left hand side, the reaction must be energetically favourable.

Eyre & Bullough show further that as the $\frac{1}{2}a\langle 111 \rangle$ loop grows bigger it is energetically favourable for it to rotate from the $\{110\}$ onto the $\{111\}$ plane.

The fact that the number of clustered point defects remains roughly constant through the annealing suggests that the loops grow in size by migrating and coalescing with opposite types avoiding one another, rather than by breaking up and reforming. In the latter case one would expect to have lost some defects during the process. The long range interaction force between loops is capable of drawing loops together, provided they are sufficiently mobile. As mentioned above, mobility could be provided either by slip or by self-climb. Several of the larger loops in plate XXV appear very irregular and this strongly supports the model of loop migration.

The main difficulty of these results is in understanding the apparent lack of effect during the anneal through 170 °C, where physical property changes indicate that a point defect becomes mobile (see § 6.4.2). One possibility is that the effect exists but goes undetected because of the inaccuracies inherent in loop counting. Downey and Eyre suggest that vacancies become mobile at 170 °C and either condense together locally, near the site of the collision cascade, to form sub-microscopic clusters, or that they annihilate impurity trapped interstitials rather than clustered interstitials. The relatively high impurity concentration in Mo make this fairly reasonable and it is even conceivable that vacancies are repelled by the small interstitial clusters existing at this temperature.

The field-ion microscope can add something here, for W is one of the easiest metals to study in this way. Ralph, Fortes & Bowkett (1966) conducted experiments similar to those on iridium described in §7.1.2 and find essentially similar results with sponge-like vacancy clusters. In W, however, there is less tendency to collapse into dislocation loops.

Annealing of these W specimens showed that during Stage III there was a reduction in the proportion of single vacancies observed on the surface. This confirms the earlier suggestion that vacancies migrate in Stage III.

Iron is another b.c.c. metal that has received considerable attention. Masters (1964) bombarded thin foils with 10^4 eV Fe^+ ions at $550\,^{\circ}C$ in order to simulate primary recoil atoms directly, and found interstitial loops with $\mathbf{b} = a\langle 100\rangle$. Eyre & Bartlett (1965) and Meakin (1964) irradiated with neutrons at $60\,^{\circ}C$ and found both this type of loop, and the $\frac{1}{2}a\langle 111\rangle$ type found in Mo. It is suggested by Eyre & Bullough that the nucleation is exactly the same as in Mo, with $\frac{1}{2}a\langle 110\rangle$ loops forming on the $\{110\}$ planes. If these are swept by a different type of partial with $\mathbf{b} = \frac{1}{2}a\langle 110\rangle$ the stacking fault can also be removed giving a Burger's vector $a\langle 100\rangle$, as observed.

$$\tfrac{1}{2}a[110] + \tfrac{1}{2}a[\bar{1}10] \rightarrow a[010].$$

Again this can be seen to have an energy balance in its favour when the stacking fault is taken into account.

7.2. Radiation growth and cluster formation in anisotropic metals

Because of its importance as a reactor fuel, uranium has been better studied than any other anisotropic metal. This study has been made all the more urgent by the appearance of an unusual phenomenon known as *radiation growth*. In this, individual crystals undergo a large distortion without any appreciable volume change. For many years this was a baffling phenomenon, but it now proves to be an interesting manifestation of cluster behaviour.

Below 660 °C uranium exists in the α-phase with an orthohombic structure. The lattice cell has three unequal orthogonal edges denoted by the vectors $a\langle 100\rangle$, $b\langle 010\rangle$ and $c\langle 001\rangle$. Kittel and colleagues (1955) found that when a single crystal is irradiated with neutrons, inducing fission, growth occurs along the $\langle 010\rangle$ axis, shrinkage along the $\langle 100\rangle$ and no change along the $\langle 001\rangle$. Near room temperature the magnitude is 400 % when 1 % of the uranium atoms have undergone fission. The *growth factor G* is defined by

$$\frac{l-l_0}{l_0} = GB,$$

where l_0 is the initial length, B is the *burn up*, or fraction of uranium atoms fissioned, given by:

$$B = \sigma_t \Phi t.$$

Thus $G = 400$ at room temperature.

G is a decreasing function of temperature becoming zero somewhere between 400 and 500 °C. As the temperature is lowered, G increases to about 10^4 at 77 °K and double this at 20 °K (Buckley, 1961; Queré, 1963).

The effect is not unique to α-uranium. Buckley has found it in certain other anistropic metals when bombarded with fission fragments. For Cd, Zr, Zn and Ti at 77 °K, G is 70, 60, 20 and 15 respectively. The effect is not found in any cubic metals nor in any uranium alloys with cubic structure, though in its anisotropic alloys it is generally present. There is therefore strong evidence to link radiation growth with anistropy. But this in itself is not enough, for fast neutron or proton irradiation will not cause growth of this large magnitude, even though comparable concentrations of atoms have been displaced (Buckley, 1961; Thompson, 1961). The effect must therefore be associated with fission fragment damage.

It is worth emphasizing the differences between radiation growth and the dimensional changes found in cubic metals and discussed in chapter 2 and 6. In the first place the effect can be very large ($\sim$ 1000 %) and does not saturate. Secondly it is not recoverable by thermal annealing. Thirdly it is only found after fission fragment bombardment. It is very difficult to explain these facts in terms of accumulated point defects distributed throughout the crystal.

Buckley suggested that growth could result from clusters of point defects forming preferentially on certain crystallographic planes. Consider the effect of N interstitial loops, each of area A and with Burgers vector $\mathbf{b}$, in a crystal that was initially a cube with edges of unit length parallel to axes Ox, Oy and Oz. Suppose that they all lie on (100) planes, parallel to Oyz, and that $\mathbf{b}$ is parallel to Ox. Neglecting any elastic relaxation, the volume of interstitials forming the loops is NAb (see the definition $\mathbf{b}$ in Chapter 3). This volume can be represented as a slab of new material on an Oyz face of the crystal, of thickness NAb. If the interstitials were formed by taking equal numbers of atoms from each of the three faces Oyz, Oxy, Oxz, a thickness $\frac{1}{3}NAb$ is removed from each of these. Then the net expansion along Ox is

$$\frac{\Delta x}{x} = + \tfrac{2}{3}NAb$$

and the net contraction along Oy and Oz is given by

$$\frac{\Delta y}{y} = \frac{\Delta z}{z} = -\tfrac{1}{3}NAb.$$

If we had chosen vacancy loops for this example, the signs would be reversed in these expressions.

In a cubic crystal one would always expect equal numbers of loops on (001), (010) and (100) planes and the net effect would then be zero since each dimension would increase by $\frac{2}{3}Nab$ but decrease by two terms of $-\frac{1}{3}Nab$. Even if more complicated planes were chosen, the high symmetry of the cubic crystal, that leads to isotropy in so many bulk properties, would result in cancellation of the effect.

However, in anisotropic metals there is no reason to assume that interstitial or vacancy loops will be distributed amongst the crystal planes in such a way that their effect cancels. Buckley's hypothesis was that interstitial loops would lie on planes giving growth along (010) and that vacancy loops would lie on other planes that allowed a contraction along (100). If the vacancy and interstitial loops contained equal numbers of point defects, and their growth had no component along ⟨100⟩, then this axis should remain unchanged.

A systematic study of loops in fission-irradiated uranium was

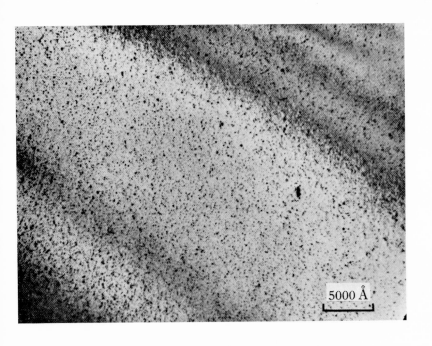

Plate XXVII Random array of small dislocation loops in U after 1.5×10^{17} n.cm^{-2} at -196 °C. (From Hudson, 1964.)

(*Facing p.* 334)

5000 Å

Plate XXVIII Zone refined uranium irradiated to ∼ 1·5 × 10¹⁷ n.cm⁻² at 80 °C, showing the two rows of loops at right angles. The vertical rows represent the intersection of the foil with a sheet of loops on (010), the horizontal rows a sheet on sites (100). (From Hudson, 1964.)

therefore carried out by Hudson, Westmacott & Makin (1962) and Hudson (1964). Loops were found that lay on either $\langle 010 \rangle$ or $\langle 100 \rangle$ planes with $\mathbf{b} = \frac{1}{2}\sqrt{(a^2+b^2)}\langle 110 \rangle$ and $a\langle 100 \rangle$ respectively.

The former class of Burger's vectors consists of two sets with $\mathbf{b}$ parallel to either (110) or $(\bar{1}10)$. If these loops are interstitial their lateral effects will cancel and there will be a growth along (010). If the $a(100)$ loops are vacancy, contraction will occur along (100). Making these assumptions the observed number of loops per unit volume is sufficient to explain the magnitude of radiation growth. In effect, one is removing atoms from $[100]$ planes and replacing them onto $[010]$, the actual redistribution taking place by interstitial and vacancy migration. By condensing point defects into loops the number of defective sites in the crystal is reduced, and since these sites are concentrated around the periphery of the loops the interstitials forming the centre of a large loop have been incorporated into perfect lattice. One can therefore understand how growth can continue indefinitely without saturation.

The relative arrangement of the loops is very interesting. After irradiation at $77\,^{\circ}\mathrm{K}$ they appear randomly distributed as in plate XXVII. But plate XXVIII shows that higher temperatures lead to alignment into rows. In fact these rows correspond to the intersection of sheets of loops with the thin foil specimen. One set of sheets contain the $\frac{1}{2}\sqrt{(a^2+b^2)}(010)$ loops and are parallel to $[010]$. The other contains the $a(100)$ loops and are parallel to $[100]$ planes, thus in both cases the loop plane is coplanar with the sheet.

This is almost certainly another manifestation of loop interaction forces. Using the calculations of Foreman & Eshelby (1962) as a basis, Hudson (1964) was able to demonstrate that similar loops should be attracted into sheets. At low temperature lack of mobility prevents alignment but, at room temperature, Hudson suggests that the interloop forces are enough to move the loops by slip. At high temperatures a combination of slip and self-climb by pipe diffusion seems probable.

The origin of temperature dependence in growth is not certain but it is likely to be associated with the nucleation of clusters. For instance, it could be the case the nuclei dissociate under thermal activation and would persist longer at low temperature, thus having a better chance of growing into a cluster. The presence of growth at

20 °K seems to imply that long range migration of at least one point defect is possible at these temperatures in uranium. It would be interesting to test this point by other techniques, such as the recovery of resistivity after low temperature bombardment.

The fact that radiation growth only occurs with fission fragments is also attributed to nucleation, it being suggested that only the fission fragment can produce a high enough defect density to form nuclei.

7.3. Some effects of clusters on plastic deformation

7.3.1. *Irradiation hardening in Cu.* In chapters 6 and 7 we saw how the elastic modulus and the internal friction are affected by the presence of point defects. These properties are determined by small movements of the dislocation lines, a few Å in general. We shall now consider deformation processes which take the crystal beyond the limit of elasticity, causing dislocation lines to sweep through large distances across their slip planes. It will become apparent that here it is the small clusters, and not the individual point defects, that have the greatest influence.

The situation in f.c.c. metals will be considered first, taking Cu as the example, since it is on this metal that most experiments have been performed. Consider the stress *versus* strain curves shown in fig. 146, which demonstrate some of the phenomena encountered when deforming irradiated single crystals. The initial near-vertical portion is the reversible elastic region, which terminates abruptly in a yield point. The most obvious effect of irradiation is the raising of the stress at the yield point, and is loosely referred to as *hardening*. Figure 146 shows how this gradually recovers with annealing. With a single crystal sample, the load at the yield point gives a valuable parameter for comparison with theoretical calculations; the resolved *critical shear stress* σ_{css}. This is found by converting the applied load into a shear stress across the slip plane operating in the particular sample (this would generally be a $\{111\}$ plane in f.c.c. crystals).

Beyond the yield point the irradiated specimen sometimes requires a slightly smaller load to maintain plastic deformation, and this feature of the curve is the yield drop. Next, there is often a flat region where deformation proceeds with very little extra load being required. In the single crystal this is often accompanied by the progressive spread of slip from an initiating slip plane right through the crystal, forming a so called *Luder's band*, a visible region of striated roughening on the sides of the crystal. Elongation in the Luder's region is generally jerky as the slip spreads from one set of slip planes to the next. Next comes a smoothly rising portion where *work-hardening* is said to occur. This ends when a necking down of the sample raises the stress locally and leads to fracture.

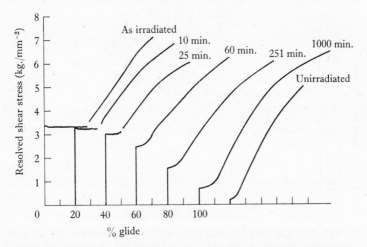

Fig. 146. Typical stress–strain curves at 20 °C showing the effect of annealing at 336 °C. Curves displaced horizontally for clarity. (From Makin & Manthorpe, 1964.)

The explanation of such phenomena as these, in both irradiated and unirradiated crystals, is the subject of intensive research. The theoretical models are still developing and it cannot be said that any features of the stress–strain curve are entirely understood. However, by concentrating mainly on the effect of irradiation on the critical shear stress we shall be able to utilize the better established models.

Many experiments have been performed with polycrystalline samples and until quite recently it was thought that these could be analysed to give σ_{css}. However, there are current doubts about this procedure for f.c.c. metals (Johnson, Sargent & Wronski, 1964) and it seems that values of σ_{css} deduced from the polycrystalline data, though potentially valuable, must be used with some caution.

The effect of neutron dose Φt on σ_{css} is shown in fig. 147. The irradiation was at 30 °C and testing was at three temperatures; 4, 78 and 300 °K. Two things are clear. First, σ_{css} increases linearly with $(\Phi t)^{\frac{1}{3}}$ over the whole range of observation from 10^{16} to 10^{20} neutron.cm^{-2}.

Figure 148 shows the dependence of σ_{css} on testing temperature T_t in more detail. There appears to be a roughly linear relationship between $\sigma_{css}^{\frac{2}{3}}$ and $T_t^{\frac{2}{3}}$. This is found at all neutron doses so far investigated. Taking this fact with the observation that three testing

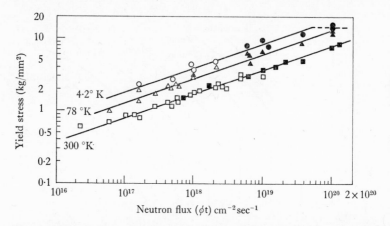

Fig. 147. Effect of neutron irradiation on the yield stress of Cu at various temperatures. The log of the yield stress is plotted as a function of the log of the dose. A straight line of slope $\frac{1}{3}$ is drawn through each set of points. Open symbols: bombarded in graphite reactor; closed symbols: bombarded in LITR. (From Blewitt, 1962.)

temperatures in fig. 147 give parallel lines means that σ_{css} must be a function of the form:

$$\sigma_{css} = F(\Phi t)\,\Theta(T_t) \qquad (7.1)$$

in which the variables are separable.

The effect of irradiation temperature T_i is not very marked. Blewitt (1962) found that the increase in σ_{css} following irradiation at 4.2 °K is only about 10 % greater than that following irradiation at 300 °K ($T_t = 4$ °K in each case). This was confirmed by annealing a sample irradiated at low temperature (Blewitt, 1962, Diehl, Leitz & Schilling, 1963) when it was found that σ_{css} decreased by 5 % in Stage I, 10 % in Stage II, and 10 % in Stage III. This is shown in fig. 149.

It has been found by many workers that the major recovery occurs in the range 300 to 400 °C, sometimes referred to as Stage V. It is not a uniquely activated process, the activation energy varying around 2 eV.

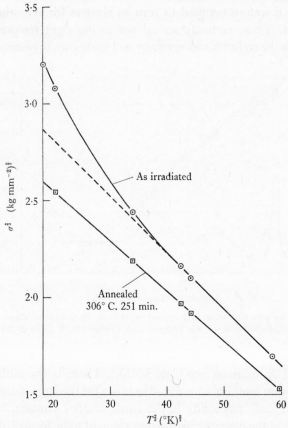

Fig. 148. $\sigma^{\frac{2}{3}}$ as a function of $T^{\frac{2}{3}}$ for irradiated crystals showing much better agreement after annealing with Seeger's theory, which predicts that $\sigma^{\frac{2}{3}} = A - BT^{\frac{2}{3}}$. (From Makin, 1963.)

These facts suggest that irradiation hardening is not directly associated with the presence of individual point defects for the following reasons:

(1) During the anneal from low temperature σ_{css} always decreases, if point defects were the impediment to dislocation

movement one might expect some increases, similar to those seen in internal friction, in temperature ranges where defect migration occurs. Certainly one would expect much bigger effects.

(2) When hardening does anneal it is at a temperature at least 200 °C higher than the highest point defect migration temperatures.

One is therefore tempted to turn to clusters for the origin of hardening. These certainly anneal out in the right temperature range with the right activation energy and are known to obstruct the

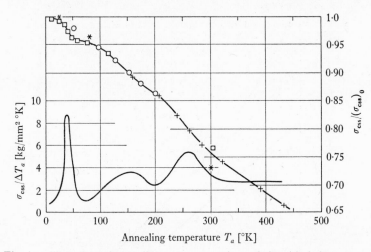

Fig. 149. The effect of annealing copper crystals on their critical shear stress, following irradiation at very low temperature. (From Diehl, Leitz & Schilling, 1963.)

motion of dislocations (see plate XXIX). There is the difficulty, however, that mobility is required in order for the large clusters to form. The only possibility is the small vacancy clusters, barely resolvable in the electron microscope, thought to be formed almost instantaneously in the cascade, perhaps assisted by activation in the thermal spike. We have already considered the strong evidence for such clusters in §7.1.2 where we found them behaving as nuclei for larger clusters.

In 1958, when only a limited selection of the experimental data was available, A. Seeger (1958) proposed a model for the irradiation hardening effect, which has proved remarkably successful. He suggested that zones of defective lattice are produced by the collision

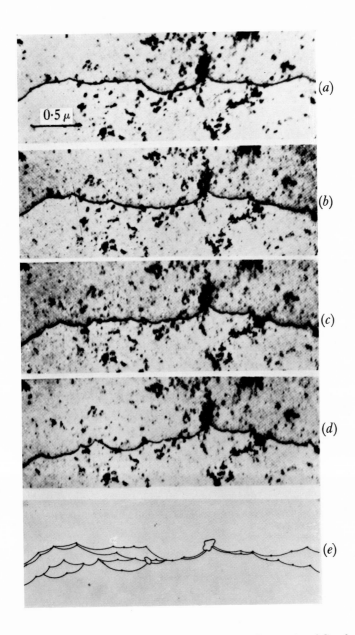

Plate XXIX A sequence of four transmission electron micrographs of Cu after it has been bombarded with 3×10^{17} 38 MeV α-particles cm^{-2} showing in (a), (b), (c) and (d) successive positions of a dislocation line which moves downwards, (e) showing these positions superimposed. (From Barnes, 1964.)

(*Facing p.* 340)

cascade, and these prevent the free movement of dislocations. In order for a dislocation to cut through such a barrier either one must increase the external stress to provide the necessary energy, or thermal activation must be relied upon. Thus, for a given rate of surmounting obstacles (i.e. a given strain rate), the external stress will decrease with increasing temperature.

The simplest version of the model assumes all the barriers to be identical, requiring an activation energy U for a dislocation to pass

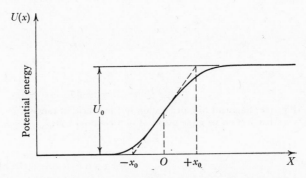

Fig. 150. Energy profile for the cutting of a dislocation through one of the 'zones' in neutron-irradiated Cu in the absence of applied stress. (From Seeger, 1958.)

through. The detailed behaviour is illustrated in fig. 150 showing how the energy of the crystal increases progressively as the zone is cut. Seeger assumed that a reasonable function to describe this behaviour would be

$$U(x) = U_0\left[1 - 1\Big/\left(1 + \exp\frac{x}{x_0}\right)\right]. \qquad (7.2)$$

In the light of experiments since 1958 it is probable that many barriers are attractive centres with $U(x)$ some sort of symmetrical potential well. This should not affect the results of the Seeger analysis provided one remembers that U_0 may represent the depth of an attractive potential well with a half-width of roughly $2x_0$.

Under stress the energy of the complete system (i.e. crystal + tensometer) must include the work done by the stress. If the barriers along the dislocation line are separated by a distance l, as shown in fig. 151, a movement forward at one barrier by x causes the dislocation to sweep an area of roughly lx. The swept area slips by b, therefore the shear strain is blx and the work done is $\sigma_{\text{css}}blx$. Then the energy relation is

$$U(x) = U_0\left[1 - 1\bigg/\left(1 + \exp\frac{x}{x_0}\right)\right] - \sigma_{\text{css}}blx \qquad (7.3)$$

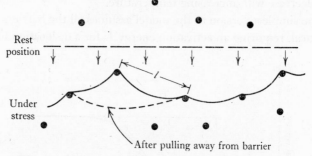

Fig. 151. Segment of a dislocation line in a field of random barriers on its glide plane. (From Holmes, 1964.)

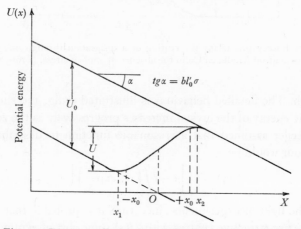

Fig. 152. Energy profile for the cutting through the 'zones' in the presence of applied shear stress. (From Seeger, 1958.)

and this is shown in fig. 152. The required activation energy is clearly $U(x_2) - U(x_1)$ and for reasonable values of U_0 and x_0 this is given approximately from (7.3) as

$$U = U_0[1 - 4blx_0\,\sigma_{\text{css}}/U_0]^{\frac{3}{2}}. \qquad (7.4)$$

The number of barriers per unit length of dislocation is roughly

$1/l$. With a dislocation density ρ_d there is a total length ρ_d of dislocation per unit volume, and therefore a total number of barriers on dislocation lines of ρ_d/l. The rate of jumping these barriers is $(\nu\rho_d/l)\exp(-U/kT_t)$, by (2.9), where ν is the frequency factor defined in §2.2. Each barrier jumped allows the dislocation to bow out onto the next and sweep out an area of the order of l^2, producing a strain of bl^2. See the derivation of (3.4). Therefore the strain per unit time is given by

$$\dot{\epsilon} = \rho_d\, blv \exp(-U/kT_t). \tag{7.5}$$

This can be rewritten as:

$$U = kT_t \log(\rho_d\, blv/\dot{\epsilon}). \tag{7.6}$$

Eliminating U from (7.4) and (7.6) one obtains

$$\sigma_{css} = \frac{U_0}{4blx_0}\left\{1 - \left[\frac{kT_t}{U_0}\log(\rho_d\, blv/\dot{\epsilon})\right]^{\frac{2}{3}}\right\}.$$

The distance l must now be expressed in terms of ρ_b, the volume density of barriers. To calculate the number that influence unit area of slip plane one takes a slab, containing the slip plane, and $2x_0$ in thickness and assume that this contains all the effective barriers. Then there are $2x_0\rho_b$ barriers per unit area and the mean distance between them is therefore $1/\sqrt{(2x_0\rho_b)}$. As a first approximation one could take

$$l \simeq \frac{1}{\sqrt{(2x_0\rho_b)}}.$$

However, this is not correct because a completely straight line would intersect no barriers and therefore one might expect $l \to \infty$ as $\sigma_{css} \to 0$. The greater the stress the more dislocations are forced to bulge through the barriers, intersecting new ones on the other side. Friedel (1956) gives the following expression for l under moderate stress,

$$l = \left(\frac{\mu b}{2x_0\rho_b\,\sigma}\right)^{\frac{1}{3}}. \tag{7.7}$$

Since $\mu b/\sigma = R$ by (3.5) it follows that (7.7) agrees with the approximate expression when $R \sim l$, which is the case for moderate stresses.

In the expression giving σ_{css} we now introduce (7.7), using the approximate form in the logarithm, since this is insensitive, then,

$$\sigma_{\text{css}}^{\frac{2}{3}} = \frac{U_0}{4bx_0}\left[\frac{2x_0\rho_b}{\mu b}\right]^{\frac{1}{3}}\left\{1 - \left[\frac{kT_t}{U_0}\log\frac{\rho_d\,bv}{\dot{\epsilon}\sqrt{(2x_0\rho_b)}}\right]^{\frac{2}{3}}\right\} \qquad (7.8)$$

The linear relationship between $\sigma_{\text{css}}^{\frac{2}{3}}$ and $T_t^{\frac{2}{3}}$ agrees with the observed temperature dependence of σ_{css}. The dose dependent quantity is ρ_b and this should be proportional to Φt, hence, if one neglects the slight influence of the logarithmic term, σ_{css} is proportional to $(\Phi t)^{\frac{1}{2}}$, which is in fair agreement with the observations. If one again forgets about the insensitive logarithmic term, the function (7.8) satisfies the condition of (7.1) that dose and T_t should be separable variables.

Several suggestions have been made to remove the discrepancy in the dose dependence, one being a saturation effect in ρ_b caused either by the overlap of zones or the weakening of zones by dynamic interstitials thrown out from nearby cascades. The possibility that normal point defect migration is responsible for weakening the zones should be ruled out by the observation by Diehl, Leitz & Schilling, during a 4 °K irradiation, of a $(\sigma t)^{0.4}$ relationship. The saturation model has serious objections raised against it by Blewitt who shows that the power law holds over too large a range of Φt. In view of the approximations involved in Seeger's model there seems little point in pursuing this any further. The prediction of $(\Phi t)^{\frac{1}{2}}$ seems remarkably close to experiments and the direct observation of dislocation behaviour, to be described below, will show that Seeger's model cannot be correct in detail.

Makin, Whapham & Minter (1961) have shown a correlation between the hypothetical barriers and the clusters observed in the electron microscope. Taking the density of clusters ρ_c, as described in §7.1.1, the graph of $\sqrt{\rho_c}$ *versus* σ_{css} was plotted as in fig. 153. For the smallest loops (< 25 Å) a direct proportionality is found, for the larger loops a curve is obtained. Equation (7.8) shows σ_{css} proportional to $\sqrt{\rho_b}$ and the observation therefore suggests that the barriers are the small vacancy clusters, discussed in §7.1.1, some of which are visible in the electron microscope.

The effect of annealing on the relation between σ_{css} and T_t throws further light on the nature of the barriers. One sees from fig. 148

that (7.8) is obeyed more closely by annealed specimens than those in the irradiated state, where the hardening at low temperatures is greater than the model predicts. Makin (1963), who made these observations, suggests that instead of single type of barrier one must have a size distribution with a corresponding range of energies U_0. Then at low temperatures of testing the effective density of

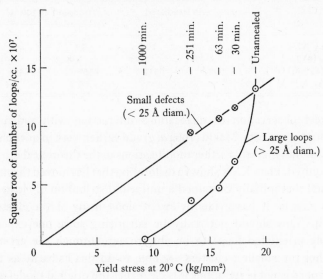

Fig. 153. The (defect density)$^{\frac{1}{2}}$ as a function of the yield stress during annealing for both large and small defects. (From Makin, Whapham & Minter, 1961.)

barriers would be higher than at high temperatures, where the smaller ones would be transparent to dislocations. After annealing, the smaller clusters would either dissociate and disappear or else grow larger, giving a more similar set of barriers, and hence behaviour according to (7.8).

By studying σ_{css} as a function both of $\dot{\epsilon}$ and T_t and analysing the results with a modified version of (7.8) Makin (1963) deduced values for U_0 and the quantity $bl(x_2 - x_1)$. This has the dimensions (length)^3 and is the *activation volume* for the process.

Assuming that in the annealed state the small clusters observed in the microscope to be the only barriers, ρ_b can be measured directly. Then assuming a value for b the effective width of the barriers $(x_2 - x_1)$, as experienced by dislocations, may be deduced.

The results of Makin's analysis are summarized in table XXXI and these are clearly consistent with the picture presented above. After irradiation the effective barriers at low T_t are smaller than at high T_t. After some annealing the barriers are nearer the same size, irrespective of T_t.

TABLE XXXI

	As irradiated		Annealed at 306 °C	
T_t (°K)	77	293	77	293
U_0 (eV)	0·87	2·88	2·00	3·01
$bl(x_2-x_1)$ (Å³)	993	8510	2550	7740
(x_2-x_1) (Å)	—	—	0·7	2·2

Direct observation of dislocations interacting with radiation produced clusters by Makin (1964 a, b) and earlier work in quenched metals shows that still further modifications to the theoretical model are required. Plate XXX shows a dislocation that has moved through a crystal that initially contained a uniform distribution of clusters. In its passage it has evidently swept along many of the clusters with it. This should not really be surprising since one expects suitably orientated loops to be able to slip through the crystal. Whether the smallest clusters are able to do this as easily as the larger ones is not certain, but it is clear that the original model of a dislocation cutting rigid barriers must be modified somewhat. Further, our interpretation of U_0 and (x_2-x_1) may require some alteration.

The sweeping effect means that subsequent dislocations will have an easier passage across the same slip plane. It is suggested that this is responsible for the sudden increases in local slip that are found in the jerky-flow of the Luders region.

So far we have considered only the effect of neutron irradiation. The comparison by Makin & Blewitt (1961, 1962) with electron irradiation provides some interesting further information. The behaviour of σ_{css} with electron dose at various temperatures T_i is shown in fig. 154. There is a strong dependence on T_i, in contrast with the neutron case, and a virtual saturation after a dose producing a concentration of displaced atoms $C_d < 10^{-4}$. No such saturation is found after neutron irradiations giving $C_d \sim 10^{-2}$. Secondly,

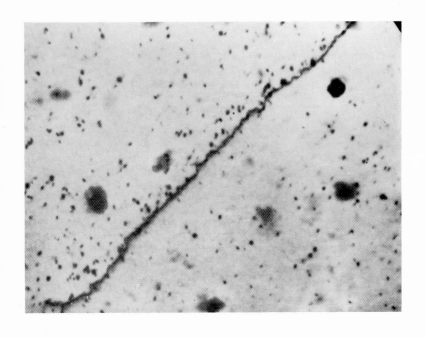

Plate XXX The sweeping up of loops in neutron irradiated Cu by a moving dislocation. × 40,000. (From Makin, 1964.)

fig. 155 shows that much of the hardening anneals at relatively low temperatures, for below Stage V.

The doses in this experiment were too small to invoke the athermal cluster formation by statistical fluctuation, discussed in §5.8.1. It is suggested by Makin & Blewitt that the electron irradiation can only form clusters by homogeneous nucleation and their density would not vary in proportion to dose. As shown in §7.1.1 such

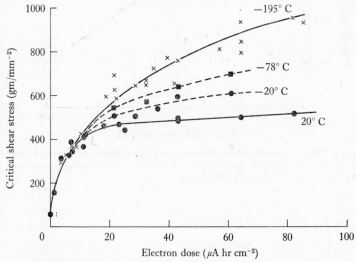

Fig. 154. The critical shear stress of Cu single crystals as a function of electron dose at various temperatures. (From Makin & Blewitt, 1961.)

clusters must be very small and their numbers must saturate rather rapidly at room temperature. Annealing would cause dissociation of the smaller clusters at relatively low temperatures.

7.3.2. *The b.c.c. transition metals.* Deformation of the b.c.c. transition metals such as Fe, Nb, Mo, and W after neutron irradiation show that there is an increase in hardness similar to that in Cu, with the increase roughly proportional to $(\Phi t)^{\frac{1}{3}}$. It is also found that σ_{css} varies with strain rate in a way that allows activation volumes to be deduced. (See Johnson, Sargent & Wronski, 1964, for a review.)

The striking difference between Cu and these b.c.c. metals is in the annealing behaviour, illustrated for the case of Nb in fig. 156. An increase of about 100 % in hardening is found on annealing through

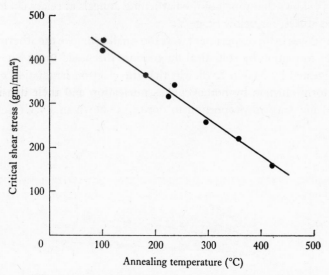

Fig. 155. The decrease in critical shear stress on annealing of crystals irradiated at 20 °C. $\Phi = 31$ μ.amp. h^{-1}. cm^{-2}; Annealing time 30 min; testing temperature -195 °C. (From Makin & Blewitt, 1961.)

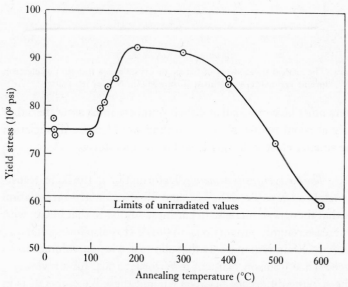

Fig. 156. The effect of annealing irradiated Nb on the yield stress. (From Makin & Makin, 1959.)

a narrow temperature range near $0.2T_m$. This is the range where experiments described in §6.4 showed a point defect to be mobile. Activation energy measurements on this hardening effect give the same result as the resistivity experiments. Final recovery of the hardening occurs over a wide range of temperatures, with a corresponding range of activation energies, just as in the case of Cu.

Attempts to correlate hardening with defects visible in the electron microscope (see §7.2.4) have proved difficult because no clusters are visible until high doses $> 10^{19}$ neutron.cm^{-2} are attained. From the observations it appears that sub-microscopic defects, possibly even point defects in part, are responsible for irradiation hardening in b.c.c. metals. The exact mechanisms have yet to be worked out but Downey & Eyre (1965) suggest that some vacancy clusters are formed during irradiation and that these are responsible for hardening. During the annealing peak at $0.2T_m$ they suppose vacancies to become mobile causing a strengthening of existing clusters and perhaps the formation of some new ones. Presumably the final recovery occurs by dislocation of the smallest clusters and growth of the larger ones beyond the point where they are effective as barriers.

An important feature of these metals is the brittle fracture that occurs when they are stressed at low temperature. This occurs for a given strain rate when T_t is below T_{db} the critical *ductile-brittle transition temperature*, and T_{db} increases with strain rate. The effect of irradiation is always to raise T_{db} and in Mo it changes from -136 to -73 °C for 5×10^{19} neutrons.cm^{-2} (Makin & Gillies, 1957). The mechanisms that lead to brittle fracture are not well established and it is unfortunately not possible to give an explanation of the *irradiation embrittlement* effect at present.

CHAPTER 8

IMPURITY DAMAGE

8.1. Introduction

So far we have only considered effects that originate from the displacement of atoms in the collision cascade. These are often referred to as *displacement damage*. An entirely different type of damage is associated with impurity atoms introduced either by transmutation of the target nucleus or when the bombarding ions come to rest in the specimen. This is known as *impurity damage*.

In the transmutation case, the concentration C_{imp} of impurities is calculated as follows. Suppose a pure element is bombarded with a dose Φt particles per unit area which induce transmutation with an effective cross-section σ_{trans}. Where the element consists of several isotopes, only one of which undergoes the reaction, the isotopic cross-section must be weighted by its fractional abundance to give σ_{trans}. Then if each reaction produces x atoms of a particular impurity, its concentration is given by

$$C_{\text{imp}} = x\sigma_{\text{trans}}\Phi t. \qquad (8.1)$$

Equation (8.1) is only true for small enough doses that the concentration of target nuclei remains essentially constant. If we take the case of uranium fission as an example, a wide selection of product atoms are possible, distributed according to the well-known curve giving x as a function of atomic number and shown in fig. 157. Of special interest is the fact that the inert gases Xe and Kr are amongst the most probable products.

An alternative origin of impurity damage is found in irradiation with accelerated ions. When these come to rest in the solid they become impurity atoms. For instance protons become hydrogen and α-particles helium. To calculate concentrations in such cases one needs to know that $1\,\mu\text{A}.\text{cm}^{-2}$ is 6×10^{12} singly-charged particles per cm^2 and, knowing the range and straggling data, an estimate can be made of the concentration distribution. In the

U.S.A.E.C. Nuclear Data Tables, Part 3, will be found a useful collection of range and straggling data for various particles.*

Of course, the concentration of impurities introduced by irradiation bears no relation to the equilibrium concentration and in some

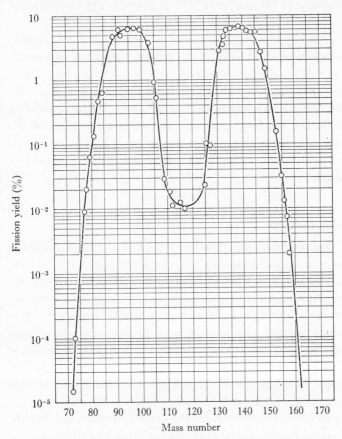

Fig. 157. Fission yields of products of various mass numbers. [(From Gladstone & Edland, 1952.)

cases will greatly exceed it. In many ways such cases are the most interesting scientifically, for there is often no other way of preparing such over-saturated solid solutions. An example of extreme insolubility is Pb in Al and if Pb^+ ions are injected at about 50 keV

* These tables are published by National Academy of Sciences, Washington, D.C., U.S.A.

into an Al foil at sufficiently high temperatures for diffusion to occur, precipitate particles of Pb form, as illustrated in plate XXXI. Another example referred to already in chapter 2 are the inert gases. These are also very important in reactor technology because of their origin in ^{10}B (n, α) and ^{235}U $(n,$ fission$)$. The remainder of this chapter will be devoted to these gases.

Other cases of importance are in semiconductors, where impurity concentrations as small as 10^{-10} can influence electrical properties. In the irradiation damage of metals, such as Au, with a large trans-mutation cross-section with neutrons, the induced impurities can trap interstitials and vacancies and influence the accumulation process. In fact Martin (1962 b) studied the trapping of interstitials on Hg atoms in Au by this method and Barnes & Mazey (1961) used He in Cu to decide between interstitial and vacancy loops (see Chapter 7).

8.2. Gas bubbles

8.2.1. *Introduction.* The practical problems of impurity damage first arose in materials of nuclear reactors. It was found, for instance, that metallic uranium, when irradiated above about 500 °C, suffered a large volume increase. One might expect a volume increase in any fissile material simply because each atom that undergoes fission is replaced by *two* new ones, hence when 1 % of the atoms have fissed the total number of atoms is increased by 1 % and the volume should increase by about 1 %. However one should also allow for the fact that most fission products have a greater atomic radius than uranium and then one predicts about 2·5 % volume increase.

Near room temperature this is the magnitude observed, but the effects above 500 °C can be as big as 100 % and clearly cannot be explained by simply multiplying the atoms. In such cases, metallo-graphic examination revealed the presence of small holes in the metal, the number and size of which would account for most of the volume increase. Similar effects were found in boron-containing control materials and in beryllium moderators. A glance at table V will show that all these materials have in common a neutron-induced reaction which forms an inert gas. We have seen in chapter 2 that such gases are likely to have large positive energies of solution. It

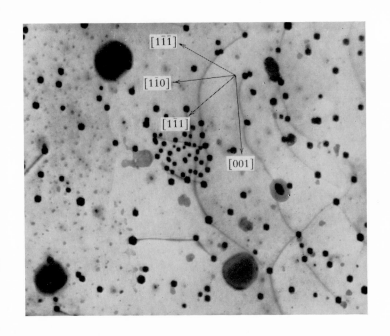

Plate XXXI A precipitate of Pb particles formed as impurity damage in Al by bombarding a thin foil with 50 keV Pb$^+$ ions (by courtesy of P. A. Thakery and R. S. Nelson).

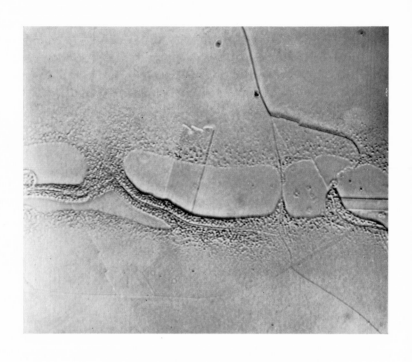

Plate XXXII He bubbles formed in α-irradiated Cu by heating at 150 °C for 1 h
(by courtesy of R. S. Barnes).

was therefore suggested that the holes were actually bubbles of this gas which, due to extreme insolubility, had precipitated out of solution. Support for this idea came from the observation that externally applied pressure would reduce the volume increase (Churchman, Barnes & Cottrell, 1958). Further, when the holes grew large enough to link up and connect with the outer surface, large quantities of the inert gas were released into the irradiation capsule.

Experiments on reactor irradiated materials proved very tedious because of the high levels of radioactivity in the samples. An alternative method of studying gas bubbles was used with great success by R. S. Barnes and his colleagues at Harwell. They used α-particles from an accelerator to inject He into metal samples. In one series of experiments they injected 38 MeV α's into Cu and Be targets producing a 1 % concentration of He in a layer about 30μ thick, at a depth of about 1 mm below the surface. Metallographic examination of a section through the sample, perpendicular to the bombarded surface, showed no effect immediately after irradiation. But after heating to 750 °C the Cu samples showed bubbles in the He-rich layer. In the first stages of anneal the bubbles were not uniformly distributed but were confined to the edges of the layer or to regions where grain boundaries crossed the layer. Plate XXXII illustrates this effect. After long periods of anneal the bubbles were large and filled the whole layer.

From this observation it was deduced that the helium alone is not able to form bubbles and that vacancies are required to provide the necessary space for expansion. It was inferred that vacancies came mainly from grain boundaries or free surfaces, being formed by thermal activation (Barnes, Redding & Cottrell 1958). Where they entered the He-rich zone, bubbles were able to form. In the early stages the available vacancies were all absorbed in the outer regions of the layer and unable to penetrate into the centre.

Apart from its interest in radiation damage this experiment showed that vacancies are mainly produced by thermal activation at surfaces or grain boundaries. The more perfect types of interface such as twin boundaries did not behave in this way.

Further light was shed on the precipitation of He in Cu by the work of Russell and his colleagues (Russell & Hastings, 1965; Vela &

Russell, 1966). By measuring the electrical resistivity and the lattice spacing as well as conducting microscopical examinations they were able to follow the behaviour of the He whilst it was still in the point defect state.

Rather than use an external source of alpha particles, Russell prepared a dilute alloy of 0·25 % B in Cu, the B being finely dispersed in small particles and not in solid solution. Irradiating this

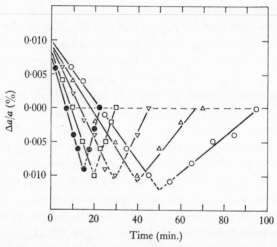

Fig. 158. The effect of time upon $\Delta a/a$ % for a range of temperatures. O, 550 °C; △, 575 °C; ▽, 600 °C, □, 625 °C; ● 650 °C. (From Russell & Hastings, 1965.)

with thermal neutrons in a reactor then caused the $^{10}B\,(n,\alpha)\,^{7}Li$ reaction to produce He.

Immediately following irradiation at 100 °C, both the lattice spacing and the resistivity had increased. Annealing at various temperatures between 300 and 800 °C produced the effect shown in fig. 158 where the lattice spacing first decreases and becomes negative, then increases to the normal value for Cu.

Remembering the discussion in § 2.3.5 the initial increase is consistent with the presence of He in interstitial positions, causing outward dilatation of the lattice. The decrease during the first stage of annealing denotes the combination of each He interstitial with perhaps two vacancies, eventually giving a decrease in lattice spacing due to an inward dilatation around the cavity. Finally, the complexes (He + 2v) migrate and coalesce with further vacancies to form bubbles, and when these become big enough the dilatation of the lattice becomes negligible (see (2.29) and the text immediately following it).

By analysing the curves shown in fig. 158 activation energies for the two stages were deduced as 0·76 and 0·97 eV. The first of these probably corresponds to the migration energy of single vacancies and the second to the migration energy of the complex (He + 2v).

8.2.2. *The behaviour of bubbles.* Consider the equilibrium of a gas-filled bubble in a solid containing mobile vacancies. There will be a continual arrival of vacancies, tending to make it grow, and also a departure, tending to shrink the bubble. The equilibrium size will be reached when the two rates exactly balance. The criterion for equilibrium is somewhat analogous to the bubble in a liquid.

Suppose the bubble has radius r and contains gas at a pressure p. The volume $v = \frac{4}{3}\pi r^3$ and the surface area $a = 4\pi r^2$. If the bubble radius changes by dr then

$$dv = 4\pi r^2 \, dr \quad \text{and} \quad da = 8\pi r \, dr.$$

During the expansion the pressure does work $p \, dv$ and the surface energy increases by $\gamma \, da$, where γ is the surface energy per unit area. For a bubble or equilibrium size these two must be equal in magnitude, i.e.

$$p \cdot 4\pi r^2 \, dr = \gamma 8\pi r \, dr,$$

hence
$$p = \frac{2\gamma}{r}. \tag{8.2}$$

The formation energy of the bubble is given by adding the surface energy of the cavity to the work done to compress gas into a volume v at pressure p, i.e.
$$U_{\mathrm{f}}^{\mathrm{b}} = 4\pi r^2 \gamma + pv$$

and substituting for p from (8.2) gives

$$U_{\mathrm{f}}^{\mathrm{b}} = \tfrac{20}{3}\pi r^2 \gamma. \tag{8.3}$$

This treatment assumes, first, that γ is isotropic and secondly that the gas obeys the perfect gas laws. We shall examine later the consequences of γ's anisotropy in crystals. Using the perfect gas laws again gives
$$pv = \tfrac{3}{2}mkT.$$

With $p = 2\gamma/r$ and $v = \frac{4}{3}\pi r^3$ this gives for m, the number of gas atoms:

$$m = \frac{8\pi\gamma r^2}{3kT}. \tag{8.4}$$

Good evidence that these relations are correct comes from the observation in the electron microscope of He bubbles coalescing in Cu. It was found by Barnes & Mazey (1963) that when two or more join together the (radius)^2 of the large bubble equals the sum of the squared radii of the individual bubbles.

Plate XXXIII shows a typical sequence of micrographs. If the nth small bubble contains m_n atoms and the large bubble contains M, and none are lost, then

$$M = \Sigma m_n$$

then using (8.4) to give M and m_n in terms of R and r_n we obtain

$$\Sigma r_n^2 = R^2. \tag{8.5}$$

Hence one sees that the observed relationship is a good test of the theoretical model.

The bubble can be imagined as a vacancy cluster containing gas atoms. It is interesting to compare the number of vacancies n with the associated number of gas atoms m. Using the notation of §2.3.2

$$\frac{n}{m} = \frac{4\pi r^2}{3\Omega} \cdot \frac{3kT}{8\pi\gamma r^2},$$

$$\frac{n}{m} = \frac{kT}{2\Omega\gamma} r.$$

At 1000 °K, taking $\gamma \sim$ 0·1 eV/Å^2 and $\Omega = 3$ Å^3 we find

$$\frac{n}{m} = \frac{r}{3} \quad \text{(with r in Å).}$$

Hence when $r = $ 100 Å vacancies outnumber gas atoms by 30 to 1. When $r = $ 10^5 Å the ratio is 30,000 to 1. This clearly explains the observation that vast numbers of vacancies are needed to grow large gas bubbles.

Now look at the magnitude of the pressures involved, this time taking γ in c.g.s. units as $\sim$ 10^3 erg. cm^{-2}. For 100 Å bubbles $p \sim$ 10^3 atm. and for 10^5 Å $p \sim$ 1 atm. This is immediately useful in relating the bubble size to volume increase, or swelling, of the specimen.

Plate XXXIII A sequence of six electron micrographs of a thin foil of Cu containing o·1 atomic % ⁴He illustrating the mobility of bubbles. (From Barnes & Mazey, 1963.)

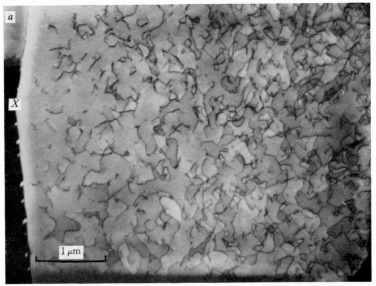

(*a*) Showing a dislocation network (resulting from the irradiation) upon which the bubbles have formed as a result of pulse heating to 800 °C.

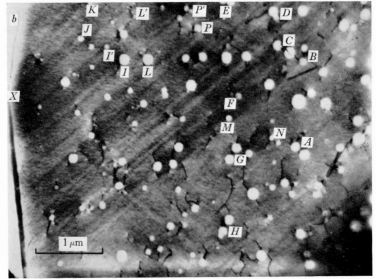

(*b*) The same area after three heating pulses, illustrating the redistribution and coarsening of the bubbles. The bubbles seen in the grain boundary in (*c*) have all left the foil.

(*Facing p.* 356)

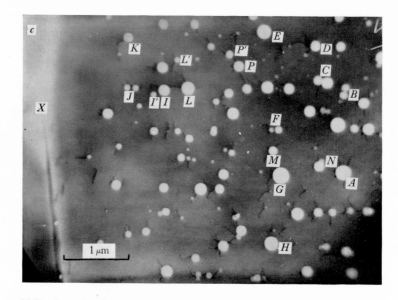

(c) Further migration of bubbles following another pulse, many have coalesced, e.g. the groups *A, D, E, K, H*, and the group of three at position *G* in (*b*) have each formed one composite bubble.

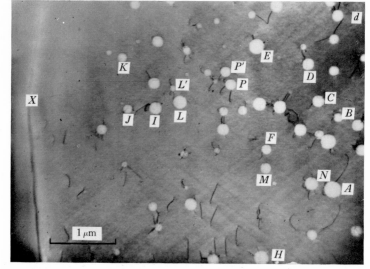

(*d*) After another pulse showing more migration, coalescence, and loss of bubbles to the surface, e.g. the large bubble at *G*. Short lengths of dislocations are seen to connect some bubbles (*A, E, C* and *H*), to the foil surface.

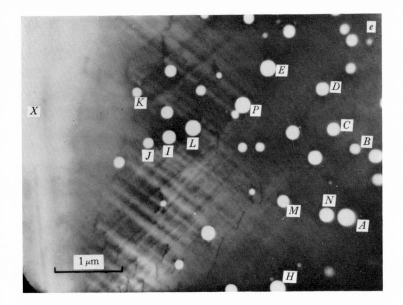

(e) After two more pulse anneals. Note the bubble at *P* has a small satellite.

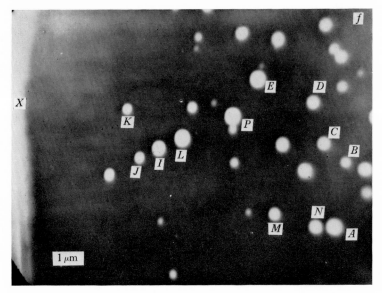

(f) After another, the tenth pulse, the small satellite at *P* has rotated around the larger bubble, and many more bubbles have been lost to the foil surface.

Plate XXXIV Electron microscope replica of U 14 wt % Mo alloy irradiated at 800 °C to 0·22 % burnup, swelling 1 %, spacing 10^{-4} cm, radius 10^{-5} cm. The bubbles have white 'shadows' behind them (by courtesy of V. J. Haddrell).

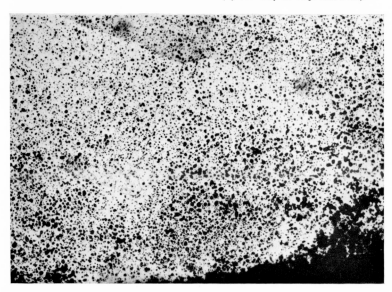

Plate XXXV Metallographic section of U 4 wt % Mo, irradiated at 700 °C to 0·48 % burnup, swelling 32 %, spacing 10^{-3} cm. (by courtesy of G. W. Greenwood and D. C. V. Jones).

Obviously, gas that is stored under high pressure requires less volume and small bubbles will therefore bring a small volume increase. To make this quantitative, suppose that unit volume of solid contains gas atoms equivalent to a volume w at NTP. (In uranium $w = 4.7$ c.c. for 1% burnup). If the fractional volume increase, or *swelling*, is (dV/V) then in unit volume, w is compressed into a volume (dV/V) at temperature T. Assuming all the gas to be contained in bubbles of the same radius r, and that the Gas Laws can be applied:

$$p = \frac{Tw}{273(dV/V)}. \tag{8.6}$$

From (8.2) and (8.6) it follows that

$$\frac{dV}{V} = \frac{Twr}{546\gamma} \tag{8.7}$$

which expresses the relation between swelling and bubble radius.

Another important parameter is the spacing between bubbles and if we idealize further and assume a uniform spacing d, then each bubble of volume $\frac{4}{3}\pi r^3$ is contained in a volume of approximately d^3. Hence:

$$\frac{dV}{V} = \frac{\frac{4}{3}\pi r^3}{d^3}. \tag{8.8}$$

Eliminating p and r from (8.3), (8.6) and (8.8) gives

$$\frac{dV}{V} = \sqrt{\left(\frac{3}{4\pi}\right)}\left[\frac{Twd}{546\gamma}\right]^{\frac{3}{2}}. \tag{8.9}$$

Thus, once the bubble spacing is set, the swelling is immediately defined. This is an important result technologically, for if a material can be prepared in which bubbles remain finely dispersed, they will always be small, contain their gas at high pressure, and produce little swelling (Greenwood, Foreman & Rimmer, 1959).

Plates XXXIV and XXXV show metallographic sections through two alloys of uranium after similar irradiations. In the first case the bubbles are on a fine scale, $d \sim 10^{-4}$ cm and swelling is only 1% for 0.22% burnup. With the second $d \sim 10^{-3}$ cm and swelling is 32% for 0.48% burnup.

Equation (8.3) was derived assuming that bubbles are spherical, and when observed by transmission electron microscopy some of them certainly take this shape. However, in a crystal the surfaces with lowest free energy are those which expose closely packed planes, and we might expect a bubble that was truly in equilibrium to take up a polyhedral form, just as a crystal grown from solution develops with low index facets. Several workers have found that such polyhedral bubbles exist, but only when they have been held for some time at a temperature where there was a plentiful supply of vacancies. Plate XXXVI a shows a typical example of helium bubbles in Cu soon after formation. The effect of destroying equilibrium is shown by comparing plate XXXVI a with XXXVI b in which the sample had been annealed at 700 °C for 15 min. During this period sufficient vacancies have been assimilated to get the bubbles to an equilibrium size and their shape is clearly polyhedral.

Studies of these polyhedral bubbles have proved fruitful in the field of Surface Physics. Because the bubbles are in the interior of a solid and contain only an inert gas one may expect their surfaces to be free from chemisorped layers. Hence the surface energies γ^{hkl} of the polyhedral facets should be of fundamental interest. From measurements of the shape of bubbles it is possible to deduce the ratios of surface energies between their facet planes.

The method of deducing such ratios, or knowing them, of deducing the shape of bubbles will now be dealt with in outline. According to Herring (1953) a hole in thermal equilibrium should have flat facets of low surface energy, joined in some cases by smoothly rounded edges and corners. Thus the cavity need not be strictly polyhedral but could be roughly spherical with some flattened surfaces. The determining criterion is the polar graph of γ as a function of direction in the crystal. Such a graph is shown schematically in fig. 159. The radius vector gives the magnitude of γ for the plane to which it is normal. Cusps on the plot indicate planes of low surface energy. The equilibrium bubble shape is found graphically by fitting a polyhedron inside the polar plot, as shown. The polyhedron must be contained inside the γ plot, otherwise the corners will be rounded. This is intuitively reasonable, but the reader is referred to Herring's paper for a rigorous proof.

In practice the shape of the polar graph is seldom known, indeed bubble measurements are proving invaluable in providing data. As an example, suppose we have bubbles shaped like those in fig. 159, and that observation has enabled the ratio b/a to be determined. Then it is a straightforward matter of geometry to show that the ratio $\gamma^{100}/\gamma^{110}$ is given by

$$\frac{\gamma^{100}}{\gamma^{110}} = \sqrt{2} - \frac{b}{a}.$$

Bubbles like this, formed of {100} and {110} facets, are found in Cu bombarded with α's and annealed for several hours at about

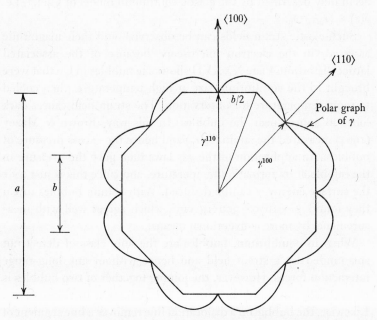

Fig. 159. The shape of a void in relation to the polar graph of γ.

800 °C. From measurements on them it was deduced that the ratio $\gamma^{100}/\gamma^{110} = 1 \cdot 2$ (Nelson, Mazey & Barnes, 1965). In Al at 550 °C bubbles formed from {100}, {110} and {111} planes were measured and it was deduced that $\gamma^{100}/\gamma^{110} = 0 \cdot 98$, and $\gamma^{100}/\gamma^{111} = 1 \cdot 03$.

If the equilibrium of bubbles is suddenly disturbed, by either heating or cooling, the internal pressure will be either too large or too small for it to be balanced by surface tension. The surrounding medium is then stressed. For plastic flow to occur, the stress must occur near a segment of dislocation line and extend over a region of comparable dimensions to the segment length. This could only occur for relatively large bubbles ($r \sim 10^{-4}$ cm) where pressures and therefore stresses are unlikely to exceed 10 atmospheres. Relatively low pressures, such as these, should not produce much plastic deformation.

Small bubbles ($r \sim 10^{-6}$ cm) could produce very high stresses but do not extend far enough to make dislocations operate. Instead they will be surrounded by a region of elastic strain, which can be accurately described by the elastic continuum model of §2.3.3, i.e. $\delta(r) = (r_0/r)^2 \delta_0$.

Such elastic strain fields can be observed, and their magnitude assessed, in the electron microscope because of the associated lattice distortion. Plate XXXVII shows He bubbles in Cu that were brought to the equilibrium size at high temperature, then cooled to room temperature for observation. The strain field causes dark lobes to appear near the bubbles. In this way, Brown & Mazey (1964) measured the strain (δ_0/r_0) and hence the excess pressure of bubbles. An application of the gas laws then gave the pressure in the bubble at its formation temperature, and since this is just $2\gamma/r$ the surface energy γ can be deduced. With helium bubbles in Cu they found $\gamma = 1670 \pm 300$ erg.cm^2, which agrees well with measurements by more conventional means.

When in equilibrium, bubbles are the only class of defect not surrounded by a strain field and hence without any long-range interaction forces. However, the joining together of two bubbles is energetically favoured, because of the reduction in surface area. Likewise, the bubble on a dislocation line removes a line segment of length $2r$, with formation energy $2\mu b^2 r$. This energy must be supplied by external work if the dislocation is to be pulled away from the bubble, hence bubbles will be strongly attached to dislocations and can influence mechanical properties. In this role they are particularly important at high temperatures where displacement damage has all recovered. Examples of bubbles on dislocation lines are seen in plate XXXVIII.

Out of equilibrium, the strain field around bubbles makes them interact with one another, or with dislocations, over long distances. In the next section we shall see that this has an important bearing on bubble movement.

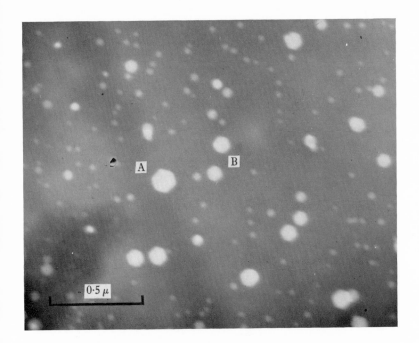

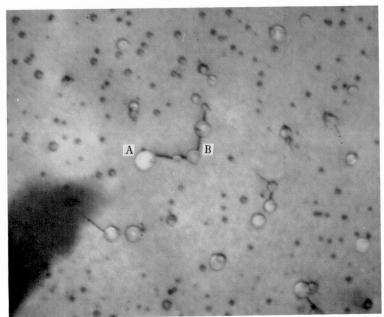

Plate XXXVI Transmission electron micrographs of a thin film of Cu containing (a) spherical helium bubbles which have formed during a pulse anneal to ~ 900 °C in the electron microscope, and (b) the same area after an anneal for 15 min, *in vacuo* at 700 °C, demonstrating the development of the crystallographic shape, e.g. at A and B. (From Nelson, Mazey & Barnes, 1965.)

(*Facing p.* 360)

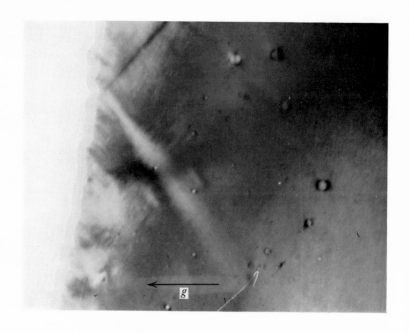

Plate XXXVII Strain fields of He bubbles in Cu cooled from 700 °C. Shown by images at the periphery of the bubbles, each with a line of no contrast normal to g; g here is a high order reflexion, (311). (From Brown & Mazey, 1964.)

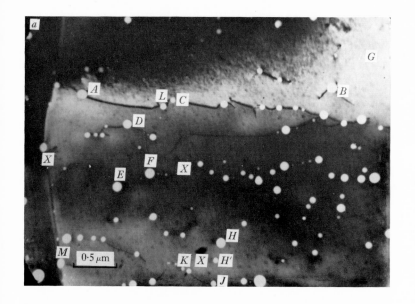

Plate XXXVIII Helium bubbles (some lying on dislocation lines) in a thin foil after several heating pulses. Reference points are the grain boundary, and surface impurities marked X. (From Barnes & Mazey, 1963.)

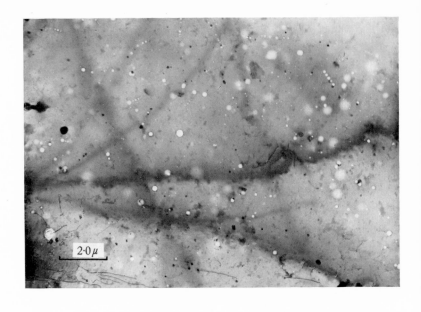

Plate XXXIX Helium bubbles in aluminium containing iron, showing the tendency for the bubbles to be attached to the precipitates. (From Barnes & Mazey, 1963.)

In two-phase alloys containing the second phase as a finely dispersed precipitate of small solid particles, bubbles are often found in association with the precipitate particles. An example is shown in plate XXXIX. This could arise from the strain field which often surrounds the precipitate particle in which the bubble strain energy could be lowered, if the dilatations were of opposite sign (see §2.3.3), or from favourable changes in surface energies at the precipitate boundary when a bubble is attached. This type of inter-action has an important application in reducing the swelling of reactor fuel. If a suitable precipitate can be found, and the disper-sion of particles made very fine, the spacing between bubbles d will also be correspondingly small. Equation (8.8) then shows that the volume increase will be small. Dilute alloys of Al and Fe in U were found to behave in this way and are used in the fuel elements of the first generation of British power station reactors (Bellamy 1962).

If the scale of a bubble precipitate is to coarsen, and we have plenty of evidence that this does occur during swelling, then there are two possible explanations. First, the gas in the bubbles may go back into solution, re-emerging preferentially into the larger bubbles where the pressure is lowest. Secondly, the bubbles themselves may migrate as entities, collision between bubbles leading to coalescence and hence coarsening.

There is certainly no evidence of re-solution, the first alternative, in the case of He in Cu. Barnes & Mazey (1963) did not observe any bubbles decrease in size during heating, furthermore they found that when two bubbles of dissimilar size were separated by only 100 Å for 10 sec at 800 °C no change in relative size occurred. The difficulty of re-solution lies in the high energy of solution rather than the migration energy, which must be fairly small for the bubbles to form in the first place.

With regard to the other inert gas-metal combinations this first alternative has yet to be dismissed so positively. Some interpretations of uranium swelling under neutron irradiation postulate a Xe and Kr re-solution process, but there is no *direct* evidence to support this so far, and there is no encouragement for the re-solution hypothesis from the calculated energies of solution in §2.6.3. There, we saw that U_s is expected to be positive for all inert gases, becoming larger with increasing atomic number. Thus if Xe and Kr do redissolve from bubbles in U, the energy of solution must be provided in some anomalous way and not by thermal activation.

Taking the second alternative, He bubble migration in a temperature gradient has been observed by Barnes & Mazey (1963) in thin films of Cu in the electron microscope. The effect is clearly seen in plate XXXVIII. For a 350 Å bubble the velocity can be as high as 10^3 Å/sec and there is apparently an empirical proportionality between velocity v and inverse radius $1/r$. Those bubbles which reached the surface burst and formed small craters, but these disappeared in a fraction of a second showing that surface diffusion is extremely rapid at the temperature involved. Now bubble migration requires the transfer of solid material from one side of the bubble to the other and for He in Cu it seems likely that surface diffusion around the bubble periphery is responsible. (In other materials, like UO_2, where the vapour pressure can be much higher, it appears that the larger bubbles move by evaporation and condensation across the bubble.)

The driving force for bubble migration in Barnes & Mazey's experiments was related to the temperature gradient. Other driving forces are also possible, of particular inportance is the binding energy between bubbles and dislocations. In plate XXXVIII one sees evidence of the moving bubbles dragging dislocations along with them. Conversely, if dislocations move, due to plastic deformation, or climb during annealing, we must expect them to sweep up the bubbles in their path and hence cause a coarsening.

The situation in nuclear fuels must be highly complex, for radiation growth and loop expansion is ocurring at the same time as bubble formation. Not only can the loops sweep up bubbles but the plastic strains due to differential growth in polycrystalline materials will cause glissile dislocations to do the same thing. The precise mechanisms by which uranium swelling occurs remains obscure at present owing to the complexity of the system, but bubble migration and coalescence under stress appear to be the dominant factors in the coarsening process. (See Barnes, 1965.)

BIBLIOGRAPHY

The figures in square brackets indicate the pages of the text in which the reference is quoted.

ABRAHAMSON, A. A. (1963). *Phys. Rev.* **130**, 693. [96–104]

ABRAHAMSON, A. A. (1964). *Phys. Rev.* **133**, A 490. [97, 98, 99, 101]

ADAM, J. & MARTIN, D. G. (1958). *Phil. Mag.* **3**, 1329. [304]

AIROLDI, A., BACHELLA, G. L. & GERMAGNOLI, E. (1959). *Phys. Rev. Letters* **2**, 145. [35, 37]

ALEXANDER, L. E. *see under* KLUG, H. P.

AMDUR, I. & BERTRAND, R. (1962). *J. Chem. Phys.* **36**, 1078. [98–103]

AMDUR, I. & MASON, E. A. (1954). *J. Chem. Phys.* **22**, 670. [98–103]

AMELINCKX, S. (1962). *Radiation Damage in Solids*, ed. Billington, p. 82, (London: Academic Press). [67]

AMELINCKX, S. (1964). *Interaction of Radiation with Solids*, ed. R. Strumane *et al.* (North Holland Publishing Co.). [78]

ASHKIN, J. *see under* BETHE, H.

ASKILL, J. & TOMLIN, D. (1963). *Phil. Mag.* **8**, 997. [38]

AUGUSTINIAK, W. *see under* McREYNOLDS, A. W.

BACHELLA, G. L. *see under* AIROLDI, A.

BAKER, C. & KELLY, A. (1962). *Nature, Lond.* **193**, 235. [38]

BALLUFFI, R. W. *see under* SIMMONS, R. O.; VENABLES, J. A.

BARNES, R. S. (1964). UKAEA, unclassified report, AERE–R–4655. Plate XXIX

BARNES, R. S. (1965). A.S.T.M. Technical publication **380**, p. 40. [362]

BARNES, R. S. & MAZEY, D. J. (1961). *Disc. Farad. Soc.* **31**, 38. [322, 330, 352, 356, 361, 362]

BARNES, R. S. & MAZEY, D. J. (1963). *Proc. Roy. Soc.* A **275**, 47. [316, 353]

BARNES, R. S., REDDING, G. B. & COTTRELL, A. H. (1958). *Phil. Mag.* **3**, 25, 97. [353]

BARNES, R. S. *see also under* CHURCHMAN, A. T.; MAZEY, D. J.; NELSON, R. S.; SILK, E.; WESTMACOTT, K. H.

BARRETT, C. S. (1945). *The Structure of Metals* (New York: McGraw-Hill). [169]

BARSCHALL, H. H. *see under* WALT, M.

BARTLETT, A. F. *see under* EYRE, D. L.

BARTLETT, J. M. & DIENES, G. J. (1953). *Phys. Rev.* **89**, 848. [20, 37]

BAUER, W., DE FORD, J. W., KOEHLER, J. S. & KAUFFMAN, J. W. (1962). *Phys. Rev.* **128**, 1497. [271, 273]

BAUER, W., SEEGER, A. & SOSIN, A. (1967). *Phys. Lett.* **24**A, 195. [269, 271]

BAUER, W. & SOSIN, A. (1964a). *Phys. Rev.* **135**, A 521. [271]

BAUER, W. & SOSIN, A. (1964b). *J. Appl. Phys.* **35**, 703. [271]

BAUER, W. & SOSIN, A. (1964c). *Phys. Rev.* **136**, 255. [271].

BAUERLE, J. E. & KOEHLER, J. S. (1957). *Phys. Rev.* **107**, 1493. [29, 30–37]

BEELER, J. E. & BESCO, D. G. (1963a). *J. Applied Phys.* **34**, 2873. [163–164, 183–188]

BEELER, J. E. & BESCO, D. G. (1963 b). *J. Phys. Soc. Jap.* **18**, 159. [163–164, 183–188]

BEELER, J. R. (1963). GEC Report TM–63–10–13. [164, 188]

BEELER, J. R. (1964 a). ASTM Symposium, Chicago, 1964. [163, 186]

BEELER, J. R. (1964 b). Symposium on Atomic Collision Cascades UKAEA unclassified report AERE–R–4694. [186, 243]

BEELER, J. R. (1964 c). *J. Appl. Phys.* **35**, 2226. [163–164, 183–188]

BEELER, J. R. & BESCO, D. G. (1964). *Phys. Rev.* **134**, A 530. [163–164, 183–188]

BEEVERS, C. J. & MAZEY, D. J. (1962). *Phil. Mag.* **7**, 1061. [329]

BEEVERS, C. J. & NELSON, R. S. (1961). *Properties of Reactor Materials and the Effects of Radiation Damage*, p. 127 (London: Butterworths). [317, 328]

BEEVERS, C. J. & NELSON, R. S. (1963). *Phil. Mag.* **8**, 1189. [328]

BELLAMY, R. G. (1962). Proc. Inst. Met., Symposium on Uranium and Graphite, p. 53. [361]

BENNEMAN, K. H. (1961 a). *Phys. Rev.* **124**, 669. [37, 44]

BENNEMAN, K. H. (1961 b). *Z. Physik* **165**, 445. [37, 44]

BERRY, H. W. (1955). *Phys. Rev.* **99**, 553. [99, 100, 104]

BERTRAND, R. *see under* AMDUR, I.

BESCO, D. G. *see under* BEELER, J. R.

BETHE, H. A. & ASHKIN, J. (1953). *Experimental Nuclear Physics* **1**, 166. [118, 154, 260, 277]

BLATT, F. J. (1957). *Solid State Physics* **4**, 322. [37]

BLEICK, W. E. & MAYER, J. E. (1934). *J. Chem. Phys.* **2**, 252. [96]

BLEWITT, T. H. (1962). *Radiation Damage in Solids*, ed. Billington (London: Academic Press). [282, 283, 287–289, 297–300, 303, 306–308, 338]

BLEWITT, T. H. & COLTMAN, R. R. (1961). *Experimental Cryophysics*, ed. Hoare, Jackson and Kurti (London: Butterworths). [288]

BLEWITT, T. H. *see also under* LUCAS, M. E.; MAKIN, M. J.; THOMPSON, D. O.

BOHR, N. (1948). *Matt. Fys. Medd.* **18**, 8. [97, 154]

BORN, M. & MAYER, J. E. (1932). *Z. Physik* **75**, 1. [96]

BOWDEN, P. & BRANDON, D. G. (1963). *J. Nuc. Mat.* **9**, 348. [325, 327, 329]

BOWDEN, F. P. & CHADDERTON, L. T. (1961). *Nature, Lond.* **192**, 31. [Plate XII]

BOWKETT, K. *see under* RALPH, B.

BRADSHAW, F. J. & PEARSON, S. (1957). *Phil. Mag.* **2**, 570. [37]

BRANDON, D. G., SOUTHON, M. & WALD, M. (1962). *Reactor Materials and the Effects of Radiation Damage*, ed. Littler, p. 113 (London: Butterworth). [225, 226]

BRANDON, D. G. *see also under* BOWDEN, P.

BRINKMAN, J. A. (1954). *J. App. Phys.* **25**, 961. [161]

BRINKMAN, J. A. (1962 a). Int. Conf. on Crystal Defects in Kyoto, Japan. [120]

BRINKMAN, J. A. (1962 b). Radiation Damage in Solids, ed. D. S. pp. 830–60 (London: Academic Press). [97, 99, 101, 109, 119, 120, 199, 222]

BRINKMAN, J. A. *see also under* MEECHAN, C. J.

BROWN, E. & GOEDECKE, G. H. (1960). *J. App. Phys.* **31**, 932. [146]

BROWN, E. *see also under* JOHNSON, R. A.

BROWN, F. *see under* PIERCY, G. R.

BROWN, L. M. & MAZEY, D. J. (1964). *Phil. Mag.* **10**, 1081. [360]

BUCKLEY, S. N. (1961). Proceedings of Int. Conf. on Reactor Materials and Effects of Radiation Damage, p. 413 (London: Butterworth). [333]

BUDIN, C., DENAYRON, F., LUCASSON, A. & LUCASSON, P. G. (1963). *C.R. Acad. Sci.* **256**, 1518. [37]

BUDIN, C. *see also under* LUCASSON, P. G.

BULLOUGH, R. & FOREMAN, A. J. E. (1964). *Phil. Mag.* **9**, 98, 315. [80]

BULLOUGH, R. *see also under* EYRE, B. L.

BURGER, G., MEISSNER, H. & SCHILLING, W. (1964). *Physica Status Solidi*, **4**, 281. [292]

CHADDERTON, L. T. *see under* BOWDEN, F. P.

CHAPMAN, G. E. & KELLY, J. C. (1967). *Aust. J. Phys.* **20**, 3. [239]

CHARPNAU, H. P. *see under* FEDER, R.

CHIK, K. P. *see under* SEEGER, A.

CHURCHMAN, A. T., BARNES, R. S. & COTTRELL, A. H. (1958). *J. Nuc. Mat.* **7**, 88. [353]

COLTMAN, R. R., KLABUNDE, C. E., McDONALD, D. L. & REDMAN, J. K. (1962). *J. App. Phys.* **33**, 3509. [137, 290, 291, 295, 305]

COLTMAN, R. R. *see also under* BLEWITT, T. H.

COOPER, H. G., KOEHLER, J. S. & MARX, J. W. (1955). *Phys. Rev.* **97**, 599. [277, 279, 284, 286, 299, 308]

CORBETT, J. W., DENNEY, J. M., FISKE, M. D. & WALKER, R. M. (1957). *Phys. Rev.* **108**, 4, 954. [260, 266]

CORBETT, J. W., SMITH, R. B. & WALKER, R. M. (1959a). *Phys. Rev.* **114**, 1452. [260]

CORBETT, J. W., SMITH, R. B. & WALKER, R. M. (1959b). *Phys. Rev.* **114**, 1460. [260]

CORBETT, J. W. & WALKER, R. M. (1958). *Phys. Rev.* **110**, 767. [260]

CORBETT, J. W. & WALKER, R. M. (1959). *Phys. Rev.* **115**, 67. [260]

COTTERILL, R. M. J. & JONES, M. W. (1962). *J. Phys. Soc. Jap.* **18**, Supp. III, 158. [325]

COTTRELL, A. H. (1953). *Dislocations and Plastic Flow in Crystals* (Oxford University Press). [52, 56]

COTTRELL, A. H. *see also under* BARNES, R. S.; CHURCHMAN, A. T.; RIMMER, D. E.

CUDDY, L. J. & MACHLIN, E. S. (1962). *Phil. Mag.* **7**, 745. [37]

CUSHING, R. L. *see under* DAVIES, J. A.

DANNEBURG, W. (1961). *Metall.* **15**, 977. [38]

DAVIES, J. A., FRIESEN, J. & McINTYRE, J. D. (1960). *Can. J. Chem.* **38**, 1526. [155, 165]

DAVIES, J. A., FRIESEN, J., McINTYRE, J. D., CUSHING, R. L. & LOUNS-BURY, M. (1960). *Can. J. Chem.* **38**, 1535. [155, 165]

DAVIES, J. A. & SIMS, G. A. (1961). *Can. J. Chem.* **39**, 601. [155, 165]

DAVIES, J. A. *see also under* PIERCY, G. R. [155, 165]

DEDERICHS, P. H. & LEIBFRIED, G. (1962). *Zeits. f. Physik*, **170**, 320. [219]

DE FORD, J. W. *see under* BAUER, W.

DE JONG, M. & KOEHLER, J. S. (1963). *Rev. Phys.* **129**, 40. [37]

DEKKER, A. (1958). *Solid State Physics* (London: Macmillan). [7]

DENAYRON, F. *see under* BUDIN, C.; LUCASSON, P. G.

DENNEY, J. M. *see under* CORBETT, J. W.; SIMON, G. W.

DE SORBO, W. & TURNBULL, D. (1959a). *Acta Met.* **7**, 83. [37]

DE SORBO, W. & TURNBULL, D. (1959b). *Phys. Rev.* **115**, 560. [37]

DIEHL, J., LEITZ, C. & SCHILLING, W. (1963). *Phys. Lett.* **4**, 236. [338, 340, 344]

DIENES, G. J. *see under* BARTLETT, J. M.

DOWNEY, M. E. & EYRE, B. L. (1965). *Phil. Mag.* **11**, 53. [330, 349]

DOWNING, R. G. *see under* SIMON, G. W.

DOYAMA, M. & KOEHLER, J. S. (1960). *Phys. Rev.* **119**, 939. [37]

DOYAMA, M. & KOEHLER, J. S. (1962). *Phys. Rev.* **127**, 21. [37]

DUCKWORTH, H. E. *see under* ORMROD, J. H.

DWORSCHAK, F., HERSCHBACH, K. & KOEHLER, J. S. (1964). *Phys. Rev.* **133**, 293. [285, 308]

DWORSCHAK, F., SCHUSTER, H., WURM, J. & WOLLENBURGER, H. (1967). Private communication. [264, 325]

EDLUND, M. C. *see under* GLASSTONE, S.

EDMONDSON, B. & WILLIAMSON, G. K. (1964). *Phil. Mag.* **9**, 277. [85]

ELLIOTT, R. O. *see under* OLSEN, C. E.

ENGLERT, A. *see under* ERGINSOY, C.

ERGINSOY, C. (1964). *Interaction of Radiation with Solids*, ed. R. Strumane *et al.* (North Holland Publishing Co.). [101, 189, 194, 195, 225]

ERGINSOY, C. & THOMPSON, M. W. (1964). Unpublished work. [223]

ERGINSOY, C., VINEYARD, G. H. & ENGLERT, A. (1964). *Phys. Rev.* **133**, A 595. [23, 189, 225, 331]

ERGINSOY, C. *see also under* VINEYARD, G. H.

ESHELBY, J. D. (1954). *J. App. Phys.* **25**, 255. [18, 24]

ESHELBY, J. D. *see also under* FOREMAN, A. J. E.

EVANS, E. L. *see under* KUBASCHEWSKI, O.

EVERHART, E. & LANE, G. H. (1960). *Phys. Rev.* **120**, 2064, also earlier papers. [100]

EVERHART, E. *see also under* MORGAN, G. H.

EYRE, B. L. & BARTLETT, A. F. (1965). *Phil. Mag.* **12**, 261. [332]

EYRE, B. L. & BULLOUGH, R. (1965). *Phil. Mag.* **12**, 31. [331, 332]

EYRE, B. L. *see also under* DOWNEY, M. E.

FARMERY, B. W. & THOMPSON, M. W. (1968). *Phil. Mag.* **18**, 415. [231, 236]

FARMERY, B. W. *see also under* THOMPSON, M. W.

FEDER, R. & CHARPNAU, H. P. (1966). *Phys. Rev.* **149**, 464. [38]

FEDER, R. & NOWICK, A. S. (1958). *Phys. Rev.* **109**, 1959. [26, 37]

FERMI, E. & TELLER, E. (1947). *Phys. Rev.* **72**, 399. [152]

FESHBACH, H. *see under* McKINLEY, W. A.

FIRSOV, O. B. (1957a). *Zh. eksper. teor. Fiz.* **32**, 1464. [97]

FIRSOV, O. B. (1957b). *Zh. eksper. teor. Fiz.* **33**, 696. [97]

FIRSOV, O. B. (1958). *JETP*, **6**, 534. [97]

FISKE, M. D. *see under* CORBETT, J. W.

FLUIT, J. M. *see under* SANDERS, J. B.

FOREMAN, A. J. E. & ESHELBY, J. D. (1962). UKAEA unclassified report, AERE–R–4170. [80, 81, 335]

FOREMAN, A. J. E. *see also under* BULLOUGH, R. & GREENWOOD, G. W.

FORTES, M. A. *see under* RALPH, B.

FRANK, F. C. & READ, W. T. (1950). *Phys. Rev.* **79**, 723. [68, 69]

FREDERIGHI, T. (1965). *Lattice Defects in Quenched Metals*, p. 217, (London: Academic Press). [29–38]

FREDERICHI, T. *see also under* PANSERI, C.

FRIEDEL, J. (1956). *Les Dislocations*, Gauthier-Villars, Paris (English edition, Pergamon, 1964). [343]

FRIESEN, J. *see under* DAVIES, J. A.

FUJITA, F. E. *see under* IWATA, T.

FUMI, F. G. (1955). *Phil. Mag.* **46**, 1007. [21, 38]

FUMI, F. G. & TOSI, M. P. (1962). *J. Phys. Chem. Solids*, **23**, 1671. [98–103]

GEORGE, G. G. & GUNNERSEN, E. M. (1964). 7th Int. Conf. on Physics of Semiconductors, Publ. Dunod, Paris, p. 385. [195]

GERMAGNOLI, E. (1962). *Radiation Damage in Solids*, ed. Billington, p. 318 (London: Academic Press). [34, 35]

GERMAGNOLI, E. *see also under* AIROLDI, A.

GEROTEL, V. *see under* SEEGER, A.

GIBSON, J. B., GOLAND, A. N., MILGRAM, M. & VINEYARD, G. H. (1960). *Phys. Rev.* **120**, 1229. [21, 37, 44, 45, 97–103, 189, 190, 199, 200, 215, 218, 222, 235]

GILBERT, R. W. *see under* HOWE, L.

GILLIES, E. *see under* MAKIN, M. J.

GILMAN, J. J. (1961). *Progress in Ceramic Science*, Vol. 1 (London: Pergamon Press). [68]

GILMAN, J. J. *see also under* JOHNSTON, W. G.

GIRIFALCO, L. A. & STREETMAN, J. R. (1958). *J. Phys. Chem. Solids*, **4**, 182. [38]

GLASSTONE, S. & EDLUND, M. C. (1952). *Nuclear Reactor Theory* (London: Macmillan). [351]

GLOCKLER, G. (1954). *J. Chem. Phys.* **22**, 159. [38]

GOEDECKE, G. H. *see under* BROWN, E.

GOLAND, A. N. *see under* GIBSON, J. B.

GOLDSTEIN, *see under* KALOS.

GRANATO, A. & NILAN, T. G. (1961). *Phys. Rev. Lett.* **6**, 171. [285, 286, 299]

GRAY, D. L. (1959). *Acta Met.* **7**, 431. [304]

GREENWOOD, G. W., FOREMAN, A. J. E. & RIMMER, D. E. (1959). *J. Nuc. Mat.* **4**, 305. [357]

GUGAN, D. (1966). *Phil. Mag.* **13**, 533. [38]

GUNN, J. F. *see under* HOWE, L.

GUNNERSEN, E. M. *see under* GEORGE, G. G.

HAM, R. K. *see under* BROOM, T.

HASIGUTI, R. R. (1965). *J. Phys. Soc. Jap.* **20**, 625. [48]

HASTINGS, I. J. *see under* RUSSELL, B.

368 BIBLIOGRAPHY

HEEGER, J. D., RAU, R., TISCHER, P. & WENZL, H. (1966). *Phys. Lett.* **21**, 393. [299]

HEITLER, W. (1944). *The Quantum Theory of Radiation* (Oxford University Press). [139]

HERRING, C. (1953). *Structure and Properties of Solid Surfaces*, ed. by Gomer and Smith (University of Chicago Press). [358]

HERSCHBACH, K. *see under* DWORSCHAK, F.

HESKETH, R. V. & RICKARDS, G. K. (1966). Harwell Symposium on Defect Clusters. UKAEA unclassified report, AERE–R–5269. [326]

HIRAKA, A. & SUITA, T. (1962). *J. Phys. Soc. Jap.* **17**, 408. [38]

HIRSCH, P. B. (1962). *Radiation Damage in Solids*, ed. D. S. Billington, p. 39 (London: Academic Press). [78]

HIRSCH, P. B., HOWIE, A. & WHELAN, M. J. (1960). *Phil. Trans. Roy. Soc.* A **252**, 61. [85, 88]

HIRSCH, P. B., SILCOX, J., SMALLMAN, R. E. & WESTMACOTT, K. H. (1958). *Phil. Mag.* **3**, 897. [79]

HIRSCH, P. B. [316] *see also under* SILCOX, J.

HOARE, F. E., JACKSON, L. C. & KURTI, N. (1961). *Experimental Cryophysics* (London: Butterworth). [388]

HOLMES, D. K. (1962). *Radiation Damage in Solids*, Summer School Proceedings, p. 777, (London: Academic Press). [72, 292]

HOLMES, D. K. (1964). *Interaction of Radiation with Solids* (North Holland Publishing Co.). [160, 165, 166, 264, 292, 342]

HOLMES, D. K. & LEIBFRIED, G. (1960). *J. App. Phys.* **31**, 1046. [159]

HOLMES, D. K., LEIBFRIED, G., OEN, O. S. & ROBINSON, M. T. (1962). *Radiation Damage*, p. 3, Conference Proceedings (I.A.E.A., Vienna). [165].

HOLMES, D. K. *see also under* OEN, O. S.; THOMPSON, D. O.

HOWE, L., GILBERT, R. W. & PIERCY, G. R. (1963). *App. Phys. Lett.* **3**, 125. [324]

HOWE, L. & GUNN, J. F. (1964). *App. Phys. Lett.* **4**, 99. [326]

HOWIE, A. & WHELAN, M. J. (1962). *Proc. Roy. Soc.* A **267**, 206. [85]

HOWIE, A. *see also under* HIRSCH, P. B.; MAZEY, D. J.

HUDSON, B. (1964). *Phil. Mag.* **10**, 949. [335]

HUDSON, B., WESTMACOTT, K. H. & MAKIN, M. J. (1962). *Phil. Mag.* **7**, 377. [335]

HUDSON, B. *see also under* MAKIN, M. J.

HUGHES, D. J. & SCHWARZ, R. B. (1958). USAEC unclassified report, BNL–325 (2nd ed.). [128]

HULL, D. *see under* WESTMACOTT, K. H.

HUNTINGTON, H. B. (1953). *Phys. Rev.* **91**, 1092, and earlier papers. [18]

HUNTINGTON, H. B. & SEITZ, F. (1942a). *Phys. Rev.* **61**, 315. [18, 37, 44]

HUNTINGTON, H. B. & SEITZ, F. (1942b). *Phys. Rev.* **61**, 324. [18, 37, 44]

HUTCHINSON, T. S., ROGERS, D. H. & TURKINGTON, R. L. (1964). UKAEA unclassified report AERE–R–4697. [76]

HYDER, H. R. McK. & KENWARD, C. J. (1959). UKAEA unclassified report, AERE–R–2886. [128]

ISEBECK, K., RAU, F., SCHILLING, W., SONNENBERG, K., TISCHER, P. & WENZL, H. (1966a). *Phys. Stat. Solidi*, **17**, 259. [299, 300, 308]

ISEBECK, K., MULLER, R., SCHILLING, W. & WENZL, H. (1966b). *Phys. Stat. Solidi*, **18**, 467. [299, 300, 308]

IWATA, T., FUJITA, F. E. & SUZUKI, H. (1961). *J. Phys. Soc. Jap.* **16**, 197. [44]

JACKSON, L. C. *see under* HOARE, F. E.

JOHNSON, A. A., SARGENT, G. A. & WRONSKI, A. S. (1964). A.S.T.M. Symposium on 'The flow and Fracture Properties of Metals in Nuclear Environments'. [337, 347]

JOHNSON, A. A. *see also under* PEACOCK, D. E.

JOHNSON, R. A. (1964). *Phys. Rev.* **134**, A 1329. [42–44, 331]

JOHNSON, R. A. & BROWN, E. (1962). *Phys. Rev.* **127**, 446. [18, 37, 40, 42–44

JOHNSTON, W. G. & GILMAN, J. J. (1959). *J. Appl. Phys.* **30**, 129. [62]

JONES, H. *see under* MOTT, N. F.

JONES, M. W. *see under* COTTERILL, R. M. J.

KALOS & GOLDSTEIN (1956). USAEC unclassified report, NDA–12–16. [127]

KANTER, M. A. (1957). *Phys. Rev.* **107**, 655. [38]

KAUFFMAN, J. W. *see under* BAUER, W.; WARD, J. B.

KEEFER, D. & SOSIN, A. (1964). *App. Phys. Lett.* **3**, 185. [266, 267]

KELLY, A. *see under* BAKER, C.

KELLY, J. C. *see under* CHAPMAN, G. E.

KENWARD, C. J. *see under* HYDER, H. R. McK.

KINCHIN, G. H. & PEASE, R. S. (1955). *Rep. Prog. Phys.* **18**, 1. [143, 157, 185]

KINCHIN, G. H. & THOMPSON, M. W. (1958). *J. Nuc. Energy*, **6**, 275. [304, 308, 312]

KING, E., LEE, J. A., MENDLESSOHN, K. & WIGLEY, D. A. (1965). *Proc. Roy. Soc.* A **284**, 325. [282, 283]

KING, K. *see under* SCHÜLE, W.

KIRITANI, M. (1964). *J. Phys. Soc. Jap.* **19**, 618. [78]

KIRITANI, M., SHIMOMURA, Y. & YOSHIDA, S. (1965). *J. Phys. Soc. Jap.* **19**, 1624. [78]

KIRITANI, M. & YOSHIDA, S. (1963). *J. Phys. Soc. Jap.* **18**, 915. [78]

KIRITANI, M. & YOSHIDA, S. (1964). *Jap. J. App. Phys.* **4**, 148. [78]

KIRITANI, M. *see also under* YOSHIDA, S.

KITTEL, J. H. *see under* PAINE, S. H.

KLABUNDE, C. E. *see under* COLTMAN, R. R.

KLUG, H. P. & ALEXANDER, L. E. (1954). *X-ray Diffraction Procedures* (New York: Wiley). [207]

KOEDAM, J. (1959). *Physica*, **25**, 742. [237]

KOEHLER, J. S., SEITZ, F. & BAUERLE, J. E. (1957). *Phys. Rev.* **107**, 499. [32]

KOEHLER, J. S. *see also under* BAUER, W.; BAUERLE, J. E.; COOPER, H.; DE JONG, M.; DOYAMA, M.; DWORSCHAK, F.; MAGNUSON, G. D.; PALMER, W.; SEITZ, F.

KUBASCHEWSKI, O. & EVANS, E. L. (1958). *Metallurgical Thermochemistry*. (London: Pergamon Press). [38]

KUPER, A., LETAW, H., SLIFKIN, L., SONDER, E. & TOMIZUKA, C. T. (1955). *Phys. Rev.* **98**, 1870. [37]

370 BIBLIOGRAPHY

KURTI, N. *see under* HOARE, F. E.

LANE, G. H. *see under* EVERHART, E.

LE CLAIRE, A. D. *see under* MAKIN, S. M.

LEE, J. A. *see under* KING, E.

LEHMANN, C. (1961). *Nukleonik*, **3**, 1. [146]

LEHMANN, C. & LEIBFRIED, G. (1963a). *J. App. Phys.* **34**, 2821. [172, 182]

LEHMANN, C. & LEIBFRIED, G. (1963b). *Z. für Physik*, **172**, 465. [172]

LEHMANN, C. & SIGMUND, P. (1966). *Phys. Stat. Solidi* **16**, 507. [237]

LEIBFRIED, G. (1959). *J. App. Phys.* **30**, 1388. [234]

LEIBFRIED, G. & LEHMANN, C. (1961). *Z. für Physik*, **162**, 2, 203. [199, 200] Also in English: UKAEA unclassified translation AERE–TRANS– 885.

LEIBFRIED, G. *see also under* DEDERICHS, P. H.; HOLMES, D. K.; LEHMANN, C.

LEITZ, C. *see under* DIEHL, J.

LETAW, H., PORTNYE, W. & SLIFKIN, L. (1956). *Phys. Rev.* **102**, 636. [38]

LETAW, H. *see also under* KUPER, A.

LINDHARD, J. (1965). *Matt. Fys. Medd.* **34**, 14. [178]

LINDHARD, J. & SCHARFF, G. (1961). *Phys. Rev.* **124**, 128. [153, 160]

LINDHARD, J., SCHARFF, G. & SCHIOTT, H. E. (1963). *Matt. Fys. Medd.* **33**, 14. [153, 155, 160]

LINDHARD, J. & THOMSEN, P. V. (1962). *Proc. Conf. on Radiation Damage* (Venice: IAEA (UNESCO)). [152, 153, 155]

LINDHARD, J. & WINTHER, A. (1964). *Matt. Fys. Medd.* **34**, 4. [154]

LOGAN, R. A. (1956). *Phys. Rev.* **101**, 1455. [38]

LOMER, J. N. (1963). *Phil. Mag.* **8**, 90, 951. [260, 266, 267]

LOMER, J. N. & NIBLETT, D. H. (1962). *Phil. Mag.* **7**, 79, 1211. [260, 268]

LOMER, J. N. & PEPPER, M. (1967). *Phil. Mag.* **16**, 1119. [195]

LOMER, J. N. & TAYLOR, R. J. (1968). *Phil. Mag.* (In press.) [263, 277]

LOMER, W. M. (1959). *Prog. Met. Phys.* **8**, 255. [20, 37]

LOUNSBURY, M. *see under* DAVIES, J. A.

LUCAS, M. E. & BLEWITT, T. H. (1967). *Bull. Am. Physics*, **12**, 302. [296, 297

LUCASSON, A., LUCASSON, P. G. & WALKER, R. M. (1962). *Proceedings of Int. Conf. on Reactor Materials and Radiation Damage*, p. 83 (London: Butterworth). [263]

LUCASSON, A. *see also under* BUDIN, C.; LUCASSON, P. G.

LUCASSON, P. G. & LUCASSON, A. (1963). *J. Phys. Chem. Sol.* **27**, 1423. [37, 277]

LUCASSON, P. G., LUCASSON, A., BUDIN, C. & DENAYRON, F. (1963). *J. Phys. Radium*, **24**, 508. [37]

LUCASSON, P. G. & WALKER, R. M. (1962). *Phys. Rev.* **127**, 485. [195, 260, 263, 299, 304]

LUCASSON, P. G. *see also under* BUDIN, C.; LUCASSON, A.

LÜCK, G. & SIZMANN, R. (1964a) *Phys. Stat. Solidi*, **5**, 683. [254]

LÜCK, G. & SIZMANN, R. (1964b). *Phys. Stat. Solidi*, **6**, 263. [254

LUNDY, T. S. & MURDOCK, J. F. (1961). *J. Metals*, **13**, 676. [37]

LUNDY, T. S. & MURDOCK, J. F. (1962). *J. App. Phys.* **33**, 1671. 37]

LUTZ, H. & SIZMANN, R. (1963). *Phys. Lett.* **5**, 113 [182]

MACHLIN, E. S. *see under* CUDDY, L. J.

MAGNUSON, G. D., PALMER, W. & KOEHLER, J. S. (1958). *Phys. Rev.* **109**, 1990. [277, 284]

MAKIN, M. J. (1963). Unclassified UKAEA report AERE–R–4403. [339, 345]

MAKIN, M. J. (1964a). *Phil. Mag.* **9**, 81. [328, 346]

MAKIN, M. J. (1964b). *Phil. Mag.* **10**, 695. [346]

MAKIN, M. J. (1964c). Unclassified UKAEA report AERE–R–4610. [Plate XX]

MAKIN, M. J. (1964d). Private communication.

MAKIN, M. J. (1966). Symposium on the Nature of Small Defect Clusters. Unclassified UKAEA report AERE–R–5269 (H.M.S.O. price 96s.). [318]

MAKIN, M. J. & BLEWITT, T. H. (1961). AERE–R–3800. [346, 347, 348]

MAKIN, M. J. & BLEWITT, T. H. (1962). *Acta Met.* **10**, 241. [346]

MAKIN, M. J. & GILLIES, E. (1957). *J. Inst. Met.* **86**, 108. [349]

MAKIN, M. J. & HUDSON, B. (1963). *Phil. Mag.* **8**, 447. [80]

MAKIN, M. J. & MANTHORPE, E. (1964). Unclassified UKAEA report AERE–R–4538. [337]

MAKIN, M. J. & MINTER, F. J. (1959). *Acta Met.* **7**, 361. [348]

MAKIN, M. J., WHAPHAM, A. D. & MINTER, F. J. (1961). *Phil. Mag.* **6**, 465. [323]

MAKIN, M. J., WHAPHAM, A. D. & MINTER, F. J. (1962). *Phil. Mag.* **7**, 285. [317, 319, 320, 344, 345]

MAKIN, M. J. [316] *see also under* HUDSON, B.

MAKIN, S. M., ROWE, A. H. & LE CLAIRE, A. D. (1957). *Proc. Phys. Soc.* **70**B, 545. [37]

MANN, E. *see under* SEEGER, A.

MANTHORPE, E. *see under* MAKIN, M. J.

MARTIN, D. G. (1961). *Phil. Mag.* **6**, 67, 839. [266, 268].

MARTIN, D. G. (1962). *Phil. Mag.* **7**, 77, 803. [299–302, 308, 352]

MARTIN, D. G. (1963). *Phil. Mag.* **7** 1721. [299, 300]

MARTIN, D. G. *see also under* ADAM, J.

MARX, J. W. *see under* COOPER, H. G.

MASON, E. A. *see under* AMDUR, I.

MASSEY, H. W. *see under* MOTT, N. F.

MASTERS, B. C. (1964). Central Electricity Generating Board (U.K.) reports, RD/B/N245 and RD/B/N321. [332]

MAYBURG, S. (1954). *Phys. Rev.* **95**, 38. [38]

MAYER, J. E. *see under* BLEICK, W. E.; BORN, M.

MAZEY, D. J., BARNES, R. S. & HOWIE, A. (1963). *Phil. Mag.* **7**, 1861. [328]

MAZEY, D. J. [316, 353] *see also under* BARNES, R. S.; BEEVERS, C. J.; BROWN, L. M.; NELSON, R. S.

MAZUR, P. & MONTROLL, E. (1960). *J. Math. Phys.* **1**, 70. [206]

McCARGO, M. *see under* PIERCY, G. R.

McDONALD, D. L. *see under* COLTMAN, R. R.

McINTYRE, J. D. *see under* DAVIES, J. A.

McKEOWN, M. *see under* McREYNOLDS, A. W.

McKinley, W. A. & Feshbach, H. (1948). *Phys. Rev.* **74**, 1759. [124]

McReynolds, A. W., Augustiniak, W. & McKeown, M. (1955). *Phys. Rev.* **98**, 418. [308]

Meakin, J. D. (1964). *Nature, Lond.* **201**, 915.   [332]

Meechan, C. J. & Brinkman, J. A. (1956). *Phys. Rev.* **103**, 33, 1193. [266, 308]

Meechan, C. J. & Sosin, A. (1959*a*). *Phys. Rev.* **113**, 422. [260, 299]

Meechan, C. J. & Sosin, A. (1959*b*). *Phys. Rev.* **113**, 424. [274, 276, 299]

Meechan, C. J., Sosin, A. & Brinkman, J. A. (1960). *Phys. Rev.* **120**, 411. [266, 269]

Meissner, H. *see under* Burger, G.

Mendlessohn, K. *see under* King, E.

Menter, J. W., Pashley, D. S. & Presland, A. E. B. (1962). *Properties of Reactor Materials and the Effects of Radiation Damage* (London: Butterworth). [324]

Merkle, K. (1966). *Phys. Stat. Solidi*, **18**, 173. [226]

Merkle, K. (1966). *Symposium on Defect Clusters*, UKAEA unclassified report AERE–R–5269. [325]

Milgram, M. *see under* Gibson, J. B.

Minter, F. J. *see under* Makin, M. J.

Montgomery, H. *see under* Nelson, R. S.

Montroll, E. *see under* Mazur, P.

Morgan, G. H. & Everhart, E. (1962). *Phys. Rev.* **128**, 2, 667. [150]

Mott, N. F. & Jones, H. (1937). *The Theory of Metals and Alloys* (Oxford University Press). [18]

Mott, N. F. & Massey, H. W. (1949). *Atomic Collisions* (Oxford University Press). [126]

Muller, R. *see under* Isebeck, K.

Murdock, J. F. *see under* Lundy, T. S.

Nachtrieb, N. H. (1952). *J. Chem. Phys.* **20**, 1185. [38]

Neely, H. H. *see under* Sosin, A.

Nelson, R. S. (1963). *Phil. Mag.* **8**, 693. [223]

Nelson, R. S. (1964). *Phys. Lett.* **10**, 723. [324]

Nelson, R. S. (1965). *Phil. Mag.* **11**, 291. [251, 252]

Nelson, R. S., Mazey, D. J. & Barnes, R. S. (1965). *Phil. Mag.* **11**, 109, 91. [359]

Nelson, R. S., Thompson, M. W. & Montgomery, H. (1962). *Phil. Mag.* **7**, 1385. [207, 238]

Nelson, R. S. & Thompson, M. W. (1962*a*). *Phys. Lett.* **2**, 124. [240, 241]

Nelson, R. S. & Thompson, M. W. (1962*b*). *Phil. Mag.* **7**, 1425. [237]

Nelson, R. S. & Thompson, M. W. (1963). *Phil. Mag.* **8**, 94, 1677. [169–172]

Nelson, R. S. *see also under* Beevers, C. J.; Thompson, M. W.

Neufeld, J. *see under* Snyder, W. S.

Newson, P. A. *see under* Thompson, M. W.

Niblett, D. H. *see under* Lomer, J. N.

Nicholls, D. K. *see under* Van Lint, V. A. J.

Nielsen, K. O. (1956). *Electromagnetically Enriched Isotopes and Mass Spectrometry* (London: Butterworth). [155, 159]

NIHOUL, J. (1962). *Phys. Stat. Solidi*, **2**, 308. [304, 308]

NIHOUL, J. (1964). *Phil. Mag.* **9**, 167. [304, 308]

NILAN, T. G. *see under* GRANATO, A.

NOWICK, A. S. *see under* FEDER, R.

OEN, O. S. (1965). USAEC unclassified report ORNL–3813. [126]

OEN, O. S., HOLMES, D. K. & ROBINSON, M. T. (1963). *J. App. Phys.* **34**, 302. [184, 185]

OEN, O. S. & ROBINSON, M. T. (1963). *Appl. Phys. Lett.* **2**, 83. [184, 185]

OEN, O. S. *see also under* HOLMES, D. K.

OGILVIE, G. J. (1965). Private communication. [327]

OGILVIE, G. J., SANDERS, J. V. & THOMSON, A. A. (1963). *J. Phys. Chem. Sol.* **24**, 2, 247. [327]

OLSEN, C. E., ELLIOT, R. O. & SANDENAW, T. A. (1963). Radiation Damage in Reactor Materials, IAEA Conference Proceedings. [382]

ORFANOV, I. V. *see under* YURASOVA, V. E.

ORMROD, J. H. & DUCKWORTH, H. E. (1963). *Can. J. Phys.* **41**, 1424. [152, 153]

OVERHAUSER, A. W. (1955). *Phys. Rev.* **90**, 393. [285, 308]

PAINE, S. H. & KITTEL, J. H. (1955). 1st Geneva Conf. on Peaceful Uses of Atomic Energy, paper 745. [333]

PALMER, W. & KOEHLER, J. S. (1958). *Bull. Am. Phys. Soc.* **3**, 336. [37]

PALMER, W. *see also under* MAGNUSON, G. D.

PANSERI, C. & FREDERIGHI, T. (1958). *Phil. Mag.* **3**, 1223. [37]

PARÉ, V. K. *see under* THOMPSON, D. O.

PASHLEY, D. S. *see under* MENTER, J. W.

PEACOCK, D. E. & JOHNSON, A. A. (1963). *Phil. Mag.* **8**, 563. [308]

PEARSON, S. *see under* BRADSHAW, F. J.

PEASE, R. S. *see under* KINCHIN, G. H.

PEPPER, M. *see under* LOMER, J. N.

PIERCY, G. R. (1960). *Phil. Mag.* **5**, 201. [308]

PIERCY, G. R., BROWN, F., DAVIES, J. A. & McCARGO, M. (1963). *Phys. Rev. Lett.* **10**, 399. [167, 168]

PIERCY, G. R., McCARGO, M., BROWN, F. & DAVIES, J. A. (1964). *Can. J. Phys.* **42**, 1116. [167, 168]

PIERCY, G. R. *see also under* HOWE, L.

PLESHIVTSEV, N. V. *see under* YURASOVA, V. E.

PORAT, D. I. & RAMAVATARUM, K. (1959). *Proc. Roy. Soc.* A **252**, 394. [152, 153]

PORAT, D. I. & RAMAVATARUM, K. (1961). *Proc. Phys. Soc.* **77**, 97. [152, 153

PORTNYE, W. *see under* LETAW, H.

PRESLAND, A. E. B. *see under* MENTER, J. W.

QUÉRÉ, Y. (1960). *C.R. Acad. Sci.* **251**, 367.

QUÉRÉ, Y. (1961). *C.R. Acad. Sci.* **252**, 2399. [37]

QUÉRÉ, Y. (1963). *J. Nuc. Mat.* **9**, 3, 290. [333]

RACHAL, L. H. *see under* SOSIN, A.

RALPH, B., FORTES, M. A. & BOWKETT, K. M. (1966). Harwell Symposium on Defect Clusters. UKAEA unclassified report AERE–R–5269. [322, 323, 332]

RAMAVATARUM, K. *see under* PORAT, D. I.

RAMSTEINER, F. *see under* SCHÜLE, W.

RAU, F. *see under* HEEGER, J. D.; ISEBECK, K.

READ, W. T. (1953). *Dislocations in Crystals* (New York: McGraw-Hill). [52, 56]

READ, W. T. *see also under* FRANK, F. C.

REDDING, G. B. [353] *see under* BARNES, R. S.

REDMAN, J. K. *see under* COLTMAN, R. R.

RIGGAUER, M., SCHILLING, W., VÖLKL, J. & WENZL, H. (1967). To be published. [285]

RICKARDS, G. K. *see under* HESKETH, R. V.

RIMMER, D. E. & COTTRELL, A. H. (1957). *Phil. Mag.* **2**, 1345. [49, 50, 52]

RIMMER, D. E. *see also under* GREENWOOD, G. W.

ROBERTS, A. C. *see under* WESTMACOTT, K. H.

ROBINSON, M. T. (1965). *Phil. Mag.* **12**, 741. [146]

ROBINSON, M. T. & OEN, O. S. (1963a). *Phys. Rev.* **132**, 2385. [165, 167, 172, 183]

ROBINSON, M. T. & OEN, O. S. (1963b). Referred to by Piercy *et al.* (1963) as a private communication. [167]

ROBINSON, M. T. *see also under* HOLMES, D. K.; OEN, O. S.

ROGERS, D. H. *see under* HUTCHISON, T. S.

ROWE, A. H. *see under* MAKIN, S. M.

RÜHLE, M. *see under* SEEGER, A.

RUSSELL, B. & HASTINGS, I. J. (1965). *J. Nuc. Mat.* **17**, 30. [353, 354]

RUSSELL, B. *see also under* VELA, P.

SANDENAW, T. A. *see under* OLSEN, C. E.

SANDERS, J. B. (1967). Thesis, University of Leiden. [146]

SANDERS, J. B. & FLUIT, J. M. (1964) *Physica* **30**, 129. [206, 211]

SANDERS, J. B. & THOMPSON, M. W. (1967). To be published. [234]

SANDERS, J. V. *see under* OGILVIE, G. J.

SARGENT, G. A. *see under* JOHNSON, A. A.

SCHARFF, G. *see under* LINDHARD, J.

SCHIFF, L. I. (1955). *Quantum Mechanics*, ch. XI, 2nd ed. (New York: McGraw-Hill). [97]

SCHILLING, W. (1967). To be published. [301]

SCHILLING, W. *see also under* BURGER, G.; DIEHL, J.; ISEBECK, K.; RIGGAUER, M.

SCHIOTT, H. E. *see under* LINDHARD, J.

SCHMID, G. *see under* SCHOTTKY, G.

SCHMITT, R. A. & SHARP, R. A. (1958). *Phys. Rev. Lett.* **1**, 12. [160]

SCHOLZ, A. (1963). *Physica Status Solidi*, **3**, 42. [38]

SCHOTTKY, G., SEEGER, A. & SCHMID, G. (1964a). *Phys. Sol. Stat.* **4**, 419. [37]

SCHOTTKY, G., SEEGER, A. & SCHMID, G. (1964b). *Phys. Sol. Stat.* **4**, 439. [37]

SCHÜLE, W., SEEGER, A., RAMSTEINER, F., SCHUMACHER, D. & KING, K. (1961). *Zeit. Nat.* **16**A, 323. [37]

SCHULTZ, H. (1965). *Lattice Defects in Quenched Metals*. (London: Academic Press). [38]

SCHUMAKER, D. *see under* SCHÜLE, W.; SEEGER, A.

SCHUSTER, H. *see under* DWORSCHAK, F.

SCHWARZ, R. B. *see under* HUGHES, D. J.

SEEGER, A. (1958). 2nd Geneva Conference on Peaceful Uses of Atomic Energy, PUAE, **6**, 250. [269, 340–342]

SEEGER, A. (1962). *Radiation Damage in Solids*, ed. D. S. Billington (London: Academic Press). [76]

SEEGER, A., GEROTEL, V., CHIK, K. P. & RÜHLE, M. (1963). *Phys. Lett.* **5**, 107. [37]

SEEGER, A. & MANN, E. (1960). *Journ. Phys. Chem. Solids*, **12**, 326. [18, 37, 44]

SEEGER, A. & SCHUMAKER, D. (1965). *Lattice Defects in Quenched Metals*, ed. Cotterill *et al.* p. 15 (London: Academic Press). [37]

SEEGER, A. *see also under* BAUER, W.; SCHOTTKY, G.; SCHÜLE, W.

SEITZ, F. (1940). *Modern Theory of Solids*, p. 176 (New York: McGraw-Hill). [27]

SEITZ, F. (1954). *Rev. Mod. Phys.* **26**, 7. [18, 37]

SEITZ, F. & KOEHLER, J. S. (1956). *Solid State Physics*, **2**, 327. [246]

SEITZ, F. *see also under* HUNTINGTON, H. B.

SHARP, R. A. *see under* SCHMITT, R. A.

SHEWMAN, P. (1962). *Diffusion in Solids* (New York: McGraw-Hill). [36]

SHIMOMURA, Y. *see under* KIRITANI, M.; YOSHIDA, S.

SIGMUND, P. (1964). *Phys. Lett.* **6**, 251 [184, 185]

SIGMUND, P. *see under* LEHMANN, C.

SILCOX, J. & HIRSCH, P. B. (1959). *Phil. Mag.* **4**, 72. [82, 316]

SILCOX, J. *see also under* HIRSCH, P. B.

SILK, E. & BARNES, R. S. (1959). *Phil. Mag.* **4**, 44, 970. [117]

SILSBEE, R. H. (1957). *J. App. Phys.* **28**, 1246. [197]

SIMMONS, R. O. (1962). *Radiation Damage in Solids*, ed. D. S. Billington (London: Academic Press). [281, 286]

SIMMONS, R. O. & BALLUFFI, R. W. (1958). *Phys. Rev.* **109**, 1142. [281, 286, 299]

SIMMONS, R. O. & BALLUFFI, R. W. (1960*a*). *Phys. Rev.* **117**, 52. [26, 37]

SIMMONS, R. O. & BALLUFFI, R. W. (1960*b*). *Phys. Rev.* **117**, 62. [37]

SIMMONS, R. O. & BALLUFFI, R. W. (1960*c*). *Phys. Rev.* **119**, 600. [37]

SIMMONS, R. O. & BALLUFFI, R. W. (1962*a*). *Phys. Rev.* **125**, 862. [26, 37]

SIMMONS, R. O. & BALLUFFI, R. W. (1962*b*). *Bull. Am. Phys. Soc.* **7**, 233. [26, 37]

SIMON, G. W., DENNEY, J. M. & DOWNING, R. G. (1963). *Phys. Rev.* **129**, 6, 2454. [136]

SIMS, G. A. *see under* DAVIES, J. A.

SIZMANN, R. *see under* LÜCK, G.

SLATER, J. C. (1928). *Phys. Rev.* **32**, 339. [96]

SLICHTER, C. P. *see under* SPOKAS, J. J.

SLIFKIN, L. *see under* KUPER, A.; LETAW, H.

SMALLMAN, R. E. & WESTMACOTT, K. H. (1959). *J. App. Phys.* **30**, 603. [317]

SMALLMAN, R. E. *see also under* HIRSCH, P. B.; WESTMACOTT, K. H. [316]

SMITH, R. B. *see under* CORBETT, J. W.

SMOLUCHOWSKI, R. (1956). Proc. Int. Conf. on Peaceful Uses of Atomic Energy, 7, 676. (U.N. Publication.) [136]

SNYDER, W. S. & NEUFELD, J. (1955). *Phys. Rev.* 97, 6, 1636. [145, 185]

SONDER, E. *see under* KUPER, A.; TOMIZUKA, C. T.

SONNENBERG, K. *see under* ISEBECK, K.

SOSIN, A. (1962a). *Phys. Rev.* 126, 5, 1968. [260, 263, 266, 267, 299]

SOSIN, A. (1962b). *J. Appl. Phys.* 33, 11, 3373. [260, 266]

SOSIN, A. & NEELY, H. H. (1961). *Rev. Sci. Inst.* 32, 922. [258, 259, 261]

SOSIN, A. & NEELY, H. H. (1962). *Phys. Rev.* 127, 1465. [260, 266, 267]

SOSIN, A. & RACHAL, L. H. (1963). *Phys. Rev.* 130, 2238. [260, 271, 272]

SOSIN, A. *see also under* BAUER, W.; KEEFER, D.; MEECHAN, C. J.

SOUTHON, M. J. *see under* BRANDON, D. G.

SPOKAS, J. J. & SLICHTER, C. P. (1959). *Phys. Rev.* 113, 1462. [37]

STREETMAN, J. R. *see under* GIRIFALCO, L. A.

SUITA, T. *see under* HIRAKA, A.

SUZUKI, H. *see under* IWATA, T.

SWALIN, R. A. (1961). *J. Phys. Chem. Sol.* 18, 290. [38]

TAKAMURA, J. (1961). *Acta Met.* 9, 547. [37]

TAKAMURA, J. (1965). *Lattice Defects in Quenched Metals*, ed. R. Cotterill *et al.*, p. 521 (London: Academic Press). [48]

TAYLOR, R. J. *see under* LOMER, J. N.

TELLER, E. *see under* FERMI, E.

TEWORDT, L. (1958). *Phys. Rev.* 109, 61. [18, 19, 37, 44]

THOMAS, G. (1963). *Transmission Electron Microscopy of Metals* (New York: Wiley). [78]

THOMPSON, D. O., BLEWITT, T. H. & HOLMES, D. K. (1957). *J. App. Phys.* 28, 742. [296]

THOMPSON, D. O. & PARÉ, V. K. (1964). *Physical Acoustics*, Vol. III, chapter 7, ed. W. P. Mason (London: Academic Press). [72, 292–294]

THOMPSON, M. W. (1958). *Phil. Mag.* 3, 421. [308]

THOMPSON, M. W. (1959). *Phil. Mag.* 4, 139. [237, 304]

THOMPSON, M. W. (1960). *Phil. Mag.* 5, 278. [304, 308, 309]

THOMPSON, M. W. (1961). *J. Nuc. Mats.* [333]

THOMPSON, M. W. (1962). *Radiation Damage in Solids*, p. 753, ed. D. S. Billington (London: Academic Press). [303, 304]

THOMPSON, M. W. (1963). *Physics Letters*, 6, 124. [227]

THOMPSON, M. W. (1968). *Phil. Mag.* 18, 377. [98, 101, 199, 229, 230, 231, 235, 236]

THOMPSON, M. W. & FARMERY, B. W. (1968). *Phil. Mag.* [231, 236]

THOMPSON, M. W., FARMERY, B. W. & NEWSON, P. A. (1968). *Phil. Mag.* 18, 261. [227]

THOMPSON, M. W. & NELSON, R. S. (1961). *Proc. Roy. Soc.* A 259, 458. [216, 219, 237]

THOMPSON, M. W. & NELSON, R. S. (1962). *Phil. Mag.* 7, 2015. [250]

THOMPSON, M. W. & WRIGHT, S. B. (1965). *J. Nuclear Materials*, 16, 146. [129]

THOMPSON, M. W. *see also under* ERGINSOY, C.; FARMERY, B. W.; KINCHIN, G. H.; NELSON, R. S.; SANDERS, J. B.

THOMSEN, P. V. *see under* LINDHARD, J.

THOMSON, A. A. *see under* OGILVIE, G. J.

TISCHER, P. *see under* HEEGER, J. D.; ISEBECK, K.

TOMIZUKA, C. T. & SONDER, E. (1956). *Phys. Rev.* **103**, 1182. [37]

TOMIZUKA, C. T. *see also under* KUPER, A.

TOMLIN, D. *see under* ASKILL, J.

TOZI, M. P. *see under* FUMI, F. G.

TURKINGTON, R. L. *see under* HUTCHINSON, T. S.

TURNBULL, D. *see under* DE SORBO, W.

VAN LINT. V. A. J. & NICHOLLS, D. K. (1966). *Sol. State Physics*, **18**, 1. [160]

VARLEY, J. H. O. (1961). Private communication referred to in Makin, Whapham & Minter (1962). [321]

VARLEY, J. H. O. (1962). *Phil. Mag.* **7**, 301. [321]

VELA, P. & RUSSELL, B. (1966). *J. Nuc. Mat.* **19**, 312. [353, 354]

VENABLES, J. A. & BALLUFFI, R. W. (1965*a*). *Phil. Mag.* **11**, 1021. [326]

VENABLES, J. A. & BALLUFFI, R. W. (1965*b*). *Phil. Mag.* **11**, 1039. [326]

VINEYARD, G. H. (1961). *Disc. Farad. Soc.* **31**, 7. [189, 248, 249]

VINEYARD, G. H. (1962). *Radiation Damage in Solids*, ed. D. S. Billington (London: Academic Press). [189, 192, 193]

VINEYARD, G. H. & ERGINSOY, C. (1962). Proc. of Conference on Lattice Defects in Kyoto, Japan. [189, 214]

VINEYARD, G. H. *see also under* ERGINSOY, C.; GIBSON, J. B.

VÖLTL, J. *see under* RIGGAUER, M.

VOOK, R. & WERT, C. (1958). *Phys. Rev.* **109**, 1529. [280, 281, 286, 299]

WALD, M. *see under* BRANDON, D. G.

WALKER, R. M. (1962). *Radiation Damage in Solids*, p. 594 (London: Academic Press). [260, 262, 265, 266, 270, 309, 331]

WALKER, R. M. *see also under* CORBETT, J. W.; LUCASSON, A.; LUCASSON, P. G.

WALT, M. & BARSCHALL, H. H. (1954). *Phys. Rev.* **93**, 1962. [126]

WARD, J. B. & KAUFFMAN, J. W. (1961). *Phys. Rev.* **123**, 90. [271]

WEHNER, G. K. (1956). *Phys. Rev.* **102**, 690. [237]

WEIJSENFELD, C. (1965). Thesis, University of Utrecht. [219, 220]

WENZL, H. *see under* HEEGER, J. D.; ISEBECK, K.; RIGGAUER, M.

WERT, C. *see under* VOOK, R.

WESTMACOTT, K. H., BARNES, R. S., HULL, D. & SMALLMAN, R. E. (1961). AERE unclassified report R. 3617. [Plate X]

WESTMACOTT, K. H., ROBERTS, A. C. & BARNES, R. S. (1962). UKAEA, unclassified report AERE–R–4096. [329]

WESTMACOTT, K. H. *see also under* HIRSCH, P. B.; HUDSON, B.; SMALLMAN, R. E.

WHAPHAM, A. D., *see also under* MAKIN, M. J.

WHELAN, M. J. *see under* HIRSCH, P. B.; HOWIE, A.

WIGLEY, D. A. (1965). *Proc. Roy. Soc*.A **284**, 344. [282]

WIGLEY, D. A. *see also under* KING, E.

WILLIAMSON, G. K. *see under* EDMONDSON, B.

WINTHER, A. *see under* LINDHARD, J.

WITT, H. *see under* KETTING, H.

WOLLENBURGER, H. *see under* DWORSCHAK, F.

WRIGHT, S. B. (1962). UKAEA unclassified report, AERE-R-4080. [129]

WRIGHT, S. B. *see also under* THOMPSON, M. W.

WRONSKI, A. S. *see under* JOHNSON, A. A.

WURM, J. *see under* DWORSCHAK, F.

YOSHIDA, S. (1961). *J. Phys. Soc. Jap.* **16**, 44. [163]

YOSHIDA, S., KIRITANI, M., SHIMOMURA, Y. & YOSHINAKA, A. (1965 *a*). *J. Phys. Soc. Jap.* **20**, 628. [78]

YOSHIDA, S., KIRITANI, M. & SHIMOMURA, Y. (1965 *b*). *Lattice Defects in Quenched Metals*, ed. R. Cotterill *et al.*, pp. 713–37 (London: Academic Press). [78]

YOSHIDA, S. *see also under* KIRITANI, M.

YOSHINAKA, A. *see under* YOSHIDA, S.

YURASOVA, V. E., PLESHIVTSEV, N. V. & ORFANOV, I. V. (1959). *JETP* **37**, 966. [237]

INDEX